State-of-the-Art Environmental Microbiology in China (2023–2024)

State-of-the-Art Environmental Microbiology in China (2023–2024)

Guest Editor

Zhiyong Li

MDPI

Basel • Beijing • Wuhan • Barcelona • Belgrade • Novi Sad • Cluj • Manchester

Guest Editor
Zhiyong Li
State Key Laboratory of
Microbial Metabolism
Shanghai Jiao Tong
University
Shanghai
China

Editorial Office
MDPI AG
Grosspeteranlage 5
4052 Basel, Switzerland

This is a reprint of the Special Issue, published open access by the journal *Microorganisms* (ISSN 2076-2607), freely accessible at: https://www.mdpi.com/journal/microorganisms/special_issues/75S4KF6K98.

For citation purposes, cite each article independently as indicated on the article page online and as indicated below:

Lastname, A.A.; Lastname, B.B. Article Title. *Journal Name* **Year**, *Volume Number*, Page Range.

ISBN 978-3-7258-4472-2 (Hbk)
ISBN 978-3-7258-4471-5 (PDF)
https://doi.org/10.3390/books978-3-7258-4471-5

Contents

About the Editor

Zhiyong Li

Dr. Li is a tenured professor of Shanghai Jiao Tong University (SJTU) and the State Key Laboratory of Microbial Metabolism, China, and the chair of the marine microbiology branch of the shanghai microbiology association. Dr. Li's research interests are marine microbial symbioses, with a main focus on sponge/coral–microbe symbioses, including the sponge/coral microbiome and community structure and functions, the association of sponge/coral–microbes, interactions between the host and its microbial symbionts, marine natural products from marine microbial symbionts, and the response and adaption of sponge holobionts and coral holobionts to environment and climate changes.

Preface

Microorganisms are responsible for energy and nutrient cycling in biosystems and are massively involved in the planet's sustainability. This reprint focuses on the microbial processes in the environment, microbial communities and microbial interactions, including omics technologies and cross-disciplinary studies dedicated to basic and/or applied research in China. As a Special Issue, "State-of-the-Art Environmental Microbiology in China (2023–2024)", in the *Microorganisms* journal of MDPI, this reprint consists of nine articles, one communication and one review from Peking University, Shanghai Jiao Tong University, Xiamen University, Jilin University, China University of Mining and Technology, Beijing Forestry University, Hohai University, Qingdao University of Technology, Heilongjiang University, Yunnan Academy of Agricultural Sciences, etc. This reprint highlights the latest advancements in the field of environmental microbiology in China, including bacterial, archaeal and fungal community structure and function, which will be of particular interest to professionals in the field of microbial ecology and microbial biotechnology. Thanks go to all the authors' contributions to this reprint.

Zhiyong Li
Guest Editor

microorganisms

MDPI

Article

Biodegradation of Crude Oil by Nitrate-Reducing, Sulfate-Reducing, and Methanogenic Microbial Communities under High-Pressure Conditions

Lu Wang [1,2], Yong Nie [3], Xinglong Chen [1,2], Jinbo Xu [3], Zemin Ji [1,2], Wenfeng Song [2], Xiaofang Wei [1,2], Xinmin Song [1,2,*] and Xiao-Lei Wu [3,4,*]

1 State Key Laboratory of Enhanced Oil & Gas Recovery, Beijing 100083, China; luwangmoon@petrochina.com.cn (L.W.); chxlhdpu@petrochina.com.cn (X.C.); jizemin@petrochina.com.cn (Z.J.); xiaofangwei@petrochina.com.cn (X.W.)

2 Research Institute of Petroleum Exploration & Development, Beijing 100083, China; songwf@petrochina.com.cn

3 College of Engineering, Peking University, Beijing 100083, China; nieyong@pku.edu.cn (Y.N.); 1801110631@pku.edu.cn (J.X.)

4 Institute of Ecology, Peking University, Beijing 100083, China

* Correspondence: sxm@petrochina.com.cn (X.S.); xiaolei_wu@pku.edu.cn (X.-L.W.)

Abstract: Carbon capture, utilization, and storage (CCUS) is an important component in many national net-zero strategies, and ensuring that CO_2 can be safely and economically stored in geological systems is critical. Recent discoveries have shown that microbial processes (e.g., methanogenesis) can modify fluid composition and fluid dynamics within the storage reservoir. Oil reservoirs are under high pressure, but the influence of pressure on the petroleum microbial community has been previously overlooked. To better understand microbial community dynamics in deep oil reservoirs, we designed an experiment to examine the effect of high pressure (12 megapascals [MPa], 60 °C) on nitrate-reducing, sulfate-reducing, and methanogenic enrichment cultures. Cultures were exposed to these conditions for 90 d and compared with a control exposed to atmospheric pressure (0.1 MPa, 60 °C). The degradation characteristic oil compounds were confirmed by thin-layer analysis of oil SARA (saturates, aromatics, resins, and asphaltenes) family component rods. We found that the asphaltene component in crude oil was biodegraded under high pressure, but the concentration of asphaltenes increased under atmospheric pressure. Gas chromatography analyses of saturates showed that short-chain saturates (C8–C12) were biodegraded under high and atmospheric pressure, especially in the methanogenic enrichment culture under high pressure (the ratio of change was −81%), resulting in an increased relative abundance of medium- and long-chain saturates. In the nitrate-reducing and sulfate-reducing enrichment cultures, long-chain saturates (C22–C32) were biodegraded in cultures exposed to high-pressure and anaerobic conditions, with a ratio of change of −8.0% and −2.3%, respectively. However, the relative proportion of long-chain saturates (C22–C32) increased under atmospheric pressure. Gas Chromatography Mass Spectrometry analyses of aromatics showed that several naphthalene series compounds (naphthalene, C1-naphthalene, and C2-naphthalene) were biodegraded in the sulfate-reducing enrichment under both atmospheric pressure and high pressure. Our study has discerned the linkages between the biodegradation characteristics of crude oil and pressures, which is important for the future application of bioenergy with CCUS (bio-CCUS).

Keywords: oil reservoir; anaerobic biodegradation; pressure; nitrate-reducing; sulfate-reducing; methanogenesis

1. Introduction

The increase in atmospheric greenhouse gas concentrations, primarily caused by the use of fossil fuels, has created serious risks for humanity [1]. Carbon capture, utilization, and storage (CCUS) is one way to achieve low-carbon use of high-carbon fuels, and it is recognized as an important technology package for climate change mitigation [2]. CCUS is expected to reduce emissions by approximately 600–1400 Mt CO_2 by 2050 and by 1000–1800 Mt CO_2 by 2060. One-third of this reduction is expected to result from the use of bioenergy with CCUS by 2060 [3,4]. CO_2 can be stored in oil reservoirs and change the mobility and future trapping systematics of the evolved supercritical fluid. Recent discoveries have shown that microbial processes (e.g., methanogenesis) may modify the fluid composition and fluid dynamics within a storage reservoir and could reduce the volume of injected CO_2 required [5]. In subsurface deposits, CO_2 utilization and crude oil biodegradation is coupled with the microbial community [6–10]. Recently, researchers discovered that stochastic assembly processes are critical in shaping the groundwater microbial community structure; however, the relative importance of these processes decreased as environmental stress increased [11]. In the subsurface, environmental stressors on the microbial community include temperature [12], salinity [13,14], pH [15], water content [16], electron acceptors [17–21], and the composition of crude oil [15,22]. Oil reservoirs are extreme environments with high temperature and high pressure [23], with all reservoir fluids existing under pressure. High pressure is a key characteristic of oil reservoirs, but its influence on the microbial community has been previously overlooked.

Pressure exists in a reservoir for the same reason that pressure exists at the bottom of the ocean. The Deepwater Horizon oil spill was one of the largest and deepest oil spills recorded, and it has been shown that pressure (0.1, 15, and 30 MPa) acts synergistically with low temperature to slow microbial growth and oil degradation in deep-sea environments [24]. Pressure has also been reported to restructure deep-sea hydrocarbon-degrading microbial communities [23,25,26]. Using a high-temperature and high-pressure incubation system (55 °C, 5 MPa), microbial communities capable of methanogenic crude oil degradation were obtained from the Yabase oil reservoir [27] and Yamagata oil reservoir [28] in Japan. Mild hydrostatic pressure (15 MPa) shaped the assemblage of oil-degrading communities [29]. Recently, petroleum-degrading microbial communities incubated under high-pressure conditions (12 MPa) were metabolically profiled with metagenomics, which also defined microbial community interactions and the exchange of amino acids and cofactors among members [23]. In addition, pressurized anaerobic digestion has gained increasing interest in recent years. It is a valuable process that allows the production of biogas with high methane content, reducing the energy costs for the biogas to upgrade and inject into the distribution grid. A modified Anaerobic Model Digestion n.1 showed that the higher the pressure, the higher the volumetric mass transfer coefficient [30]. The research results of Merkle et al. [31] showed that methane content had increased from 79.08% at 10 bar to 90.45% at 50 bar. Siciliano's research showed that as the pressure increased, the quality of the biogas was enhanced, while the overall amount of methane lowered [32].

These studies enriched our current knowledge of microbial communities in petroleum reservoirs under high-pressure conditions.

Biodegradation in shallow subsurface petroleum reservoirs has been attributed to aerobic bacterial hydrocarbon degradation stimulated by surface recharge of oxygen-bearing meteoric waters. However, anaerobic degradation processes dominate in subsurface sedimentary environments [33]. Approximately half of the world's in-place oil and bitumen has experienced biodegradation, which is believed to largely have occurred through anaerobic methanogenesis. The presence of secondary microbial methane is apparent in twenty-two basins, probable in twelve basins, and possible in six basins worldwide [34]. Therefore, research on anaerobic degradation of oil is important. Crude oil can contain thousands or even tens of thousands of hydrocarbon compounds with highly variable composition [35,36]. Hydrocarbons are divided into four parts based on their polarizability and polarity (saturates, aromatics, resins, and asphaltenes; SARA), with each fraction having a different composition [37]. Saturated hydrocarbons are more easily degraded than aromatic hydrocarbons. Aromatic hydrocarbons with one to three aromatic rings are also efficiently biodegraded. The asphaltene fraction contains higher-molecular-weight compounds with complex chemical structures in the range of 600 to 2,000,000 Da [38,39].

In the subsurface, the oil biodegradation rate is not limited by the supply of electron donors (i.e., hydrocarbons) but rather by the supply of nutrients or electron acceptors to the site of degradation [40]. For hydrocarbon or polycyclic aromatic hydrocarbon (PAH) degradation, NO_3^-, Fe^{3+}, SO_4^{2-}, and HCO_3^- are typical terminal electron acceptors (TEAs), which are linked to four typical reducing conditions (i.e., nitrate-reducing, ferric-reducing, sulfate-reducing, and methanogenic conditions, respectively) [18,41–44]. The standard Gibbs free energy values for NO_3^-, SO_4^{2-}, and CO_2 are −163.2, −152.2, and −62.8 kJ·mol^{-1}, respectively [45]. Previous studies have indicated that the biodegradation of PAHs and the mechanisms affecting bioremediation in PAH-polluted marine sediment may vary under different TEA-reducing conditions [44,46]. Chen et al. [47] reviewed the recent advances in the biodegradation of PAHs under anoxic conditions and provided mechanistic insights into metabolic pathways and functional genes. Fumarate addition is also an important initial activation mechanism for anaerobic alkane degradation and methanogenic alkane degradation, resulting in the generation of 1-methylalkyl succinic acids; this is followed by the C-skeleton rearrangement reaction, oxidization to fatty acids, and further conversion to methane and carbon dioxide [48–51]. In the presence of nitrate, CO_2 can improve the anaerobic biodegradation efficiency of the resins and asphaltenes in heavy oil, particularly the biodegradation selectivity of polar heterocyclic compounds by the newly isolated *Klebsiella michiganensis* [52]. However, there have been limited studies on anaerobic biodegradation using different electron acceptors under high-pressure conditions.

In the present study, we conducted laboratory-scale high-pressure incubations of production water from Jilin Oilfield under nitrate-reducing, sulfate-reducing, and methanogenic conditions. We detected and analyzed the composition of crude oil after incubation. The study aimed to characterize the role of anaerobic biodegradation at pressure under various conditions.

2. Materials and Methods

2.1. High-Pressure and Atmospheric-Pressure Incubations

Crude oil and production water were collected from Jilin Oilfield in China. The temperature of the oil reservoir was 60 °C. After the oil production well was selected, dead oil was discarded before sampling. Culture vessels (or the pressure container) were quickly filled with oil production water and sealed, and then transported to the local research institute. Solution gases were released, and the vessels were replenished with oil production water. The pressure of the samples was increased and they were stored in stainless steel reactors (1 L) at 5 MPa. The samples were transported to the laboratory within 48 h, at which point the pressure was immediately increased to 12 MPa. For the three stainless steel reactors (1 L), the first one was added by fumarate (final concentration of 10 mM), the second by nitrate (final concentration of 10 mM), and the third by sulfate (final concentration of 10 mM). All of the three high-pressure reactors were incubated statically for 90 days at 60 °C. The setup diagram is shown in Figure 1a. Samples that were intended to be cultured at atmospheric pressure were transported from Jilin Oilfield to the laboratory under atmospheric pressure. Our atmospheric pressure setup is the same as the one for conventional anaerobic experiments using Hungate technology. The setup diagram at atmospheric pressure is shown in Figure 1b. Hungate serum bottles with a total volume of 120 mL were cleaned and dried. We filled a bottle with 14 g of 40–70 mesh quartz sand (about 10 mL in volume) (Tianjin Quartz Clock Factory Bazhou Chemical Plant, analytically pure), and placed it in 180 °C oven for 3 h for sterilization. We turned off the oven and waited until the temperature dropped below 60 °C. After having opened the door of the oven, we used tin foil to seal the bottle immediately. We filled the bottle with 50 mL of oil production water (incubation broth) and 2 g of crude oil, and the headspace volume was 60 mL. Fumarate, nitrate, and sulfate were added to each of the three groups cultured under atmospheric pressure, respectively. The final concentration of additives in each group was the same, 10 mM. After having added indicator resazurin into the sample bottles, the anaerobic reduction copper column and air nitrogen replacement device were used to inject high-purity nitrogen through the injection needles for about 15–20 min until the color of the liquid in the bottles changed from red to colorless. Finally, the mixed solution of L-Cysteine HCl·H_2O (3 mM) and $Na_2S\cdot 9H_2O$ (2 mM) was quickly added into each serum bottle. We sealed these bottles immediately with a rubber plug (SANSHIN, Okinawa, Japan) and finally with an aluminum seal cap (Chemglass, Vineland, NJ, USA). Then, these bottles were incubated statically for 90 d at 60 °C.

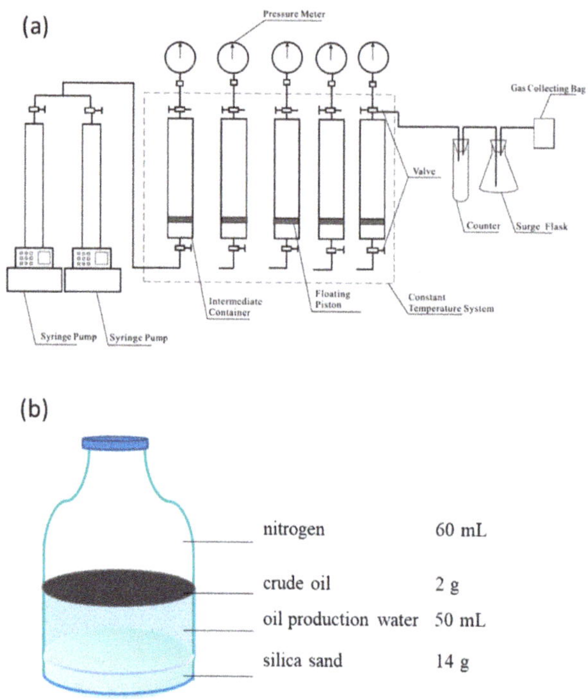

Figure 1. The setup diagram of experiment. (**a**) high-pressure group; (**b**) atmospheric pressure group.

The culture starting point sample was labeled S0. After 90 days of incubation, the samples incubated under atmospheric pressure were labeled as "additive-90", e.g., NO_3-90, SO_4-90, and Fuma-90; the samples incubated under high pressure were labeled as "additive-P90", e.g., NO_3-P90, SO_4-P90, and Fuma-P90.

After 90 days of incubation, the total microbial DNA was extracted from the incubation water, and the bacteria were amplified (16S rRNA gene), purified, and sequenced using the HiSeq platform. For each sample, 200–250 mL of water sample in the culture system was collected using a filter membrane. The filter membrane was cut into 1 mm^2 fragments and transferred to a Lysing Matrix E tube with glass beads. Amounts of 978 μL Sodium Phosphate Buffer and 122 μL MT Buffer were added. After homogenization, the bacteria were sonicated at 16 °C for 30 min, and the genomic DNA of the bacteria was extracted according to the instructions of the FastDNA® Spin Kit for soil kit (MP Biomedicals, Santa Ana, CA, USA). High-throughput sequencing of the 16S rRNA gene was conducted by Beijing Novogene Biotech Co., Ltd. (Beijing, China). The abundance of Operational Taxonomic Units (OTUs for short) at the level of the genus was labeled "g-". The OTUs of samples incubated for 90 d under atmospheric pressure were labeled as "g-additive-90" ("-1" and "-2" stand for two parallel samples), e.g., g-NO_3-90-1, g-NO_3-90-2, g-SO_4-90-

1, g-SO_4-90-2, g-Fuma-90-1, and g-Fuma-90-2. (g-NO_3-90 means the average value of g-NO_3-90-1 and g-NO_3-90-2, etc.).

The OTUs of samples incubated for 90 d under high pressure were labeled as "g-additive-P90", e.g., g-NO_3-P90, g-SO_4-P90, and g-Fuma-P90.

After 90 days of incubation, the change ratio of each *n*-alkane relative to 0 d equals the ratio of C_i at 90 d ($C_{i\text{-}90}$) in a sample minus the ratio of C_i at 0 d ($C_{i\text{-}0}$), divided by ($C_{i\text{-}0}$), i.e., ratio of change = ($C_{i\text{-}90}$ − $C_{i\text{-}0}$) × 100%/$C_{i\text{-}0}$.

2.2. Analysis of the Family Composition of Crude Oil

A group-type analysis can be used to define the relationship between the structures and properties of an oil because it is impossible to identify all of the individual components in oil [53]. The procedure to separate the family composition into saturates, aromatics, resins, and asphaltenes (SARA procedure) was conducted in accordance with the oil and gas industry standard of the People's Republic of China SY/T 5119-2016 [54], which is an analysis method for the family composition of rock extracts and crude oil. Testing was carried out by the Central Laboratory of Geological Sciences, Research Institute of Petroleum Exploration & Development, PetroChina.

The steps of using the rod thin layer flame ionization detection method to conduct analysis are as follows: (a) Take a specific amount of soluble organic matter from the rock or purified crude oil sample, dissolve it in chloroform, and prepare a solution with a concentration ranging from 10 mg/mL to 20 mg/mL. (b) Extract 0.5 μL to 1.0 μL of the sample solution using a microsyringe and apply it to an activated silica gel chromatographic rod, approximately 0.5 cm from one end, and repeat this process five–six times. (c) Place the silica gel chromatographic rod in a constant humidity cylinder for 10 min. (d) Insert the silica gel chromatographic rod into a chromatographic cylinder filled with *n*-hexane and unfold it so that the solvent rises between 8 cm and 9 cm. (e) Allow the silica gel chromatographic rod to sit in volatile solvent at room temperature for 2 min before placing it back in the constant humidity cylinder for another 10 min. (f) Subsequently, place it into a chromatographic cylinder containing a mixed solvent of dichloromethane and *n*-hexane (volume ratio of 1:1), unfold it until the solvent rises between 4 cm and 5 cm. (g) Repeat step e), successively. Finally, insert it into a chromatographic cylinder containing a mixture of *n*-hexane and isoamyl alcohol (volume ratio of 90:10); unfold it until the solvent rises between 1.5 cm and 2.0 cm. The separated sample on the silica gel chromatographic rod should be placed in volatile solvent at room temperature for 2 min. (h) Turn on the main power switch and other switches of the rod thin layer flame ionization analyzer, and adjust the instrument parameters according to the regulations (room temperature: 20 °C–30 °C; relative humidity: less than or equals to 65%; airflow rate: 2000 mL/min; hydrogen flow rate: 160 mL/min–180 mL/min; and scanning speed: 30 s each time). The separated sample on the silica gel thin layer rod should be inserted into the instrument for testing.

2.3. Gas Chromatography Analysis of Saturates

The procedure to separate saturated hydrocarbon was conducted in accordance with the oil and gas industry standard of the People's Republic of China SY/T 5779-2008 [55], which is an analytical method for hydrocarbons in petroleum and sediment by gas chromatography. Testing was conducted by the Central Laboratory of Geological Sciences, Research Institute of Petroleum Exploration & Development, PetroChina. The test method was based on Part 6 of *SY/T 5779-2008: Analysis of saturated hydrocarbons in rock chloroform extracts and crude oil.* 1. Sample preparation. (1) The sample should be fractionated according to SY/T 5119 to isolate the saturated hydrocarbon fractions, which should then be concentrated and transferred into a sealed sample bottle for storage in a refrigerated environment prior to testing. (2) When a splitter is used during sampling, the saturated hydrocarbon should be diluted with an appropriate amount of *n*-hexane; when sampling without a splitter, the saturated hydrocarbon should be diluted with an appropriate amount of isooctane. 2. Measurement procedure. (1) Activate the gas path system of the chromatograph to eliminate any blockages or leaks. (2) Initiate the instrument following the specific operating procedures for different types of chromatographs and adjust it to achieve optimal operating conditions. (3) Ignite the flame; after having stabilized the programmed temperature's chromatographic baseline, select either the shunt or non-shunt injection method, and determine the injection volume based on the sample size. (4) Inject the sample using a microinjector, and activate both the programmed temperature settings and chromatographic workstation.

2.4. Gas Chromatography Mass Spectrometry Analysis of Aromatics

The aromatic hydrocarbon samples were analyzed using the Thermo-Trace GC Ultra-DSQ II (Model: 0807173, made in Thermo Fisher Scientific, Waltham, MA, USA). The oven was heated to 100 °C for 5 min, and the temperature was increased at 3 °C/min until it reached 320 °C and then held for 20 min. To avoid the influence of a solvent peak, the filament was opened after 8 min. The separation column was an HP-5MS elastic quartz capillary column (60 m × 0.25 mm × 0.25 mm). High-purity nitrogen (99.999%) was used as a carrier gas at a flow rate of 1 mL/min. The injector temperature was 300 °C and the transmission line temperature was 300 °C.

The mass spectrometer was operated in the electron impact mode using 70 eV ionization voltage. The ion source temperature was 220 °C and the GC/MS interface was set to 250 °C. Testing was conducted by the Central Laboratory of Geological Sciences, Research Institute of Petroleum Exploration & Development, PetroChina.

3. Results

The effects of high pressure and atmospheric pressure on the biodegradation of crude oil were compared based on the results of family composition, saturated hydrocarbon chromatography, and aromatic hydrocarbon mass spectrometry.

3.1. Nitrate-Reducing Enrichment Group

3.1.1. Family Composition of Crude Oil

The family composition of the atmospheric pressure group and the high-pressure group at 0 d and after 90 d of nitrate-reducing cultivation is shown in Figure 2a. The relative abundance of saturates increased after 90 d of cultivation by 7.2% at atmospheric pressure and 10.0% at high pressure. The relative abundance of aromatics at atmospheric pressure and high pressure decreased by 16.2% and 20.0%, respectively, after 90 d cultivation. The relative abundance of resins at atmospheric pressure and high pressure increased by 6.8% and 11.1%, respectively, after 90 d cultivation. Compared with that at 0 d, the relative abundance of asphaltenes in the atmospheric pressure group was 2.3% higher after 90 d and was 1.1% lower in the high-pressure group.

3.1.2. Gas Chromatography of Saturates

The gas chromatography of saturates before and after microbial degradation of crude oil with nitrate is shown in Figure 3a–c and Figure S1. For short-chain alkanes (C8–C12), the relative abundance decreased by 7.0% after incubation at atmospheric pressure and 1.1% at high pressure after 90 d. For medium-chain alkanes (C13–C21), the relative abundance increased by 4.4% and 4.0% after incubation for 90 d at atmospheric and high pressure, respectively. For long-chain alkanes (C22–C32), the relative abundance increased by 2.1% after incubation under atmospheric pressure and decreased by 3.1% after incubation under high pressure (Figures 4 and S2). These results indicate that the microbial community tends to degrade long-chain alkanes under high pressure but not under atmospheric pressure.

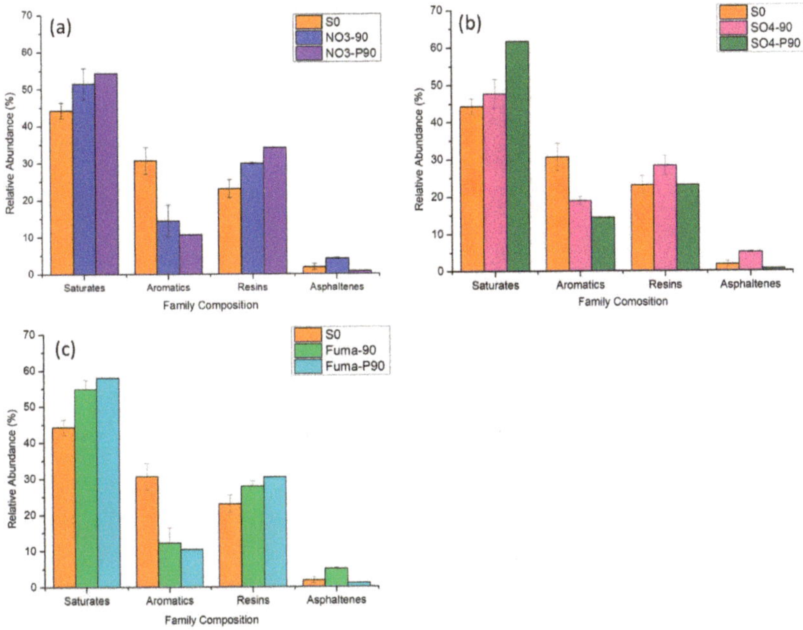

Figure 2. Relative abundance of family composition of different groups. (**a**) S0, NO_3-90, and NO_3-P90; (**b**) S0, SO_4-90, and SO_4-P90; and (**c**) S0, Fuma-90, and Fuma-P90. (The culture starting point sample was labeled S0. After 90 days of incubation, the samples incubated under atmospheric pressure were labeled as "additive-90", e.g., NO_3-90, SO_4-90, and Fuma-90, while the samples incubated under high pressure were labeled as "additive-P90", e.g., NO_3-P90, SO_4-P90, and Fuma-P90).

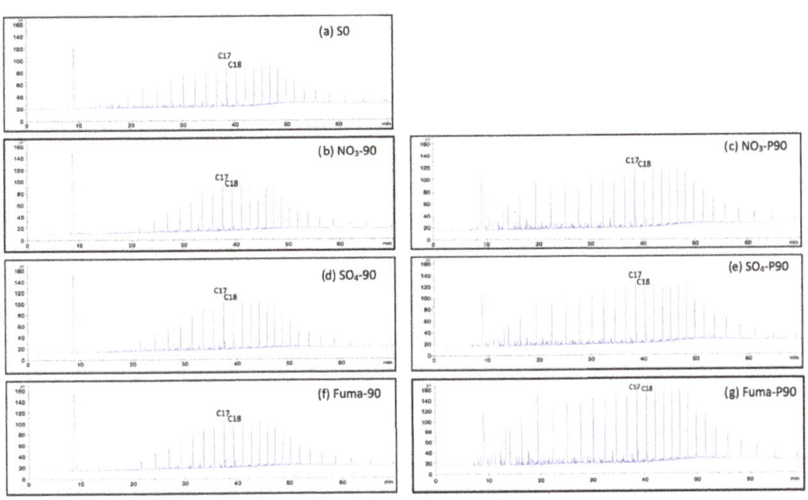

Figure 3. Gas chromatography of saturates before and after microbial degradation of crude oil. (**a**) S0, (**b**) NO_3-90, (**c**) NO_3-P90, (**d**) SO_4-90, (**e**) SO_4-P90, (**f**) Fuma-90, and (**g**) Fuma-P90. (The culture starting point sample was labeled S0. After 90 days of incubation, the samples incubated under atmospheric pressure were labeled as "additive-90", e.g., NO_3-90, SO_4-90, and Fuma-90, while the samples incubated under high pressure were labeled as "additive-P90", e.g., NO_3-P90, SO_4-P90, and Fuma-P90).

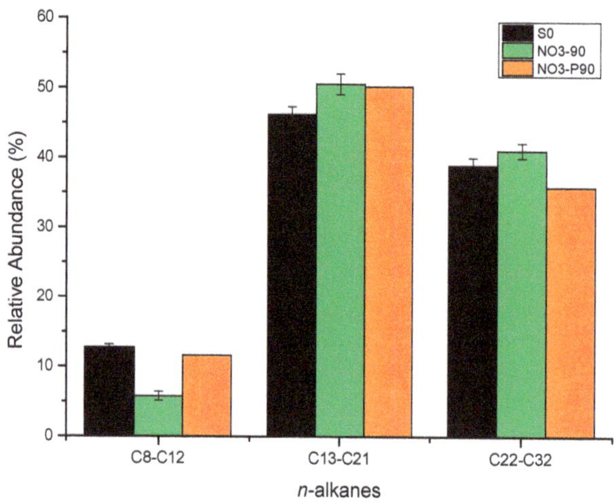

Figure 4. Relative abundance of *n*-alkanes from C8–C12, C13–C21, and C22–C32. (The culture starting point sample was labeled S0. After 90 days of incubation, the samples with NO_3^- additive incubated under atmospheric and high pressure were labeled as NO_3-90 and NO_3-P90, respectively).

3.1.3. Gas Chromatography and Mass Spectrometry of Aromatics

The results from the gas chromatography and mass spectrometry of aromatics before and after the microbial degradation of crude oil with nitrate are shown in Figure S5a–c. From these mass spectrograms, it is difficult to see a clear difference.

3.2. *Sulfate-Reducing Enrichment Group*

3.2.1. Family Composition of Crude Oil

The hydrocarbon family compositions of the atmospheric pressure group and the high-pressure group at 0 d and after 90 d of sulfate-reducing cultivation are shown in the Figure 2b. The relative abundance of saturates was 3.4% higher after 90 d at atmospheric pressure and 17.4% higher after 90 d at high pressure. The relative abundance of aromatics was 11.8% lower after 90 d at atmospheric pressure and 16.3% lower after 90 d at high pressure. The relative abundance of resins after 90 d was 5.2% lower at atmospheric pressure and 0.1% lower at high pressure. The relative abundance of asphaltenes was 3.2% higher after 90 d at atmospheric pressure and 1.2% lower after 90 d at high pressure.

3.2.2. Gas Chromatography of Saturates

The gas chromatography of saturates before and after microbial degradation of crude oil with sulfate reduction is shown in Figure S5a,d,e. For short-chain alkanes (C8–C12), the relative abundance decreased by 4.9% after incubation at atmospheric pressure and by 5.8% at high pressure. For medium-chain alkanes (C13–C21), the relative abundance increased by 3.5% after incubation at atmospheric pressure and by 6.4% after incubation at high pressure. For long-chain alkanes (C22–C32), the relative abundance increased by 1.1% after incubation at atmospheric pressure and decreased by 0.9% after incubation at high pressure (Figures 5 and S3). The results indicate that the microbial community tends to degrade long-chain alkanes at high pressure but not significantly under atmospheric pressure.

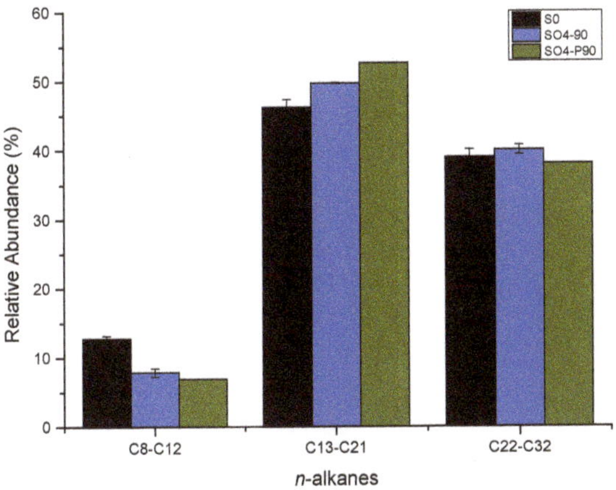

Figure 5. Relative abundance of *n*-alkanes from C8–C12, C13–C21, and C22–C32. (The culture starting point sample was labeled S0. After 90 days of incubation, the samples with SO_4^- additive incubated under atmospheric and high pressure were labeled as SO_4-90 and SO_4-P90, respectively).

3.2.3. Gas Chromatography and Mass Spectrometry of Aromatics

The results of the gas chromatography and mass spectrometry for aromatics before and after the microbial degradation of crude oil by sulfate are depicted in Figure S5a,d,e. For a comprehensive comparison of microbial degradation, Figure 6 illustrates the Total Ion Chromatography Mass Spectrometry results for aromatics over a duration ranging from 10 to 45 min. Naphthalene, C1-naphthalene, and C2-naphthalene have undergone biodegradation through sulfate reduction enrichment culture.

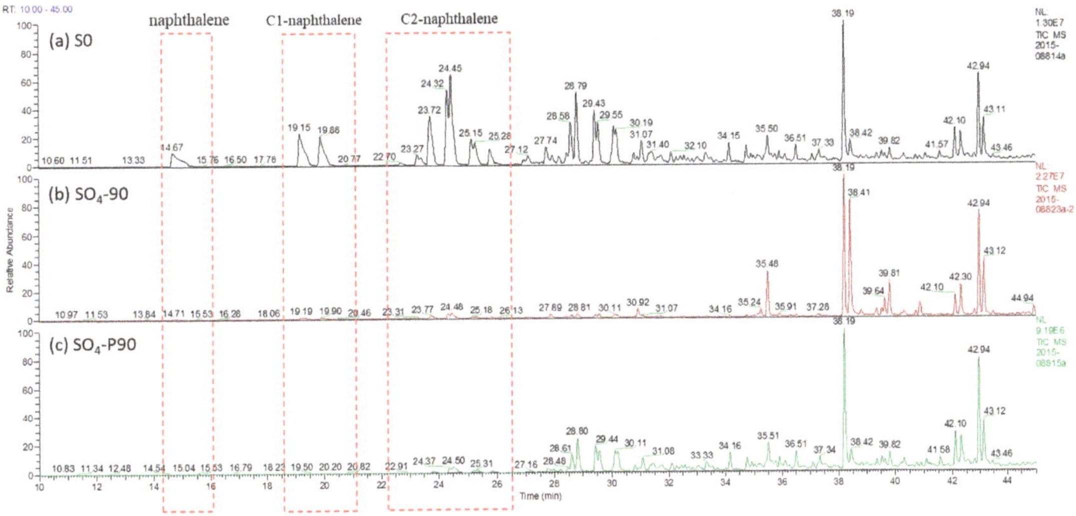

Figure 6. Total Ion Chromatography Mass Spectrometry of aromatic hydrocarbons of different samples (retention time is shown from 10 min to 45 min). (**a**) S0. (**b**) SO_4-90. (**c**) SO_4-P90. (The culture starting point sample was labeled S0. After 90 days of incubation, the samples with SO_4^- additive incubated under atmospheric and high pressure were labeled as SO_4-90 and SO_4-P90, respectively).

3.3. *Methanogenic Enrichment Group*

3.3.1. Family Composition of Crude Oil

The family compositions of hydrocarbons from the atmospheric pressure group and the high-pressure group at 0 d and after 90 d of methanogenic cultivation are shown in Figure 2c. The relative abundance of saturates was 10.5% higher at atmospheric pressure and 13.7% higher at high pressure after 90 d. The relative abundance of aromatics was 18.5% lower at atmospheric pressure and 20.3% lower at high pressure after 90 d. The relative abundance of resins at atmospheric pressure was 4.9% higher and 7.4% higher at high pressure after 90 d. The relative abundance of asphaltenes was 3.1% higher in the atmospheric pressure group and 0.8% lower in the high-pressure group.

3.3.2. Gas Chromatography of Saturates

The results from the gas chromatography of saturates before and after the microbial degradation of crude oil with sulfate are shown in Figures 3a,f,g and S1. For short-chain alkanes (C8–C12), the relative abundance was 6.1% lower after incubation at atmospheric pressure and 10.34% lower after incubation at high pressure. For medium-chain alkanes (C13–C21), the relative abundance increased by 3.1% after incubation at atmospheric pressure and by 7.6% after incubation at high pressure. For long-chain alkanes (C22–C32), the relative abundance increased by 2.7% after incubation at atmospheric pressure and by 2.2% after incubation at high pressure. These results indicate that, comparing the degradation of medium- and long-chain alkanes, the microbial community tends to degrade short-chain alkanes under high pressure as well as under atmospheric pressure (Figures 7 and S4).

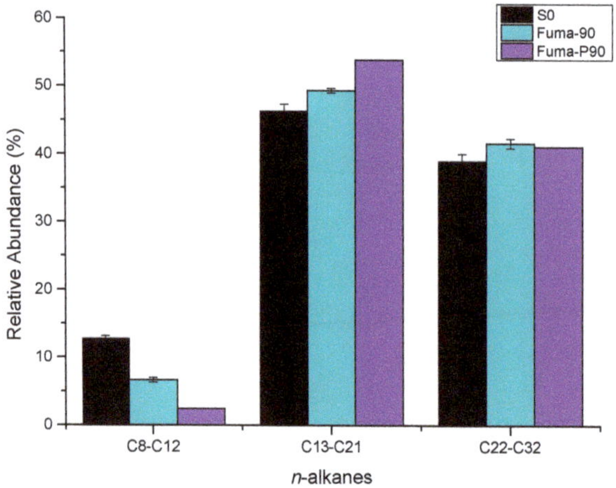

Figure 7. Relative abundance of *n*-alkanes from C8–C12, C13–C21, and C22–C32. (The culture starting point sample was labeled S0. After 90 days of incubation, the samples with fumarate additive incubated under atmospheric and high pressure were labeled as Fuma-90 and Fuma-P90, respectively).

3.3.3. Gas Chromatography and Mass Spectrometry of Aromatics

The results from the gas chromatography and mass spectrometry of aromatics before and after the microbial degradation of crude oil with nitrate are presented in Figure S5a,f,g. From these mass spectrograms, it is difficult to see a clear difference.

4. Discussion

The results from the atmospheric pressure and high-pressure incubation experiments confirm that the microbial anaerobic degradation of crude oil occurred during the 90 d incubation. However, hydrocarbon biodegradation ability under atmospheric pressure and high pressure was different.

4.1. Effect of the Biodegradation Efficiency of Oil Family

After 90 d of incubation, aromatics decreased in both the atmospheric and high-pressure groups for the nitrate-reducing, sulfate-reducing, and methanogenic enrichment cultures; the degree of decrease in the high-pressure culture group was higher than that for the atmospheric-pressure culture group. Saturates and resins increased in both the atmospheric- and high-pressure groups, except for the resins under sulfate-reducing conditions and high pressure, which increased by 0.1%. Interestingly, asphaltenes increased in the atmospheric pressure group and decreased in the high-pressure group. This indicates that asphaltenes tend to be degraded by the microbial community under high pressure. Asphaltenes are the most polar and heavy fraction of petroleum and have complex structures and toxicity [56]. Only a few microbial consortia have been reported to degrade asphaltenes. These consortia contain isolates that correspond to the genera *Rhodococcus*, *Bacillus*, *Stutzerimonas*, *Cellulosimicrobium*, *Pseudomonas*, and *Paenibacillus*, which are able to use asphaltene as a sole carbon and energy source [15,56,57]. *Bacillus*, *Pseudomonas*, and *Paenibacillus* genera were detected in the high-pressure group in this study (Table S1, sheet 1 and sheet 2). We speculated that these piezotolerant microorganisms had survived in the oil reservoir for a long time, were adapted to the high-pressure environment, and had the ability to use asphaltenes. Their ability to withstand high temperature and degrade aromatics and asphaltenes led to an increase in the saturates, subsequently improving the quality of the oil.

4.2. Effect of Oil Saturate Hydrocarbon Biodegradation Efficiency

After 90 d of incubation, the ratio of change (see Section 2.1) is shown in Figure 8. For the nitrate-reducing, sulfate-reducing, and methanogenic enrichment cultures, the *n*-alkanes C8–C12 decreased in both the atmospheric and high-pressure groups, while the *n*-alkanes C13–C21 increased in each group. The *n*-alkanes C22–C32 decreased in each group under atmospheric pressure; however, the changes in the relative abundance of this hydrocarbon fraction in each group were different under high-pressure cultivation. When nitrate and sulfate were used as electron acceptors at high pressure, the relative abundance of C22–C32 decreased. In the methanogenic enrichment cultures at high pressure, the relative abundance of the C22–C32 fraction increased by 2.2% and the ratio of change was 5.6%, while the relative abundance of the C8–C12 fraction decreased by 10.3% with a ratio of change was −81.0%. In this set of experiments, the relative abundance of short-chain alkanes may have decreased, resulting in an increase in the relative abundance of medium- and long-chain alkanes. Short-chain saturates tended to be degraded in both the atmospheric- and high-pressure groups, and long-chain saturates tended to be degraded at high pressure. Previous studies have reported the degradation of hydrocarbons by the genus *Dietzia*. *Dietzia* sp. DQ12-45-1b used C6–C40 *n*-alkanes as carbon and energy sources [58]; *Dietzia* sp. E1 used C6–C36 alkanes [59]; *D. cinnamea* P4 used C11–C36 alkanes [60]; and *Dietzia maris* AURCCBT01 used C14, C18, C20, C28, and C32 alkanes [61]. Consortia with dominant *Geobacillus*, *Parageobacillus*, and *Anoxybacillus* genera have exhibited a strong ability to degrade hydrocarbons in long-chain alkanes (C18–C40) [62]. The consortia of *Ochrobactrum* sp., *Pseudomonas aeruginosa*, and *Bacillus* sp. have the ability to use C11–C18 alkanes, with a removal rate up to 78.5%, and the removal rates of C26–C29 and C33–C35 alkanes were 36.2% and 30.5%, respectively. In addition, possible metabolic pathways of crude oil degradation were also proposed by this study, including reactions for aldehyde bond (H-C=O) and ketone bond (C-O-C) hydrodehydration, ring-opening of cyclic ether, C-C bond oxidation of benzene rings, and ring-opening hydrolysis of gentisic acid [63]. *Rhodococcus qingshengii* and *Alcanivorax venustensis* incubated in a seawater-based medium degraded 100% of C9–C12 and C16–C29 hydrocarbons and 85% of C13–C15 hydrocarbons [64]. The biodegradation of crude oil paraffin wax was demonstrated by

11 bacteria (including the genera *Geobacillus*, *Parageobacillus*, and *Anoxybacillus*) isolated from seawater and oil-contaminated soil samples; these bacteria could completely degrade C37–C40 alkanes and increase the ratio of C14–C18. In addition, enzymes associated with the biodegradation of crude oil, including alkane monooxygenase, alcohol dehydrogenase, lipase, and esterase, were detected [65]. In our study, the genera *Dietzia*, *Ochrobactrum*, *Pseudomonas*, and *Bacillus* were detected based on 16S rRNA, and their relative abundances are shown in Figure 9. Although the total relative abundance of the oil-biodegrading microorganisms incubated at atmospheric pressure was higher than that of those incubated at high pressure, the oil-degrading ability of the former was not significantly better than the latter. High pressure and high temperature in the in situ oil reservoir may be important for the function of microbial communities.

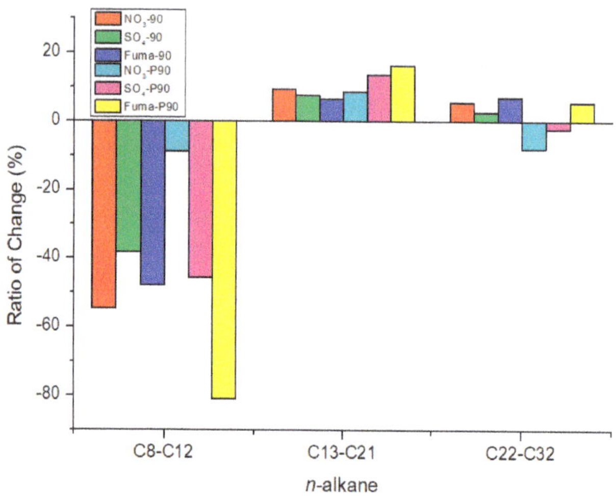

Figure 8. The ratio of change for *n*-alkanes from C8–C12, C13–C21, and C22–C32. (After 90 days of incubation, the samples incubated under atmospheric pressure were labeled as "additive-90", e.g., NO_3-90, SO_4-90, and Fuma-90, while the samples incubated under high pressure were labeled as "additive-P90", e.g., NO_3-P90, SO_4-P90, and Fuma-P90).

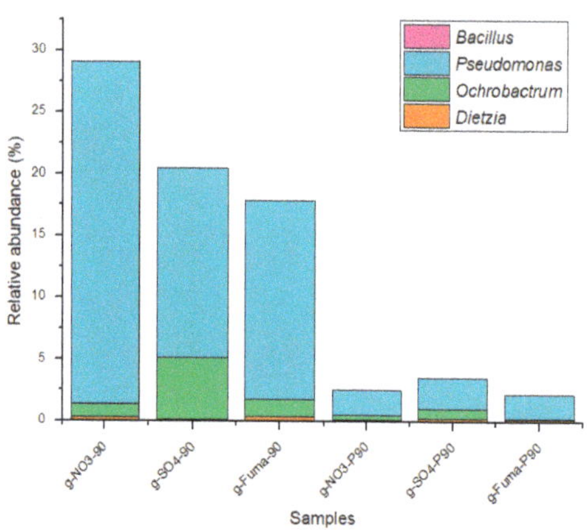

Figure 9. The composition and relative abundance of the genera *Bacillus*, *Pseudomonas*, *Ochrobactrum*, and *Dietzia* in different samples.

4.3. Effect of Oil Aromatic Hydrocarbon Biodegradation Efficiency

From the results of the Total Ion Chromatography Mass Spectrometry of aromatic hydrocarbons, degradation under atmospheric pressure and high pressure was significant when using sulfate as an electron acceptor. In Figure 6, naphthalene, C1-naphthalene (including 2-methylnaphthalene and 1-methylnaphthalene), and C2-naphthalene (including 2-ethylnaphthalene, 2,6-dimethylnaphthalene, 2,7-dimethylnaphthalene, 1,3-dimethylnaphthalene, 1,7-dimethylnaphthalene, 1,6-dimethylnaphthalene, 1,4-dimethylnaphthalene, 2,3-dimethylnaphthalene, and 1,2-dimethylnaphthalene) were largely degraded. Chen et al. [47] reported that several naphthalene-degrading consortia can use bicyclic PAHs (i.e., naphthalene) as a sole carbon source, mainly under sulfate-reducing conditions. Figure 10 presents the sulfate-reducing bacteria found in our study, including *Desulfomicrobium*, *Desulfovibrio*, *Desulfotomaculum_Desulfovirgula*, and *Thermodesulfovibrio* (Table S1, sheet1 and sheet 3). The abundance of sequences related to the genus *Thermodesulfovibrio* was 20.8% and 2.0%, respectively, under high and atmospheric pressure with sulfate as an electron acceptor. The ability of these genera to withstand high temperature and degrade bicyclic PAHs results in an improvement in the quality of the oil.

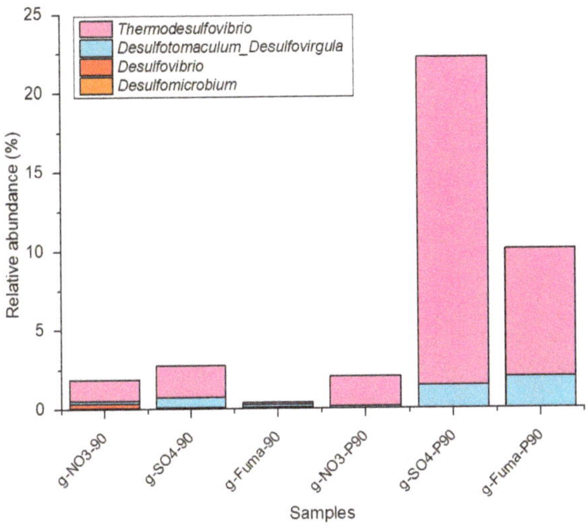

Figure 10. The composition and relative abundance of the genera *Thermodesulfovibrio*, *Desulfotomaculum_Desulfovirgula*, *Desulfovibrio*, and *Desulfomicrobium* in different samples.

5. Conclusions

In summary, we measured the change in oil composition to determine the microbial hydrocarbon degradation ability under high pressure in nitrate-reducing, sulfate-reducing, and methanogenic enrichment cultures. In the nitrate-reducing and sulfate-reducing enrichment cultures, the relative abundance of asphaltenes in the atmospheric pressure group was higher after 90 d incubation than at 0 d, while the asphaltenes in the high-pressure group decreased over the 90 d of incubation. The microbial community tended to degrade long-chain alkanes (C22–C32) under high pressure but not under atmospheric pressure. In the sulfate-reducing enrichment, the degradation of aromatics under atmospheric pressure and high pressure was significant. In the methanogenic enrichments, the relative abundance of asphaltenes in the atmospheric pressure group was higher after 90 d incubation, and was lower after 90 d at high pressure. The relative abundance of the C8–C12 alkane fraction decreased by a great degree under high-pressure incubation. This study has enriched our current knowledge of the oil-degradation profile of microbial communities under high-temperature and high-pressure conditions.

Supplementary Materials: The following supporting information can be downloaded at: https://www.mdpi.com/article/10.3390/microorganisms12081543/s1. Table S1. Composition and relative abundance of different genera in all samples; Figure S1. Gas chromatography of saturates before and after microbial degradation of crude oil; Figure S2. Relative abundance of *n*-alkanes from C8–C32 with NO_3^- additive; Figure S3. Relative abundance of *n*-alkanes from C8–C32 with SO_4^{2-} additive; Figure S4. Relative abundance of n-alkanes from C8–C32 with fumarate additive; Figure S5. Total Ion Chromatography Mass Spectrometry of aromatic hydrocarbon from 7 min to 60 min.

Author Contributions: Conceptualization, L.W., X.S. and X.W.; methodology, Y.N., X.C., J.X., Z.J., W.S. and X.W.; formal analysis, L.W.; investigation, L.W. and J.X.; writing—original draft preparation, L.W.; writing—review and editing, L.W., X.S. and X.-L.W.; Supervision, X.S. and X.-L.W. All authors have read and agreed to the published version of the manuscript.

Funding: This research was funded by Scientific Research and Technological Development Project of Research Institute of Petroleum Exploration & Development Company Limited, CNPC (Grant No. 2023ycq08); National Key Research and Development Program of China (Grants No. 2023YFF0614100 and No. 2023YFF0614101), and PetroChina Major Scientific and Technological Project (Grants No. 2021ZZ01-03 and No. 2021ZZ01-05).

Data Availability Statement: The original contributions presented in the study are included in the article, further inquiries can be directed to the corresponding authors.

Acknowledgments: The authors offer their sincere thanks to Ming Gao for funding support.

Conflicts of Interest: The authors declare no conflicts of interest.

References

1. Tsimplis, M.; Noussia, K. The Use of Ships Within a CCUS System: Regulation and Liability. *Resour. Conserv. Recycl.* **2022**, *181*, 106218. [CrossRef]
2. Zheng, Y.W.; Gao, L.; Li, S.; Wang, D. A Comprehensive Evaluation Model for Full-chain CCUS Performance Based on the Analytic Hierarchy Process Method. *Energy* **2022**, *239*, 122033. [CrossRef]
3. Chen, S.Y.; Liu, J.F.; Zhang, Q.; Teng, F.; McLellan, B.C. A Critical Review on Deployment Planning and Risk Analysis of Carbon Capture, Utilization, and Storage (CCUS) Toward Carbon Neutrality. *Renew. Sustain. Energy Rev.* **2022**, *167*, 112537. [CrossRef]
4. Cai, B.F.; Li, Q.; Zhang, X.; Cao, C.; Cao, L.B.; Chen, W.H.; Chen, Z.J.; Dong, J.C.; Fan, J.L.; Jiang, Y.; et al. *Annual Report of China's CO_2 Capture, Utilization and Storage (CCUS) (2021)—China CCUS Path Study*; Chinese Academy of Environmental Planning of the Ministry of Ecology and Environment, Institute of Rock and Soil Mechanics, Chinese Academy of Science, the Administration Center for China's Agenda 21: Beijing, China, 2021.
5. Tyne, R.L.; Barry, P.H.; Lawson, M.; Lloyd, K.G.; Giovannelli, D.; Summers, Z.M.; Ballentine, C.J. Identifying and Understanding Microbial Methanogenesis in CO_2 Storage. *Environ. Sci. Technol.* **2023**, *57*, 9459–9473. [CrossRef]
6. Mayumi, D.; Dolfing, J.; Sakata, S.; Maeda, H.; Miyagawa, Y.; Ikarashi, M.; Tamaki, H.; Takeuchi, M.; Nakatsu, C.H.; Kamagata, Y.C. Carbon Dioxide Concentration Dictates Alternative Methanogenic Pathways in Oil Reservoirs. *Nat. Commun.* **2013**, *2998*, 1998. [CrossRef]
7. Vilcaez, J.; York, J.; Youssef, N.; Elshahed, M. Stimulation of Methanogenic Crude Oil Biodegradation in Depleted Oil Reservoirs. *Fuel* **2018**, *232*, 581–590. [CrossRef]
8. Morozova, D.; Wandrey, M.; Alawi, M.; Zimmer, M.; Vieth, A.; Zettlitzer, M.; Würdemann, H. Monitoring of the Microbial Community Composition in Saline Aquifers During CO_2 Storage by Fluorescence in situ Hybridization. *Int. J. Greenh. Gas Control* **2010**, *4*, 981–989. [CrossRef]
9. Ma, L.; Liang, B.; Wang, L.Y.; Zhou, L.; Mbadinga, S.M.; Gu, J.D.; Mu, B.Z. Microbial Reduction of CO_2 from Injected $NaH^{13}CO_3$ with Degradation of *n*-hexadecane in the Enrichment Culture Derived from a Petroleum Reservoir. *Int. Biodeterior. Biodegrad.* **2018**, *127*, 192–200. [CrossRef]
10. Liang, B.; Wang, L.Y.; Zhou, Z.C.; Mbadinga, S.M.; Zhou, L.; Liu, J.F.; Yang, S.Z.; Gu, J.D.; Mu, B.Z. High Frequency of *Thermodesulfovibrio* spp. and *Anaerolineaceae* in Association with *Methanculleus* spp. In a Long-term Incubation of *n*-alkanes-degrading Methanogenic Enrichment Culture. *Front. Microbiol.* **2016**, *7*, 1431. [CrossRef]
11. Ning, D.; Wang, Y.J.; Fan, Y.P.; Wang, J.J.; Van Nostrand, J.D.; Wu, L.Y.; Zhang, P.; Curtis, D.J.; Tian, R.M.; Lui, L.; et al. Environmental Stress Mediates Groundwater Microbial Community Assembly. *Nat. Microbiol.* **2024**, *9*, 490–501. [CrossRef]
12. Rojo, F. Degradation of Alkanes by Bacteria. *Environ. Microbiol.* **2009**, *11*, 2477–2490. [CrossRef] [PubMed]
13. Cao, Y.Q.; Zhu, Z.W.; Song, X.; Cai, Q.H.; Chen, B.; Dong, G.H.; Ye, X.D. Microbial Eco-physiological Strategies for Salinity-mediated Crude Oil Biodegradation. *Sci. Total Environ.* **2020**, *727*, 138723. [CrossRef] [PubMed]
14. Chandrasekar, S.; Sorial, G.A.; Weaver, J.W. Dispersant Effectiveness on Oil Spills-Impact of Salinity. *ICES J. Mar. Sci.* **2006**, *63*, 1418–1430. [CrossRef]

15. Tavassoli, T.; Mousavi, S.M.; Shojaosadati, S.A.; Salehizadeh, H. Asphaltene Biodegradation Using Microorganisms Isolated from Oil Samples. *Fuel* **2012**, *93*, 142–148. [CrossRef]
16. Korenblum, E.; Souza, D.B.; Penna, M.; Seldin, L. Molecular Analysis of the Bacterial Communities in Crude Oil Samples from Two Brazilian Offshore Petroleum Platforms. *Int. J. Microbiol.* **2012**, *2012*, 156537. [CrossRef] [PubMed]
17. Shin, B.; Kim, M.; Zengler, K.; Chin, K.J.; Overholt, W.A.; Gieg, L.M.; Konstantinidis, K.T.; Kostka, J.E. Anaerobic Degradation of Hexadecane and Phenanthrene Coupled to Sulfate Reduction by Enriched Consortia from Northern Gulf of Mexio Seafloor Sediment. *Sci. Rep.* **2019**, *9*, 26567.
18. Zhang, K.; Hu, Z.; Zeng, F.F.; Yang, X.J.; Wang, J.J.; Jing, R.; Zhang, H.N.; Li, Y.T.; Zhang, Z. Biodegradation of Petroleum Hydrocarbons and Changes in Microbial Community Structure in Sediment Under Nitrate-, Ferric-, Sulfate-reducing and Methanogenic Conditions. *J. Environ. Manag.* **2019**, *249*, 109425. [CrossRef] [PubMed]
19. Pavlova, O.N.; Izosimova, O.N.; Chernitsyna, S.M.; Ivanov, V.G.; Pogodaeva, T.V.; Khabuev, A.V.; Gorshkov, A.G.; Zemskaya, T.I. Anaerobic Oxidation of Petroleum Hydrocarbons in Enrichment Cultures from Sediments of the Gorevoy Utes Natural Oil Seep under Methanogenic and Sulfate-reducing Conditions. *Microb. Ecol.* **2022**, *83*, 899–915. [CrossRef]
20. Ma, T.T.; Liu, L.Y.; Rui, J.P.; Yuan, Q.; Feng, D.S.; Zhou, Z.; Dai, L.R.; Zeng, W.Q.; Zhang, H.; Cheng, L. Coexistence and Competition of Sulfate-reducing and Methanogenic Populations in an Anaerobic Hexadecane-degrading Culture. *Biotechnol. Biofuels* **2017**, *10*, 207. [CrossRef]
21. Liu, J.F.; Zhang, K.; Liang, B.; Zhou, Z.C.; Yang, S.Z.; Li, W.; Hou, Z.W.; Wu, X.L.; Gu, J.D.; Mu, B.Z. Key Players in the Methanogenic Biodegradation of *n*-hexadecane Identified by DNA-Stable Isotope Probing. *Int. Biodeterior. Biodegrad.* **2019**, *143*, 104709. [CrossRef]
22. Hasinger, M.; Scherr, K.E.; Lundaa, T.; Brauer, L.; Zach, C.; Loibner, A.P. Changes in Iso- and *n*-alkane Distribution During Biodegradation of Crude Oil Under Nitrate and Sulphate Reducing Conditions. *J. Biotechnol.* **2012**, *157*, 490–498. [CrossRef] [PubMed]
23. Xu, J.B.; Lu, W.; Lv, W.F.; Song, X.M.; Nie, Y.; Wu, X.L. Metabolic Profiling of Petroleum-degrading Microbial Communities Incubated under High-pressure Conditions. *Front. Microbiol.* **2023**, *14*, 1305731. [CrossRef] [PubMed]
24. Marietou, A.; Chastain, R.; Beulig, F.; Scoma, A.; Hazen, T.C.; Bartlett, D.H. The Effect of Hydrostatic Pressure on Enrichments of Hydrocarbon Degrading Microbes from the Gulf of Mexico Following the Deepwater Horizon Oil Spill. *Front. Microbiol.* **2018**, *9*, 808.
25. Fasca, H.; Castilho, L.V.A.; Castilho, J.F.M.; Pasqualino, I.P.; Alvarez, V.M.; Azevedo Jurelevicius, D.; Seldin, L. Response of Marine Bacteria to Oil Contamination and to High Pressure and Low Temperature Deep Sea Conditions. *MicrobiologyOpen* **2018**, *7*, e00550. [CrossRef] [PubMed]
26. Calderon, L.J.P.; Gontikaki, E.; Potts, L.D.; Shaw, S.; Gallego, A.; Anderson, J.A.; Witte, U. Pressure and Temperature Effects on Deep-sea Hydrocarbon-degrading Microbial Communities in Subarctic Sediments. *MicrobiologyOpen* **2019**, *8*, e00768. [CrossRef] [PubMed]
27. Mayumi, D.; Mochimaru, H.; Yoshioka, H.; Sakata, S.; Maeda, H.; Miyagawa, Y.; Ikarashi, M.; Takeuchi, M.; Kamagata, Y. Evidence for Syntrophic Acetate Oxidation Coupled to Hydrogenotrophic Methanogenesis in the High-temperature Petroleum Reservoir of Yabase Oil Field (Japan). *Environ. Microbiol.* **2010**, *13*, 1995–2006. [CrossRef]
28. Suda, K.; Ikarashi, M.; Tamaki, H.; Tamazawa, S.; Sakata, S.; Haruo, M.; Kamagata, Y.; Kaneko, M.; Ujiie, T.; Shinotsuka, Y.; et al. Methanogenic Crude Oil Degradation Induced by an Exogenous Microbial Community and Nutrient Injections. *J. Pet. Sci. Eng.* **2021**, *201*, 108458. [CrossRef]
29. Barbato, M.; Scoma, A. Mild Hydrostatic-pressure (15 MPa) Affects the Assembly, But Not the Growth, of Oil-degrading Coastal Microbial Communities Tested under Limiting Conditions (5 °C, no added nutrients). *FEMS Microbiol. Ecol.* **2020**, *96*, fiaa160. [CrossRef]
30. Crescenzo, C.D.; Marzocchella, A.; Karatza, D.; Chianese, S.; Musmarra, D. Autogenerative High-pressure Anaerobic Digestion Modelling of Volatile Fatty Acids: Effect of Pressure Variation and Substrate Composition on Volumetric Mass Transfer Coefficients, Kinetic Parameters, and Process Performance. *Fuel* **2024**, *358*, 130144. [CrossRef]
31. Merkle, W.; Baer, K.; Lindner, J.; Zielonka, S.; Ortloff, F.; Graf, F.; Kolb, T.; Jungbluth, T.; Lemmer, A. Influence of Pressures Up to 50 Bar on Two-stage Anaerobic Digestion. *Bioresour. Technol.* **2017**, *232*, 72–78. [CrossRef]
32. Siciliano, A.; Limonti, C.; Curcio, G.M. Perfromance Evaluation of Pressurized Anaerobic Digestion (PDA) of Raw Compost Leachate. *Fermentation* **2022**, *8*, 15. [CrossRef]
33. Jones, D.M.; Head, I.M.; Gray, N.D.; Adams, J.J.; Rowan, A.K.; Aitken, C.M.; Bennett, B.; Huang, H.; Brown, A.; Bowler, B.F.J.; et al. Crude-oil Biodegradation via Methanogenesis in Subsurface Petroleum Reservoirs. *Nature* **2008**, *451*, 176–181. [CrossRef] [PubMed]
34. Milkov, A.V. Worldwide Distribution and Significance of Secondary Microbial Methane Formed During Petroleum Biodegradation in Conventional Reservoirs. *Org. Geochem.* **2011**, *42*, 184–207. [CrossRef]
35. Xue, J.L.; Yu, Y.; Bai, Y.; Wang, L.P.; Wu, Y.N. Marine Oil-degrading Microorganisms and Biodegradation Process of Petroleum Hydrocarbon in Marine Environments: A Review. *Curr. Microbiol.* **2015**, *71*, 220–228. [CrossRef] [PubMed]
36. Cheng, L.; Shi, S.; Li, Q.; Chen, J.; Zhang, H.; Lu, Y. Progressive Degradation of Crude Oil *n*-alkanes Coupled to Methane Production under Mesophilic and Thermophilic Conditions. *PLoS ONE* **2014**, *9*, e113253. [CrossRef] [PubMed]

37. Ehmedan, S.S.; Ibrahim, M.K.; Azzam, A.M.; Hamedo, H.A.; Saeed, A.M. Acceleration the Bacterial Biodegradation of Crude Oil Pollution Sing Fe_2O_3 and ZnO Nanoparticles. *Environ. Nanotechnol. Monit. Manag.* **2021**, *16*, 100613.
38. Harayama, S.; Kishira, H.; Kasai, Y.; Shutsubo, K. Petroleum Biodegradation in Marine Environments. *J. Mol. Microbiol. Biotechnol.* **1999**, *1*, 63–70. [PubMed]
39. Chuah, L.F.; Chew, K.W.; Bokhari, A.; Mubashir, M.; Show, P.L. Biodegradation of Crude Oil in Seawater by Using a Consortium of Symbiotic Bacteria. *Environ. Res.* **2022**, *213*, 113721. [CrossRef]
40. Head, I.M.; Jones, D.M.; Larter, S.R. Biological Activity in the Deep Subsurface and the Origin of Heavy Oil. *Nature* **2003**, *426*, 344–352. [CrossRef]
41. Guo, Z.Z.; Kang, Y.; Hu, Z.; Liang, S.; Xie, H.J.; Ngo, H.H.; Zhang, J. Removal Pathways of Benzofluoranthene in a Constructed Wetland Amended with Metallic Ions Embedded Carbon. *Bioresour. Technol.* **2020**, *311*, 123481. [CrossRef]
42. Han, X.K.; Wang, F.W.; Zhang, D.J.; Feng, T.; Zhang, L.L. Nitrate-assisted Biodegradation of Polycyclic Aromatic Hydrocarbons (PAHs) in the Water-level-fluctuation Zone of the Three Gorges Reservoir, China: Insights from in Situ Microbial Interaction Analyses and a Microcosmic Experiment. *Environ. Pollut.* **2021**, *268*, 115693. [CrossRef]
43. Sun, J.; Zhang, Z.T.; Wang, H.; Rogers, M.J.; Guo, H.J.; He, J.Z. Exploration of the Biotransformation of Phenanthrene Degradation Coupled with Methanogensis by Metabolites and Enzyme Analyses. *Environ. Pollut.* **2022**, *293*, 1184491. [CrossRef]
44. Bianco, F.; Monteverde, G.; Race, M.; Papirio, S.; Esposito, G. Comparing Performances, Costs and Energy Balance of Ex Situ Remediation Processes for PAH-contaminated Marine Sediments. *Environ. Sci. Pollut. Res.* **2020**, *27*, 19363–19374. [CrossRef] [PubMed]
45. Dhar, K.; Subashchandrabose, S.R.; Venkateswarlu, K.; Krishnan, K.; Megharaj, M. Anaerobic Microbial Degradation of Polycyclic Aromatic Hydrocarbons: A Comprehensive Review. In *Reviews of Environmental Contanination and Toxicology*; De Voogt, P., Ed.; Springer International Publishing: Cham, Switzerland, 2020; Volume 251, pp. 25–108.
46. Mu, J.; Chen, Y.; Song, Z.; Liu, M.; Zhu, B.K.; Tao, H.C.; Bao, M.T.; Chen, Q.G. Effect of Terminal Electron Acceptors on the Anaerobic Biodegradation of PAHs in Marine Sediments. *J. Hazard. Mater.* **2022**, *438*, 129569. [CrossRef]
47. Chen, C.; Zhang, Z.; Xu, P.; Hu, H.Y.; Tang, H.Z. Anaerobic Biodegradtion of Polycyclic Aromatic Hydrocarbons. *ACS Earth Space Chem.* **2023**, *7*, 823–837. [CrossRef]
48. Ji, J.H.; Liu, Y.F.; Zhou, L.; Mbadinga, S.M.; Pan, P.; Chen, J.; Liu, J.F.; Yang, S.Z.; Sand, W.F.; Gu, J.D.; et al. Methanogenic Degradation of Long *n*-Alkanes Requires Fumarate-dependent Activation. *Appl. Environ. Microbiol.* **2019**, *85*, e00985-19. [CrossRef] [PubMed]
49. Ji, J.H.; Liu, Y.F.; Zhou, L.; Irfan, M.; Mbadinga, S.M.; Pan, P.; Chen, J.; Liu, J.F.; Yang, S.Z.; Sand, W.; et al. Methanogenic Biodegradation of C13 and C14 *n*-alkanes Activated by Addition to Fumarate. *Int. Biodeterior. Biodegrad.* **2020**, *153*, 104994. [CrossRef]
50. Ji, J.H.; Zhou, L.; Mbadinga, S.M.; Irfan, M.; Liu, Y.F.; Pan, P.; Qi, Z.Z.; Chen, J.; Liu, J.F.; Yang, S.Z.; et al. Methanogenic Biodegradation of C9 to C12 *n*-alkanes Initiated by *Smithella* via Fumarate Addition Mechanism. *AMB Express* **2020**, *10*, 23. [CrossRef]
51. Chen, J.; Zhou, L.; Liu, Y.F.; Hou, Z.W.; Li, W.; Mbadainga, S.M.; Zhou, J.; Yang, T.; Liu, J.F.; Yang, S.Z.; et al. Synthesis and Mass Spectra of Rearrangement Bio-signature Metabolites of Anaerobic Alkane Degradation via Fumarate Addition. *Anal. Biochem.* **2020**, *600*, 113746. [CrossRef]
52. Zhang, L.; Zhou, X.Y.; Hu, C.X.; Yao, S.; Shi, L.; Niu, T.; Li, X.; Tong, L.H.; Zhang, T.; Xia, W.J. CO_2 Improves the Anaerobic Biodegradation Intensity and Selectivity of Heterocyclic Hydrocarbons in Heavy Oil. *Environ. Res.* **2023**, *224*, 115541. [CrossRef]
53. Panda, S.K.; Andersson, J.T.; Schrader, W. Mass-spectrometric Analysis of Complex Volatile and Nonvolatile Crude Oil Components: A Challenge. *Anal. Bioanal. Chem.* **2007**, *389*, 1329–1339. [CrossRef] [PubMed]
54. *SY/T 5119-2016*; Analysis Method for Family Composition of Rock Extracts and Crude Oil. National Energy Administration: Beijing, China, 2016.
55. *SY/T 5779-2008*; Analytical and Method of Hydrocarbons in Petroleum and Sediment by Gas Chromatography. National Development and Reform Commission: Beijing, China, 2008.
56. Navas-Cáceres, O.D.; Parada, M.; Zafra, G. Development of a Highly Tolerant Bacterial Consortium for Asphaltene Biodegradation in Soils. *Environ. Sci. Pollut. Res.* **2023**, *30*, 123439–123451. [CrossRef] [PubMed]
57. Das, S.; Das, N.; Choure, K.; Pandey, P. Biodegradation of Asphaltene by Lipopeptide-biosurfactant Producing Hydrocarbonoclastic, Crude Oil Degrading *Bacillus* spp. *Bioresour. Technol.* **2023**, *382*, 129198. [CrossRef] [PubMed]
58. Wang, X.B.; Chi, C.Q.; Nie, Y.; Tang, Y.Q.; Tan, Y.; Wu, G.; Wu, X.L. Degradation of Petroleum Hydrocarbons(C6–C40) and Crude oil by Novel *Dietzia* Strain. *Bioresour. Technol.* **2011**, *102*, 7755–7761. [CrossRef]
59. Bihari, Z.; Szvetnik, A.; Szabo, Z.; Blastyak, A.; Zombori, Z.; Balazs, M. Functional Analysis of Long-chain *n*-alkane Degradation by *Dietzia* spp. *FEMS Microbiol. Lett.* **2011**, *316*, 100–107. [CrossRef]
60. Weid, I.; Marques, J.M.; Cunha, C.D.; Lippi, R.K.; Santos, S.; Rosado, A.S.; Lins, U.; Seldin, L. Identification and Biodegradation Potential of a Novel Strain of *Dietzia cinnamea* Isolated from a Petroleum Contaminated Tropical Soil. *Syst. Appl. Microbiol.* **2007**, *30*, 331–339. [CrossRef]
61. Venil, C.K.; Malathi, M.; Devi, P.R. Characterization of *Dietzia maris* AURCCBT01 from Oil-contaminated Soil for Biodegradation of Crude Oil. *3 Biotech* **2021**, *11*, 291. [CrossRef]

62. Raja Abd Rahman, R.N.Z.; Latip, W.; Adlan, N.A.; Sabri, S.; Mohamad Ali, M.S. Bacteria Consortia Enhanced Hydrocarbon Degradation of Waxy Crude Oil. *Arch. Microbiol.* **2022**, *204*, 701.
63. Tang, F.; Zhang, H.; Cheng, H.; Wang, Y.R.; Liu, Q.Y.; Zhao, C.C.; Gu, Y.Y.; Wang, J.G. New Insights of Crude Oil Biodegradation Construction by Microbial Consortium B10: Responded Substrates, Genomics, Biodegradation Mechanism and Pathways. *Chem. Eng. J.* **2023**, *478*, 147143. [CrossRef]
64. Baltaci, M.O.; Omeroglu, M.A.; Ozkan, H.; Taskin, M.; Adiguzel, A. Enhanced Biodegradation of Crude Oil Contamination by Indigenous Bacterial Consortium under Real Conditions. *Biocatal. Biotransform.* **2024**, *42*, 56–67. [CrossRef]
65. Adlan, N.A.; Sabri, S.; Masomian, M.; Mohamad Ali, M.S.; Raja Abd Rahman, R.N.Z. Microbial Biodegradation of Paraffin Wax in Malaysian Crude Oil Mediated by Degradative Enzymes. *Front. Microbiol.* **2020**, *11*, 565608. [CrossRef] [PubMed]

microorganisms

MDPI

Article

Bacterial and Archaeal Communities in Erhai Lake Sediments: Abundance and Metabolic Insight into a Plateau Lake at the Edge of Eutrophication

Zhen Xie [1], Wei Li [2,3], Kaiwen Yang [1], Xinze Wang [2,3], Shunzi Xiong [2,3] and Xiaojun Zhang [1,*]

1 State Key Laboratory of Microbial Metabolism, and Joint International Research Laboratory of Metabolic & Developmental Sciences, School of Life Sciences and Biotechnology, Shanghai Jiao Tong University, Shanghai 200240, China; xiezhen07@sjtu.edu.cn (Z.X.); kevinyang97@outlook.com (K.Y.)
2 National Observation and Research Station of Erhai Lake Ecosystem in Yunnan, Dali 671000, China; andynee@foxmail.com (W.L.); xinzewang@sjtu.edu.cn (X.W.); xiongshunzi@sjtu.edu.cn (S.X.)
3 Yunnan Dali Research Institute, Shanghai Jiao Tong University, Dali 671000, China
* Correspondence: xjzhang68@sjtu.edu.cn; Tel.: +86-21-34204878

Abstract: The littoral zones of lakes are potential hotspots for local algal blooms and biogeochemical cycles; however, the microbial communities within the littoral sediments of eutrophic plateau lakes remain poorly understood. Here, we investigated the taxonomic composition, co-occurrence networks, and potential functional roles of both abundant and rare taxa within bacterial and archaeal communities, as well as physicochemical parameters, in littoral sediments from Erhai Lake, a mesotrophic lake transitioning towards eutrophy located in the Yunnan–Guizhou Plateau. 16S rRNA gene sequencing revealed that bacterial communities were dominated by Proteobacteria, Bacteroidetes, and Chloroflexi, while Euryarchaeota was the main archaeal phylum. Co-occurrence network analysis revealed that keystone taxa mainly belonged to rare species in the bacterial domain, but in the archaeal domain, over half of keystone taxa were abundant species, demonstrating their fundamental roles in network persistence. The rare bacterial taxa contributed substantially to the overall abundance (81.52%), whereas a smaller subset of abundant archaeal taxa accounted for up to 82.70% of the overall abundance. Functional predictions highlighted a divergence in metabolic potentials, with abundant bacterial sub-communities enriched in pathways for nitrogen cycling, sulfur cycling, and chlorate reduction, while rare bacterial sub-communities were linked to carbon cycling processes such as methanotrophy. Abundant archaeal sub-communities exhibited a high potential for methanogenesis, chemoheterotrophy, and dark hydrogen oxidation. Spearman correlation analysis showed that genera such as *Candidatus competibacter*, *Geobacter*, *Syntrophobacter*, *Methanocella*, and *Methanosarcina* may serve as potential indicators of eutrophication. Overall, this study provides insight into the distinct roles that rare and abundant taxa play in the littoral sediments of mesotrophic plateau lakes.

Keywords: bacterial and archaeal communities; rare and abundant taxa; metabolic potential; eutrophication indicators; mesotrophic plateau lake

1. Introduction

Eutrophication is a serious threat to the health of lacustrine ecosystems worldwide, reducing aquatic biodiversity and affecting the ecological function of lakes [1]. Microorganisms are essential components of lake ecosystems that play important roles in biogeochemical processes by participating in energy flows, materials cycling, elemental transformation, and pollutant biodegradation [2]. As a sink for many nutrients with the potential to cause secondary pollution of water bodies, lake sediment is an important pollution source that has become a key issue of concern in lakes [3]. The internal release of nitrogen (N) and phosphorus (P) from sediments, which is predominantly mediated by microbial communities, is the main nutrient source and one of the key causes of lake eutrophication [4], which emphasizes the importance of microbial communities in lake ecosystems.

The content and distribution of sediment nutrients are affected by sampling location and human disturbance. Littoral zones are transitional boundaries or ecotones between terrestrial and aquatic ecosystems that contain a variety of microbes, submerged plants, and animal communities, making them heterogeneous areas with higher niche diversity than pelagic zones [5]. However, littoral zones are potential hotspots for localized algal blooms and greenhouse gas emissions. For example, the contribution of the littoral zone to sediment methane emissions was found to be comparable to that of the profundal zone of Taihu Lake, despite being much smaller [6]. Similarly, littoral zones have been shown to contribute to as much degradation of organic matter in Lake Constance as the pelagic water column, despite the littoral zone's surface area comprising less than 10% of the lake's total surface area [7]. Therefore, understanding the littoral zone will facilitate the conservation and sustainable development of lake ecosystems. However, current knowledge regarding the microbial community of sediments in plateau lake littoral zones remains scarce.

Microorganisms are important drivers in the cycling of carbon (C), N, and P in aquatic ecosystems [8,9], and ecosystem function can be affected by the diversity, composition, and interactions of microbial communities [10]. Microbial community structure in sediments always changes with environmental variables (e.g., organic matter content, nutrient concentration, pH, salinity, etc.); thus, shifts in microbial functional groups influence elementary cycles [11,12]. Many studies have shown that microbial community structure can be used as an indicator of lake trophic status because of the quick and strong response of microbes to environmental stress [13,14]. Under normal conditions, microbial communities are expected to consist of a few abundant taxa and a large number of rare taxa in natural ecosystems [15,16]. Taxa with distinct abundance patterns are likely to have different ecological functions and environmental responses. Previous studies have demonstrated that abundant taxa drive major functions (e.g., carbon metabolism and nutrient cycling) in ecosystems because of their high numbers, whereas rare taxa tend to be involved in multi-functionality [16,17]. Moreover, recent studies have indicated that rare bacteria have broader environmental adaptability and can be considered as seed banks [18,19]. Many studies have elucidated the diversity, distribution, ecological functions, and environmental responses of abundant and rare taxa in different ecosystems, such as lakes [20], hot springs [21], shallow coastal sediments [22], and agricultural soils [23]. Notably, Zhao et al. (2022) highlighted that abundant taxa are more significantly impacted by eutrophication than their rare counterparts, which may have crucial implications for the structure and function of planktonic fungal communities in marine ecosystems [24]. Eutrophication was identified as a key driver of homogenization within bacterial and fungal communities, with generalist fungal species exerting a more pronounced effect on beta diversity across trophic gradients than specialist species [25]. This distinction underscores the importance of examining these sub-communities separately to unravel the mechanisms behind variations in microbial diversity in response to eutrophication. However, few studies of plateau lake ecosystems have focused on the roles of different taxa for both bacteria and archaea. A better understanding of the mechanisms that influence highly connected taxa composition and structure may provide insight into the underlying response of the whole community.

Plateau lakes are important aquatic ecosystems that are sensitive to anthropogenic disturbances due to their poor water exchange ability; accordingly, they are considered sensitive sentinels of global climate change [26]. Erhai Lake, which is the second largest freshwater lake in the Yunnan-Guizhou Plateau region, plays a significant role in local socio-economic development issues, including drinking water supply, irrigation, tourism, and regional climate regulation. However, the lake has undergone anthropogenic eutrophication since the rapid intensification of agricultural development and urbanization around the watershed in the last few decades. Consequently, Erhai Lake is transitioning from a mesotrophic to a eutrophic state. Previous studies indicated that N and P released from sediments were serious ecological pollution risks in Erhai Lake [27,28]. Although studies have investigated the overall composition of bacterial communities in some sites within the lake, we still lack a comprehensive understanding of the community structure of both bacteria and archaea in the littoral zone of Erhai Lake. Therefore, this study was conducted to (1) determine the composition, diversity, and interactions of archaea and bacteria in the sediment and their responses to nutrient loading; (2) identify keystone species that maintain community stability; and (3) determine the potential functions of both abundant and rare archaeal and bacterial communities in littoral sediments of a mesotrophic plateau lake.

2. Materials and Methods

2.1. Site Description and Sample Collection

Erhai Lake (25°36′–25°58′ N, 100°06′–100°18′ E), the second largest fault lake in China, is located in Dali, Yunnan Province, at an elevation of 1974 m. The covering area and watershed area are about 249.8 km^2 and 2656 km^2, respectively. The average depth is 10.5 m with a maximum depth of 20.9 m and the reservoir capacity is approximately 2.8×10^9 m^3. The annual average climate temperature is 15 °C. There are 117 tributaries that flow into the lake, but only 1 outlet.

In July 2022, a total of 34 sediment samples were collected from the littoral zone along the shore of Erhai Lake (Figure 1). From each site, three random surface sediment samples (0–5 cm) were collected using core samplers and manually mixed into one sample (200 g in total) for further use. The corresponding overlying water samples were collected using plexiglass water samplers. Samples were transported to the laboratory at 4 °C, after which the sediment samples were kept at −20 °C until processed for molecular analyses and the water samples were stored at 4 °C until their physicochemical properties were analyzed.

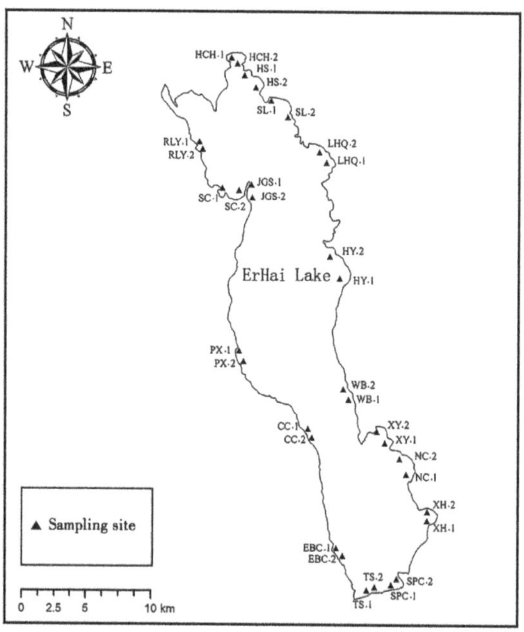

Figure 1. Sampling sites of Erhai Lake.

2.2. Environmental Factors Analysis

The pH, dissolved oxygen (DO), and temperature (WT) of the overlying water were measured in situ during sampling using a HACH HQ40D portable multimeter. Water transparency was determined based on Secchi depth (SD). Chemical oxygen demand (COD) was measured with potassium dichromate as an oxidant using a spectrophotometer (HACH DR3900, Loveland, CO, USA). The permanganate index (PI, also known as permanganate chemical oxygen demand) was determined using the potassium permanganate method (GB11892-89). Water samples for chlorophyll a (Chl-a) analysis were first filtered through 0.45 μm glass fiber membranes, after which Chl-a was extracted with 90% acetone for >2 h at 4 °C in the dark and then measured by spectrophotometry (Chinese Environmental Standard HJ 897-2017). The total phosphorus (TP) content of the overlying water samples was determined at 700 nm using the molybdate colorimetry method after potassium sulfate ($K_2S_2O_8$) digestion. Orthophosphate (PO_4^{3-}-P) concentration was determined by ion chromatography (ECO-IC, Metrohm, Herisau, Switzerland). Total nitrogen (TN) content was determined at 220–275 nm using UV spectroscopy after alkaline $K_2S_2O_8$ digestion. The total dissolved nitrogen (TDN) and ammonium nitrogen (NH_4^+-N) of the overlying water were determined by colorimetry at 220–275 nm and using Nessler's reagent spectrophotometry at 420 nm, respectively, after filtering samples through a 0.45 μm cellulose acetate membranes (Bikeman Bio, Changde, China). The ratios of total nitrogen to total phosphorus (N/P) were also calculated. To evaluate the trophic state of each site, the trophic state index (TSI) was calculated as described by [25].

2.3. DNA Extraction and PCR Amplification

Total DNA extraction from the sediment samples was performed using a FastDNA SPIN Kit for Soil (MP Biomedicals, Solon, CA, USA) according to the manufacturer's protocols. The primer set of 341F (5′-CCTACGGGNGGCWGCAG-3′)/805R (5′-GACTACHVGGG TATCTAATCC-3′) was used for PCR amplifications of the V3-V4 regions of bacterial 16S rRNA gene. The primer set of 1106F (5′-TTWAGTCAGGCAACGAGC-3′)/1378R (5′-TGTGCAAGGAGCAGGGAC-3′) was used to amplify the V9 hypervariable region of archaeal 16S rRNA gene. The amplification conditions were set as follows: for bacteria, an initial denaturation at 94 °C for 5 min, 30 cycles of 30 s at 94 °C, 30 s at 53 °C, 30 s at 72 °C, and a final extension at 72 °C for 8 min. For archaea, an initial denaturation at 95 °C for 3 min, 30 cycles of 20 s at 95 °C, 20 s at 53 °C, 30 s at 72 °C, and a final extension at 72 °C for 5 min. PCR products were purified using the EZNA® Gel Extraction Kit (Omega Bio-Tek, Winooski, VT, USA) and pooled in equimolar amounts. Then, paired-end sequencing (PE250) was performed on an Illumina Nova 6000 platform at Magigene Biotechnology Co., Ltd., Guangzhou, China, according to the standard procedures. The sequences are available as raw FASTQ files in the NCBI SRA archive under project ID PRJNA1123317.

2.4. Sequence Analysis

The sequences were processed using QIIME2 (version 2020.11.0). To ensure quality, fastp (version 0.14.1) was used for splicing and conducting quality control of the raw reads. Adapters and primers were trimmed off to get paired-end clean reads using Cutadapt (v1.9.1). Paired-end clean reads were processed by USEARCH (v10.0.240), including merging paired reads, filtering low-quality reads, and chimera removal. DADA2 was employed for denoising and identifying amplicon sequence variants (ASVs) [29]. The representative ASVs were annotated based on the SILVA 132 database to identify the taxonomy of each ASV. The ASVs assigned as chloroplasts and mitochondria and unknown were excluded from downstream analysis. Alpha diversity indices (chao1, Simpson, Shannon, ACE, and goods coverage) were calculated using the 'vegan' package in the R program. Potential metabolic functions were predicted from 16S rRNA data based on the Kyoto Encyclopedia of Genes and Genomes (KEGG) database using PICRUSt2 and the Functional Annotation tool of the Prokaryotic Taxa database (FAPROTAX).

2.5. Definition of Abundant and Rare Taxa

All ASVs were classified into the following six categories based on the relative abundance as described in previous studies [30,31]: (i) AAT (always abundant taxa, ASVs ≥ 1% in all samples); (ii) ART (always rare taxa, ASVs < 0.01% in all samples); (iii) MT (moderate taxa, ASVs between 0.01% and 1% in all samples); (iv) CRT (conditionally rare taxa, ASVs < 0.01% in some samples and <1% in all samples); (v) CAT (conditionally abundant taxa, ASVs $\geq$ 1% in some samples and >0.01% in all samples); and (vi) CRAT (conditionally rare and abundant taxa, ASVs ranging from rare (<0.01%) to abundant ($\geq$1%)). To simplify downstream analysis, CRT and ART were regarded as rare taxa, while CRAT, AAT, and CAT were considered abundant taxa. The classification of microbial taxa yielded three groups: rare taxa (CRT and ART), abundant taxa (CRAT, AAT, and CAT), and moderate taxa (MT). Further analysis was dependent on these categories and mainly focused on rare and abundant taxa.

2.6. Co-Occurrence Network Analyses and Keystone Species Definition

Network analysis and identification of potential keystone taxa in the 34 samples were conducted based on Spearman correlation coefficients. To simplify the dataset for better visualization, only ASVs that were detected in more than 50% of the samples were retained for network analysis. The Spearman correlation coefficients were calculated using the 'Hmisc' package in the R program (Version 4.2.1). A Spearman coefficient of greater than 0.6 (or less than −0.6) and a p-value less than 0.05 indicated a significant correlation. Gephi was used to visualize the networks [32]. The topological roles of each node can be evaluated based on the values of within-module connectivity (Zi) and among-module connectivity (Pi). Next, the network nodes were divided into four categories: (i) peripherals (Zi < 2.5, Pi < 0.62); (ii) module hubs (Zi > 2.5, Pi < 0.62); (iii) connectors (Zi < 2.5, Pi > 0.62); and (iv) network hubs (Zi > 2.5, Pi > 0.62). Network nodes (ASVs) identified as network hubs, module hubs, and connectors were considered keystone species, which are considered to play key roles in microbial community structure and potential functions.

2.7. Other Statistical Analyses

The functional differences between abundant and rare taxa were compared by Statistical Analysis of Metagenomic Profiles (STAMP) [33]. Spearman's correlation analysis was performed and visualized using the "Hmisc" and "corrplot" packages in the R software (Version 4.2.1). $p < 0.05$ was considered to identify statistically significant differences, and asterisks represented the significant levels at $p < 0.05$ (*), $p < 0.01$ (**), and $p < 0.001$ (***).

3. Results

3.1. Overall Microbial Community Compositions and Diversity

Overall, 2,850,609 high-quality bacterial and 1,927,903 archaeal sequences were obtained from 34 samples after amplicon sequencing, with 74,654–94,514 sequences being obtained per sample for bacteria and 42,464–63,136 sequences per sample for archaea. Good's coverage was close to 1.0 for both bacterial and archaeal communities across the 34 samples, indicating that the majority of the microbial taxa had been recovered (Table S1). To equalize sequencing depth, each sample was rarefied to the lowest sequence numbers (74,654 sequences for bacteria and 42,464 sequences for archaea) across all samples and then classified into ASVs. A total of 108,543 bacterial and 6357 archaeal ASVs were identified at 100% identity. For each sample, 1981 to 5859 bacterial ASVs and 271 to 532 archaeal ASVs were obtained (Table S1). The alpha diversity indices (Shannon and Simpson) and the richness estimators (ACE and Chao 1) for both bacteria and archaea are summarized in Table S1. Most samples showed similar bacterial alpha diversity indices and richness estimators except site HCH.1, in which the alpha diversity was relatively low and the exception of HCH.1 was not observed in the archaeal community.

Evaluation of the bacterial community composition revealed a total of 58 phyla, 125 classes, 285 orders, 350 families, 569 genera, and 438 species across all 34 samples. The three most abundant bacterial phyla were Proteobacteria (33.27–76.67%), Bacteroidetes (2.50–20.10%), and Chloroflexi (1.92–18.49%), followed by Acidobacteria (0.45–2.24%), Planctomycetes (0.67–6.03%), Verrucomicrobia (0.31–7.71%), and Nitrospirae (0.44–8.04%), except at site HCH.1, where there was a high proportion of Firmicutes (14.29%) (Figure 2a).

At the genus level, *Dechloromonas* (0.04–13.55%), *Methylobacter* (0.05–8.08%), *Methyloversatilis* (0.03–7.94%), and *Desulfatiglans* (0.33–2.93%) were abundant across most of the samples (Figure 2b). Interestingly, the proportion of unassigned genera ranged from 58.16% to 87.10% across most sampling sites, while the unassigned proportion was only 13.44% in HCH.1 due to its high level of *Acinetobacter* (67.84%) (Figure 2b).

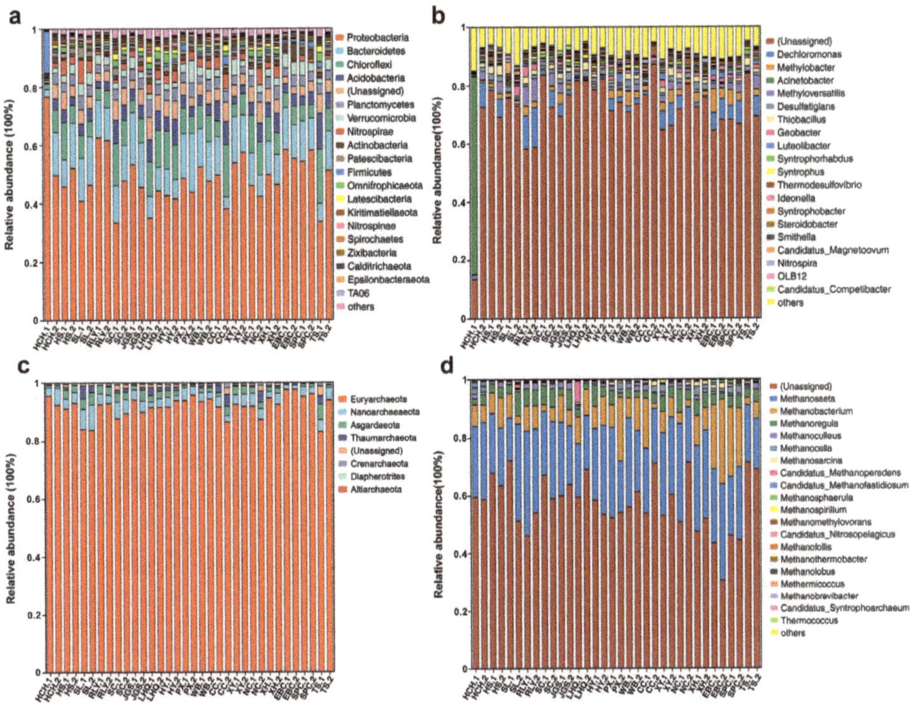

Figure 2. The relative abundance of (**a**) bacterial communities in sediments at the phylum and (**b**) genus level, and (**c**) the archaeal communities at the phylum and (**d**) genus level.

Evaluation of the archaeal community revealed 7 phyla, 15 classes, 14 orders, 19 families, 21 genera, and 19 species. The most abundant archaeal phyla among the samples was Euryarchaeota (83.13–97.63%), while Nanoarchaeaeota (1.45–8.70%), Asgardaeota (0.20–5.32%), Thaumarchaeota (0.06–4.54%), Crenarchaeota (0.03–1.66%), Diapherotrites (<0.15%), and Aliarchaeota (only at sites XY.2 and EBC.1 with a relative abundance < 0.10%) were also found (Figure 2c). The most abundant genera was *Methanosaeta* (14.76–36.18%), followed by *Methanobacterium* (1.74–29.11%), *Methanoregula* (1.53–8.33%), *Methnoculleus* (0.40–2.38%), *Methnocella* (0.13–3.21%), and others (Figure 2d).

3.2. Identification of Abundant and Rare Sub-Communities

All microbial taxa were divided into six categories and then simplified into three main categories for further analysis (Figure 3, Table 1). A summary of the classifications for the total 108,543 bacterial ASVs revealed that 10 and 30 ASVs were considered CAT and CRAT, respectively, while no ASVs were categorized as AAT. The total number of abundant taxa for ASVs was 40 with an average relative abundance of 17.63%. Only six bacterial ASVs were classified into MT, contributing 0.85% of the average relative abundance. The ASV number for rare taxa was 108,497, representing 81.52% of the average relative abundance in each sample. Specifically, ART and CRT comprised 88,143 and 20,354 ASVs, respectively, contributing 10.82% and 70.70% of the average relative abundance per sample. In contrast to the bacterial community, 64 out of 6357 archaeal ASVs affiliated as abundant taxa contributed 82.70% of the average relative abundance in each sample. Interestingly, two archaeal ASVs (ASV1 and ASV2, affiliated to Order Methanomicrobiales) belonging to AAT had a high abundance (42.85%), suggesting that potential methanogenesis was widespread in the sediment of Erhai Lake. The remaining 6293 of 6357 archaeal ASVs, which were regarded as rare, contributed only 17.30% to the average relative abundance in each sample. No ASV was divided into MT for archaea. These results implied that rare species dominated the bacterial community, while abundant taxa were predominant in the archaeal community within Erhai Lake littoral sediments.

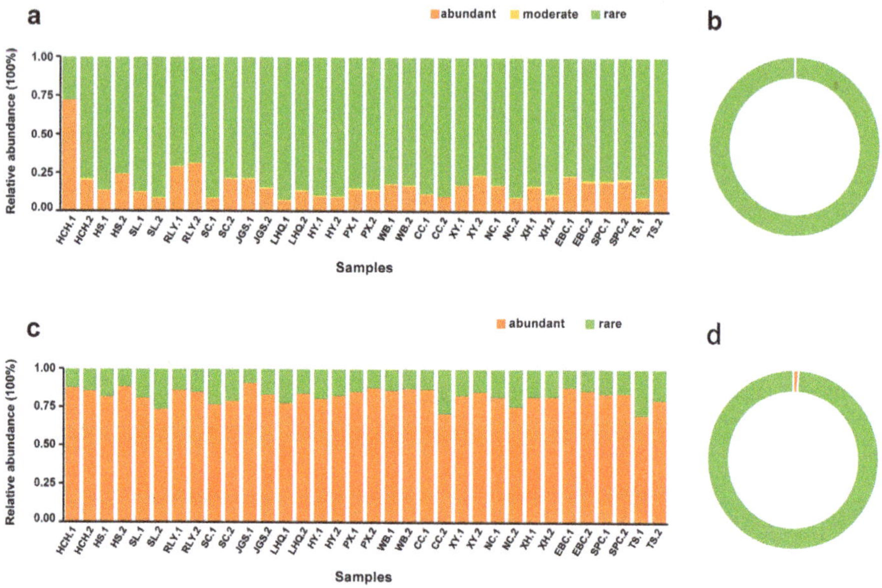

Figure 3. Relative abundance and composition of sub-communities to the overall sequences or total ASVs. (**a**) Relative abundance of bacterial sub-communities; (**b**) Composition of bacterial ASVs; (**c**) Relative abundance of archaeal sub-communities; (**d**) Composition of archaeal ASVs.

Table 1. The contribution of each taxa category to the bacterial and archaeal communities in the 34 samples at 100% identity level.

		Bacteria		Archaea	
		ASVs Number	Avg. Relative Abundance	ASVs Number	Avg. Relative Abundance
	All ASVs	108,543	100.00%	6357	100.00%
Abundant taxa	Always abundant taxa (AAT)	0	0.00%	2	42.85%
	Conditionally abundant taxa (CAT)	10	9.93%	23	29.17%
	Conditionally rare and abundant taxa (CRAT)	30	7.71%	39	10.68%
	Total	40	17.63%	64	82.70%
Rare taxa	Always rare taxa (ART)	88,143	10.82%	3565	0.58%
	Conditionally rare taxa (CRT)	20,354	70.70%	2728	16.72%
	Total	108,497	81.52%	6293	17.30%
	Moderate taxa (MT)	6	0.85%	0	0.00%

3.3. Metabolic Potential of Abundant and Rare Sub-Communities

To compare the metabolic potential of abundant and rare sub-communities, FAPROTAX and PICRUSt2 analyses were performed. The results showed that abundant bacterial sub-communities were predicted to have a higher abundance of functional pathways involved in nitrogen cycles (e.g., nitrate and nitrogen respiration), sulfur cycles (e.g., dark sulfide oxidation and dark oxidation of sulfur compounds), and chlorate reducers than other pathways (Figure 4a). Additionally, rare bacterial sub-communities had more sequences annotated to carbon cycles and other functional pathways, such as methanotrophy, hydrocarbon degradation, and sulfate respiration. For archaeal communities, methanogenesis was the main potential ecological function involving different trophic types. Archaeal-abundant sub-communities contained a much higher abundance of functional pathways associated with methanogenesis, chemoheterotrophy, and dark hydrogen oxidation than other pathways (Figure 4b). Surprisingly, some functional pathways were only assigned in rare sub-communities. For example, dark thiosulfate oxidation, dark sulfur oxidation, dark hydrogen oxidation, phototrophy, sulfite respiration, and aerobic ammonia oxidation were detected only in rare bacterial sub-communities. Additionally, nitrogen fixation, methanogenesis using formate, and methanogenesis by disproportionation of methyl groups only presented in archaeal rare sub-communities, but not in abundant sub-communities. PICRUSt2 functional prediction predicted more KEGG functions in rare sub-communities than abundant sub-communities for both bacteria and archaea (Table 2). These findings indicated that rare sub-communities were more diverse in function and different sub-communities contributed differently to specific metabolic pathways.

Table 2. Comparison of numbers of functional pathways among abundant sub-community, rare sub-community, and whole community in Erhai Lake.

	Predicted KEGG-Annotated Genes		
	Abundant Sub-Community	**Rare Sub-Community**	**Overall Community**
Bacteria	3611	7481	7481
Archaea	3701	4064	4064

3.4. Overall Microbial Network Co-Existence Patterns

The correlation-based network consisted of 635 nodes (ASVs) with 5658 edges (correlations) for bacteria and 110 nodes with 397 edges for archaea (Figure 5). Overall, taxa tended to co-occur (positive correlations, pink lines) rather than co-exclude (negative correlations, green lines). Positive correlations accounted for 83.88% and 68.01% of the potential interactions in bacterial and archaeal networks, respectively, whereas negative correlations accounted for 16.12% and 31.99% of the interactions for bacterial and archaeal co-existence patterns. Considering all correlations, the links between bacteria were more complex than those between archaea, indicating that potential interactions are stronger in bacterial networks. Nodes in the bacterial network mainly belonged to Proteobacteria (40.79%), Bacteroidetes (15.43%), Chloroflexi (10.39%), Acidobacteria (7.24%), Planctomycetes (4.09%), Verrucomicrobia (3.46%), and Nitrospirae (2.99%) (Figure 5a). Furthermore, bacterial networks were clearly parsed into 16 modules (a module is defined as a group of ASVs that are linked more tightly together). Of these, the major modules (1, 2, 3, 4, and 5) accounted for 30.24%, 24.72%, 22.20%, 14.33%, and 2.05% of the entire bacterial network, respectively (Figure 5b). The nodes in the archaeal network were primarily occupied by Euryarchaeaeota (69.09%), Nanoarchaeaeota (12.73%), and Asgardaeota (7.27%) (Figure 5c). Only six modules were clustered for archaeal networks, with major modules 1 to 5 representing 31.82%, 30.00%, 13.64%, 12.73%, and 9.09% of the entire archaeal network, respectively (Figure 5d). Among these modules, the taxonomic compositions varied, with Proteobacteria, Bacteroidetes, Chloroflexi, Nitrospirae, Acidobacteria, and Euryarchaeota being dominant for bacterial and archaeal modules, respectively (Figure S1).

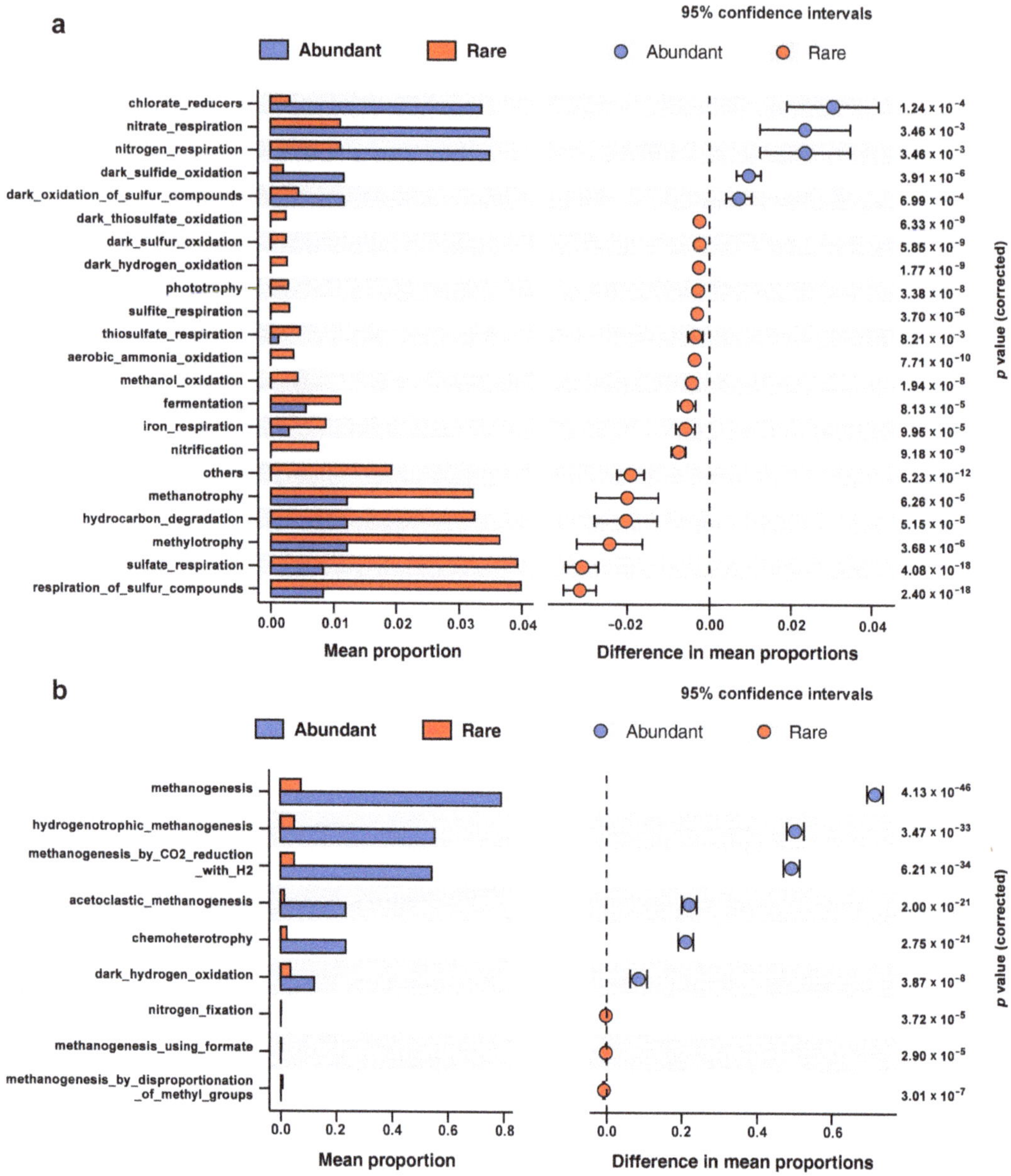

Figure 4. Difference of functional pathways between abundant and rare sub-communities in littoral sediments from Erhai Lake. (**a**) Bacteria; (**b**) Archaea.

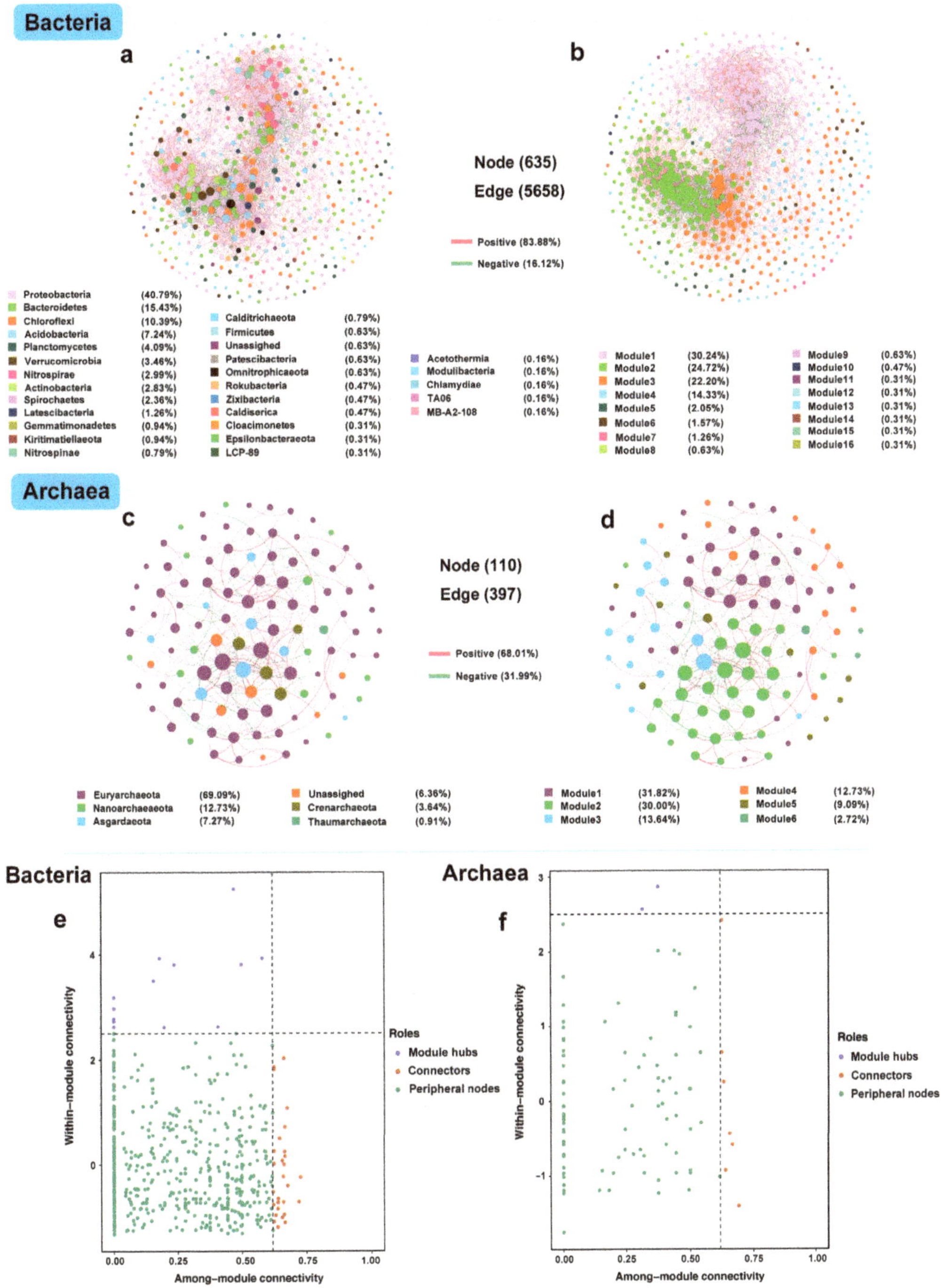

Figure 5. Overall co-occurrence networks and Zi-Pi plot of the bacterial and archaeal communities based on pairwise Spearman's correlations between ASVs. The nodes were colored according to

different types of (**a**,**c**) phylum and (**b**,**d**) modularity classes, respectively. A connection stands for a strong (Spearman r > 0.6 or r < −0.6) and significant (p value < 0.05) correlation. For each panel, the size of each node is proportional to the number of connections (i.e., degree). The green and pink edges indicate negative and positive interactions between two individual nodes, respectively. Zi-Pi plot showing the distribution of (**e**) bacterial and (**f**) archaeal ASVs based on their topological roles. Each symbol represents an ASV. The topological role of each ASV was determined according to the scatter plot of within-module connectivity (Zi) and among-module connectivity (Pi). (i) network hubs: nodes with Zi > 2.5 and Pi > 0.62; (ii) module hubs: nodes with Zi > 2.5 and Pi $\leq$ 0.62; (iii) connectors: nodes with Zi $\leq$ 2.5 and Pi > 0.62; and (iv) peripheral nodes: nodes with Zi $\leq$ 2.5 and Pi $\leq$ 0.62.

Zi-Pi analysis was performed to determine the potential keystone taxa within each network. Based on the Zi-Pi plot, 41 bacterial ASVs and 9 archaeal ASVs were defined as keystone taxa and Proteobacteria and Euryarchaeota were the most prominent module hubs and connectors for bacteria and archaea, respectively (Figures 5e,f and S2, Tables S2 and S3). The bacterial keystone taxa included taxa from the classes Gammaproteobacteria (9 ASVs), Deltaproteobacteria (8 ASVs), Bacteroidia (6 ASVs), Thermodesulfovibrionia (Nitrospirae, 1 ASV), Subgroup_6 (Acidobacteria, 1 ASVs), Subgroup_22 (Acidobacteria, 1 ASVs), Dehalococcoidia (Chloroflexi, 3 ASV), Anaerolineae (Chloroflexi, 2 ASVs), Verrucomicrobiae (3 ASVs), Planctomycetacia (2 ASVs), and Phycisphaerae (Planctomycetes, 1 ASVs). Additionally, two ASVs were assigned as Chloroflexi/LCP_89, while the other two ASVs were unassigned (Table S2, Figure S2). In the archaeal network, the keystone species included taxa from the classes Woesearchaeia (Nanoarchaeaeota, 2 ASVs), Methanomicrobia (Euryarchaeota, 4 ASVs), Thermoplasmata (Euryarchaeota, 2 ASVs), and Lokiarchaeia (Asgardaeota, 1 ASV) (Table S3). Keystone taxa spanned a range of relative abundances in the overall communities (0.02% to 0.62% for bacteria and 0.03% to 0.86% for archaea). Most of the bacterial keystone taxa (38 of 41 ASVs) were classified as rare (Table S2), while over half of the archaeal keystone taxa (5 out of 9 ASVs) were abundant (Table S3). This highlights the significant roles of rare taxa in bacterial communities and abundant taxa in archaeal communities.

3.5. Correlation between Microbial Community and Environmental Factors

The physicochemical properties of the overlying water are shown in Table S4. The TN and TP ranged from 0.526 to 0.677 mg/L and 0.024 to 0.037 mg/L, respectively. Additionally, the calculated TSI ranged from 42 to 49, indicating the water samples were mesotrophic.

The Spearman correlations between physicochemical properties and alpha diversity indices of the total microbial communities are shown in Figure S3. The ACE and Chao 1 indices of the overall bacterial community were significantly negatively correlated with TP ($p < 0.05$, Figure S3a), implying that P nutrient concentrations may affect bacterial community richness in the littoral sediments of Erhai Lake. Notably, no significant correlation was observed between the identified environmental variations and the diversity indices of the archaeal community (Figure S3b).

To further understand the response of bacteria and archaea to environmental variations, the dominant bacterial and archaeal communities were selected and their responses to environmental factors were analyzed (Figures S3c,d and 6a,b). The results showed that Proteobacteria, Verrucomicrobia, Nitrospirae, Actinobacteria, Patescibacteria, Firmicutes, Zixibacteria, Calditrichaeota, Epsilonbacteraeota, and TA06 were significantly correlated with water depth, water transparency, WT, DO, pH, NH_4^+-N, TP, N/P ratio, PO_4^{3-}-P, COD, permanganate index, phytoplankton density, and Chl-a (Figure S3c). Additionally, the NH_4^+-N and TP contents and permanganate index were positively correlated with the relative abundance of Nanoarchaeota and Asgardaeota, but negatively related with that of Euryarchaeota (Figure S3d). At the genus level, many genera responded negatively to environmental factors with *Geobacter*, *Syntrophobacter*, *Methanocella*, and *Methanosarcina* being significantly negatively correlated to TSI, while *Candidatus Competibacter* showed a positive response (Figure 6a,b).

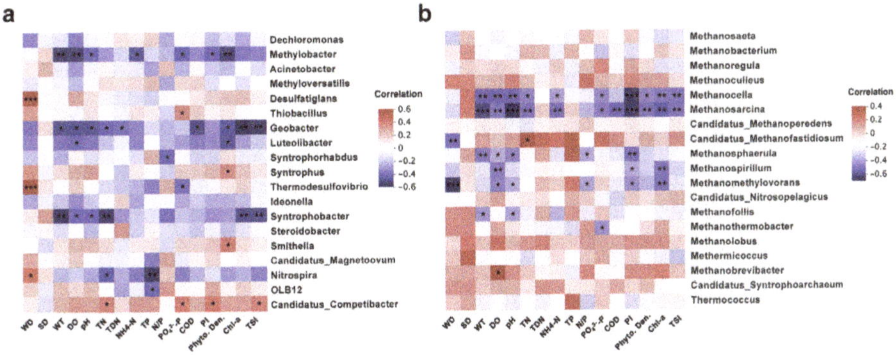

Figure 6. Heatmaps showing Spearman correlation between environmental factors and microbial species of (**a**) bacterial and (**b**) archaeal communities at genus level. Blue indicates a negative correlation, while red indicates a positive correlation. *** $p < 0.001$, ** $p < 0.01$, * $p < 0.05$.

To verify the role those environmental variables played in the abundances of keystone taxa, Spearman correlation analysis was conducted (Figure S3e,f). Generally, the results showed that ASV_97 (genera *Candidatus Competibacter*, Proteobacteria), ASV_433 (Gammaproteobacteria), ASV_496 (genera *Candidatus Competibacter*, Proteobacteria), ASV_826 (Bacteroidetes), and ASV_429 (Class Anaerolineae, Chloroflexi) were positively influenced by TN, NH_4^+-N, PI, DO, pH, COD, PO_4^{3-}-P, and Chl-a, while these environmental variables negatively affected the keystone taxa of ASV_732 (genera *Methyloparacoccus*, Proteobacteria), ASV_1179 (Verrucomicrobia), and ASV_1207 (Planctomycetes). Moreover, TSI was significantly positively related to ASV_97, ASV_433, ASV_496, ASV_429, and ASV_826, whereas it was negatively influenced by ASV_1179, implying the potential indicator roles of these keystone species in the eutrophication of Erhai Lake. Evaluation of archaeal keystone taxa revealed that most were positively correlated with environmental factors, except that ASV_47 (genera *Methanobacterium*, Euryarchaeota) was significantly negatively related to PI, DO, and WT. No significant correlation was found between ASV_21 (genera *Methanosaeta*, Euryarchaeota) and any of the detected environmental variables. Notably, only ASV_88 (genera *Methanobacterium*, Euryarchaeota) had a positive relationship with TSI. These significant correlations between keystone taxa and water nutrients or environmental factors suggest that these key species play important roles in the eutrophication of lakes.

4. Discussion

Lake sediment ecosystems harbor complex microbial communities that are involved in a variety of biogeochemical transformations. Microbial communities respond quickly to anthropogenic disturbances and are sensitive indicators of the water quality status of lakes [13,31]. Different abundance patterns exist within microbial communities, such as the presence of abundant or rare taxa, shaping complex interaction networks and playing discrepant ecological roles in sediments [34,35]. However, little is known about the littoral sediments of mesotrophic plateau lakes. In this study, we investigated the diversity, community composition, potential functions, and co-occurrence patterns of bacterial and archaeal communities, as well as the related environmental factors in the littoral sediments of mesotrophic Erhai Lake on the Yunnan–Guizhou Plateau with a high sampling coverage to draw a comprehensive picture of these ecosystems.

4.1. Characteristics of the Overall Microbial Community

The dominant phyla identified in this study have been widely recognized as prevalent in freshwater lake ecosystems worldwide, playing a pivotal role in C and N metabolism [2]. The predominance of Proteobacteria and Chloroflexi in the littoral sediments of this lake was similar to that observed in previous studies of various lakes of different trophic status [36]. At the genus level, the main bacterial genera were *Dechloromonas*, *Methylobacter*, *Acinetobacter*, *Methyloversatilis*, *Desulfatigians*, and *Thiobacillus*, which have the potential functions of nitrogen, sulfur, and aromatic pollutants metabolism as well as methane oxidation. Notably, unlike at other sites, *Acinetobacter* (affiliated with Gammaproteobacteria) occupied a relatively high abundance at site HCH.1. *Acinetobacter* is a well-known hydrocarbon degrader, especially with respect to aromatic compounds and alkanes [37]. The high proportion of *Acinetobacter* at site HCH.1 might demonstrate a unique pollution condition at this site. If so, this would indicate *Acinetobacter* could be a useful indicator. Firmicutes, which is generally ubiquitous in aquatic ecosystems, but a minor member of freshwater lake ecosystems [38], also showed a relatively high abundance at site HCH.1. Previous studies have reported that Firmicutes can grow chemo-organotrophically on a variety of organic substrates and degrade many types of organic pollutants, such as petroleum hydrocarbons [39], polychlorinated biphenyl [40], and hexahydro-1,3,5-trinitro-1,3,5-triazine [41]. The abundance of these bacteria implies that there is organic pollution near site HCH.1.

Among archaea, Euryarchaeota was identified as the most influential phylum, with a relative abundance >80%. Euryarchaeota have been found to be predominant in some lake sediments, such as those of Taihu Lake [42], Dianchi Lake [43], and Geneva Lake [44]. Some previous studies have also indicated that archaeal community diversity varies with sampling depth in lake sediments and water columns [45,46]. Archaea are widely known for their roles in numerous biogeochemical processes, including ammonia oxidation, methanogenesis, and sulfate reduction [47]. Euryarchaeotes are mainly known as methanogens, although some may also degrade hydrocarbons and be involved in sulfur, nitrogen, and iron cycling [47,48]. At the genus level, the three most abundant genera (*Methanosaeta*, *Methanobacterium*, and *Methanoregula*) were all methanogens, revealing that sediments in the littoral zones of Erhai Lake are potential hotspots for CH_4 emissions. *Methanosaeta* utilizes acetate as a substrate, while *Methanobacterium* and *Methanoregula* are hydrogenotrophic methanogens [49]. Additionally, a small number of methylotrophic methanogens (e.g., *Methanomethylovorans* and *Methanolobus*) were also detected, demonstrating a high metabolic diversity of methanogens in these sediments.

4.2. *Rare and Abundant Taxa Dominated Co-Occurrence Networks*

Network analysis has been successfully applied to explore co-occurrence patterns and interspecific interactions of microbial communities in various ecosystems. In this study, more positive correlations than negative correlations were observed in the networks of both bacteria and archaea. The enhancement of positive interactions among bacteria and archaea in this study may occur because of the higher availability of nutrients and bioavailable carbon sources. This increased availability could reduce the competitive pressure among microbial communities, potentially leading to a less stable network in this ecosystem [50,51].

The co-occurrence networks of Erhai Lake indicate that rare bacterial taxa and abundant archaeal taxa may play irreplaceable roles in shaping their respective networks within the littoral sediments of this mesotrophic plateau lake. Similar findings were reported in a recent study of oil reservoirs [52]. Previous research has primarily focused on either rare bacterial communities or rare eukaryotic plankton communities, and the results have confirmed the significance of rare species in the overall community structure and ecological functions. For instance, Xue et al. (2018) found that most of the keystone species of eukaryotic plankton communities were affiliated with rare taxa following a reservoir cyanobacterial bloom [31]. However, our study indicated that, while rare taxa played a key role in the bacterial community, the archaeal community was dominated by abundant species. Growing evidence from various ecosystems has highlighted the importance of keystone species in co-occurrence networks, in which some keystone taxa are the drivers of the structure and function of the overall microbial community and the disappearance of such keystone taxa can affect the entire community structure and functions.

4.3. Rare and Abundant Taxa Play Different Potential Functions

Typically, abundant taxa contribute the majority of the overall abundance, while rare taxa account for a smaller proportion of all species' abundance, although with a large number of ASVs. However, in our study, the contribution of rare bacterial taxa to the overall abundance was large (81.52%), which is similar to the results reported for sediments of aquaculture pond ecosystems [53]. In contrast, the archaeal community exhibited the opposite trend, with a small number of abundant taxa ASVs contributing up to 82.70% of the overall abundance. These results demonstrated that rare and abundant bacterial and archaeal sub-communities contributed differently to the overall community abundance in littoral sediments of a mesotrophic plateau lake. Specifically, abundant taxa were ubiquitous and possessed higher adaptability to environmental variations, whereas rare taxa were restricted in their distribution across samples and were stronger competitors within narrow niche breadths.

Our results showed that both rare and abundant sub-communities made different contributions to defined metabolic pathways, with rare sub-communities exhibiting greater functional diversity. Several recent studies have emphasized the paramount importance of rare taxa to ecological functions in ecosystems [18,22,53]. Rare taxa may harbor more diverse functional genes involved in C, N, P, and S cycles than abundant taxa and perform distinct or supplementary functions that contribute to the preservation of community function [16,31]. Moreover, rare taxa have been reported to be more competitive than abundant taxa within a narrow niche breadth and to preferentially metabolize some specific substrate, whereas abundant taxa have a broader niche breadth and are more adaptable to various habitats [54]. Rare taxa are important contributors to microbial diversity that increase the functional redundancy within the community [17,34]. A previous study revealed that rare sub-communities played a dominant role in shaping the complexity of bacterial ecological networks and could indirectly enhance the functionality of abundant taxa, highlighting the disproportionate roles of rare taxa in the overall communities [55].

Modularity is believed to reflect interspecies interactions and niche differentiation, with different modules potentially playing different functions. This was supported by the results of the present study, which showed the taxonomic compositions of each module differed (Figure S1). Specifically, the compositions of abundant and rare communities differed and contributed uniquely to ecological functions [21]. Moreover, keystone taxa may impact the metabolic function of entire communities, regardless of their low abundance [56]. The keystone taxa identified in our study were reportedly involved in mediating carbon, sulfur, iron, and nitrogen cycles. Consistent with previous studies, the bacterial phyla Proteobacteria (Gammaproteobacteria and Deltaproteobacteria) were predominant as keystone taxa in freshwater ecosystems. For instance, Proteobacteria are ubiquitous in aquatic habitats and play key roles in nitrogen fixation, dissimilatory nitrate reduction, denitrification, and nutrient cycling [57]. Additionally, Gammaproteobacteria and Deltaproteobacteria are recognized as dominant groups in eutrophic lakes [58,59], while Bacteroidetes have been found to be associated with cyanobacterial blooms in freshwater ecosystems [60]. Members of the phylum Verrucomicrobia act as polysaccharide degraders and play a potential role in carbon cycling in freshwater systems [61]. Chlorofexi and Nitrospirae have demonstrated diverse anaerobic metabolic capabilities and are likely involved in sulfur and nitrogen cycling [62]. The archaeal keystone ASVs affiliated with the classes Methanomicrobia, Woesearchaeia, and Thermoplasmata have the potential for carbon metabolism, sulfate reduction, and organic matter decomposition. In summary, the intricate interplay between rare and abundant taxa in the sediments of mesotrophic lakes reveals a complex ecosystem dynamic, with each group exerting unique and complementary influences on the microbial community's metabolic potential and overall biogeochemical functioning.

4.4. Effects of Environmental Factors and Potential Microbial Indicators of Lake Eutrophication

In lake ecosystems, environmental factors affect the diversity, composition, and distribution of microorganisms. In our study, Spearman correlation analysis of the bacterial community showed that richness estimators (ACE and Chao1) were significantly negatively related to TP, while no detected environmental parameters were significantly related to the diversity index of the archaeal community. A recent meta-analysis of Chinese lakes revealed that the Shannon diversity of sedimental bacterial community decreased significantly with latitude and nitrate (NO_3^--N), but increased with TN, TP, and total organic carbon (TOC) at the national scale [63]. Overall, these findings indicate that local environmental variations make greater contributions to shaping indigenous microbial communities at smaller spatial scales. For example, the dominant community in Hulun Lake was strongly affected by temperature, pH, DO, EC, NH_4^+-N, TP, TN, and COD [64]. Moreover, many low-latitude freshwater lakes such as Chaohu Lake and Dianchi Lake were influenced by nutrient physiochemical parameters (e.g., NH_4^+-N, NO_3^--N, TN, and TP) [30,59]. The key driving environmental factors of microbial communities can reflect the main nutrient status of the lake ecosystems and provide hints for controlling lake eutrophication.

Interestingly, *Candidatus Competibacter* was found to be strongly positively related to TSI, likely because it can compete with polyphosphate-accumulating organisms (PAO) for resources. This competition may reduce the capacity for P removal [65]. In contrast, *Syntrophobacter*, a member of the class Deltaproteobacteria, was negatively correlated with TSI. This genus, which is known for anaerobic respiration, can oxidize propionate using sulfate or fumarate as an electron acceptor [66]. Given that propionate is a significant precursor of methane production [66], it is speculated that *Syntrophobacter* might directly or indirectly participate in the conversion of organic matter degradation to methane emissions, thereby reducing available carbon nutrients in lakes and negatively affecting eutrophication indices. The genus *Geobacter* has been reported to be involved in iron reduction, nitrate-dependent Fe (II) oxidation, and pollutant degradation [67]. *Geobacter* can coexist with methanogens and accelerate the production of CH_4 by interspecies electron transfer [68]. However, *Geobacter* can also inhibit CH_4 production by directly competing with methanogens for acetate [69]. The specific role of *Geobacter* in Erhai Lake requires further exploration. Notably, the negative correlations between the archaeal genera *Methanocella* and *Methanosarcina* and TSI indicated these microbes may serve as potential indicators. Methanogenic microorganisms have been reported to be associated with cyanobacterial blooms [70]. The significant correlations between the identified key species and TSI or other key environmental factors demonstrate that these key species play important roles in the eutrophication processes of lakes and deserve further exploration.

5. Conclusions

We found distinct patterns of abundance and functionality among both bacterial and archaeal taxa, highlighting the complex interactions between these microorganisms and their environment. The microbial community structure in the littoral sediments of Erhai Lake was characterized by a dominance of Proteobacteria, Bacteroidetes, Chloroflexi, and Euryarchaeota. Rare bacterial taxa were often keystone species, exerting a disproportionate influence on network structure and function. In contrast, archaeal communities were primarily driven by abundant taxa, indicating a different ecological dynamic between these two domains. This divergence in communities demonstrated the importance of considering both rare and abundant taxa when studying microbial communities and their responses to environmental changes. This study shed light on the nuanced relationships between microbial communities and their environment in mesotrophic lake sediments. Identification of keystone taxa and developing an understanding of their ecosystem roles will provide valuable information that will be useful in the development of targeted strategies to mitigate eutrophication and preserve the health of these vital aquatic ecosystems.

Although microbial indicators hold potential for application in lake restoration, the vast diversity and complex interactions within microbial communities can make it challenging to predict their responses to environmental changes or interventions. Furthermore, the utilization of microbial indicators for eutrophication prediction or the ecological management of eutrophic lakes through the regulation of specific functional microbial consortia in future commercial applications must take into account the principle of adapting to local conditions. This is due to the unique structure of microbial communities in each lake, necessitating a management approach that is specific to each lake. Further research is needed to focus on key functional microbes within plateau lakes, integrating new technologies (such as metagenomic or metatranscriptomic techniques) with traditional cultivation and isolation methods to functionally validate these key microbes. It is essential to explore the metabolic coupling among the cycles of carbon, nitrogen, phosphorus, and sulfur, and to elucidate their roles in the transformation of nutrients in eutrophic lakes.

Supplementary Materials: The following supporting information can be downloaded at: https://www.mdpi.com/article/10.3390/microorganisms12081617/s1, Figure S1: Histograms show the taxonomy composition of the dominant modules at the phylum level; Figure S2: Taxonomic composition of bacterial keystone taxa at the phylum level; Figure S3: Heatmap showing Spearman correlation between the alpha diversity index, microbial communities and environmental factors; Table S1: Alpha diversity index of bacterial and archaeal communities in littoral sediments of Erhai Lake; Table S2: Lists of keystone taxa in co-occurrence network of bacteria; Table S3: Lists of keystone taxa in co-occurrence network of archaea; Table S4: Physicochemical characterization of samples in the littoral sediments of Erhai Lake.

Author Contributions: Writing—original draft, methodology, investigation, visualization, data curation, Z.X.; visualization, investigation, W.L.; visualization, data curation, K.Y.; visualization, investigation, X.W.; writing—original draft, methodology, investigation, visualization, S.X.; writing—review and editing, supervision, project administration, funding acquisition, conceptualization, X.Z. All authors have read and agreed to the published version of the manuscript.

Funding: This study was financially supported by the National Observation and Research Station of Erhai Lake Ecosystem in Yunnan, and the Natural Science Foundation of China (NSFC 31971526).

Data Availability Statement: All raw sequence data obtained in this study are available on the Sequence Read Archive database of the National Center for Biotechnology Information under accession number PRJNA1123317.

Acknowledgments: The authors are very grateful for the support of Hongyan Ren from the Mobiomed company for her help in microbial community analysis.

Conflicts of Interest: The authors declare no conflicts of interest.

References

1. Alexander, T.J.; Vonlanthen, P.; Seehausen, O. Does eutrophication-driven evolution change aquatic ecosystems? *Philos. Trans. R. Soc. B Biol. Sci.* **2017**, *372*, 20160041. [CrossRef] [PubMed]
2. Tran, P.Q.; Bachand, S.C.; McIntyre, P.B.; Kraemer, B.M.; Vadeboncoeur, Y.; Kimirei, I.A.; Tamatamah, R.; McMahon, K.D.; Anantharaman, K. Depth-discrete metagenomics reveals the roles of microbes in biogeochemical cycling in the tropical freshwater Lake Tanganyika. *ISME J.* **2021**, *15*, 1971–1986. [CrossRef] [PubMed]
3. Nowlin, W.H.; Evarts, J.L.; Vanni, M.J. Release rates and potential fates of nitrogen and phosphorus from sediments in a eutrophic reservoir. *Freshw. Biol.* **2005**, *50*, 301–322. [CrossRef]
4. Huang, W.; Dong, X.; Tu, C.; Yang, H.; Chang, Y.; Yang, X.; Chen, H.; Che, F. Response mechanism of sediment endogenous phosphorus release to functional microorganisms and its cyanobacterial growth and disappearance effects. *Sci. Total Environ.* **2024**, *906*, 167676. [CrossRef] [PubMed]
5. Cardoso, L.D.; de Faria, D.M.; Crossetti, L.O.; Marques, D.D. Phytoplankton, periphyton, and zooplankton patterns in the pelagic and littoral regions of a large subtropical shallow lake. *Hydrobiologia* **2019**, *831*, 119–132. [CrossRef]
6. Wang, H.J.; Lu, J.W.; Wang, W.D.; Yang, L.Y.; Yin, C.Q. Methane fluxes from the littoral zone of hypereutrophic Taihu Lake, China. *J. Geophys. Res.-Atmos.* **2006**, *111*, D17109. [CrossRef]
7. Sala, M.M.; Güde, H. Seasonal dynamics of pelagic and benthic (littoral and profundal) bacterial abundances and activities in a deep prealpine lake (L. Constance). *Arch. Fur Hydrobiol.* **2006**, *167*, 351–369. [CrossRef]
8. Falkowski, P.G.; Fenchel, T.; Delong, E.F. The microbial engines that drive Earth's biogeochemical cycles. *Science* **2008**, *320*, 1034–1039. [CrossRef] [PubMed]
9. Paerl, H.W.; Pinckney, J.L. A mini-review of microbial consortia: Their roles in aquatic production and biogeochemical cycling. *Microb. Ecol.* **1996**, *31*, 225–247. [CrossRef]
10. Fuhrman, J.A. Microbial community structure and its functional implications. *Nature* **2009**, *459*, 193–199. [CrossRef]

11. Chen, J.; Wu, J.L.; Liu, M.; Li, L.Q.; Zhang, W.J.; Wang, D.S.; Ma, T. Bacterial community structure in the surface sediments of different habitats of Baiyangdian Lake, Northern China: Effects of nutrient conditions. *J. Soils Sediments* **2021**, *21*, 1866–1874. [CrossRef]
12. Fagervold, S.K.; Bourgeois, S.; Pruski, A.M.; Charles, F.; Kerhervé, P.; Vétion, G.; Galand, P.E. River organic matter shapes microbial communities in the sediment of the Rhone prodelta. *ISME J.* **2014**, *8*, 2327–2338. [CrossRef] [PubMed]
13. Ji, B.; Liang, J.C.; Ma, Y.Q.; Zhu, L.; Liu, Y. Bacterial community and eutrophic index analysis of the East Lake. *Environ. Pollut.* **2019**, *252*, 682–688. [CrossRef] [PubMed]
14. Pearman, J.K.; Wood, S.A.; Vandergoes, M.J.; Atalah, J.; Waters, S.; Adamson, J.; Thomson-Laing, G.; Thompson, L.; Howarth, J.D.; Hamilton, D.P.; et al. A bacterial index to estimate lake trophic level: National scale validation. *Sci. Total Environ.* **2022**, *812*, 152385. [CrossRef] [PubMed]
15. Galand, P.E.; Casamayor, E.O.; Kirchman, D.L.; Lovejoy, C. Ecology of the rare microbial biosphere of the Arctic Ocean. *Proc. Natl. Acad. Sci. USA* **2009**, *106*, 22427–22432. [CrossRef] [PubMed]
16. Jousset, A.; Bienhold, C.; Chatzinotas, A.; Gallien, L.; Gobet, A.; Kurm, V.; Küsel, K.; Rillig, M.C.; Rivett, D.W.; Salles, J.F.; et al. Where less may be more: How the rare biosphere pulls ecosystems strings. *ISME J.* **2017**, *11*, 853–862. [CrossRef] [PubMed]
17. Pedrós-Alió, C. The rare bacterial biosphere. *Annu. Rev. Mar. Sci.* **2012**, *4*, 449–466. [CrossRef] [PubMed]
18. He, Z.B.; Liu, D.; Shi, Y.; Wu, X.J.; Dai, Y.X.; Shang, Y.W.; Peng, J.J.; Cui, Z.L. Broader environmental adaptation of rare rather than abundant bacteria in reforestation succession soil. *Sci. Total Environ.* **2022**, *828*, 154364. [CrossRef] [PubMed]
19. Sauret, C.; Séverin, T.; Vétion, G.; Guigue, C.; Goutx, M.; Pujo-Pay, M.; Conan, P.; Fagervold, S.K.; Ghiglione, J.-F. 'Rare biosphere' bacteria as key phenanthrene degraders in coastal seawaters. *Environ. Pollut.* **2014**, *194*, 246–253. [CrossRef] [PubMed]
20. Liao, J.Q.; Cao, X.F.; Wang, J.; Zhao, L.; Sun, J.H.; Jiang, D.L.; Huang, Y. Similar community assembly mechanisms underlie similar biogeography of rare and abundant bacteria in lakes on Yungui Plateau, China. *Limnol. Oceanogr.* **2017**, *62*, 723–735. [CrossRef]
21. Zhang, Y.M.; Wu, G.; Jiang, H.C.; Yang, J.; She, W.Y.; Khan, I.; Li, W.J. Abundant and rare microbial miospheres respond differently to environmental and spatial factors in Tibetan hot springs. *Front. Microbiol.* **2018**, *9*, 345048. [CrossRef]
22. Gong, X.Z.; Chen, Z.Y.; Deng, Y.; Zhao, D.; Gao, P.; Zhang, L.; Tu, Q.C.; Qu, L.Y.; Zheng, L.W.; Zhang, Y.; et al. Contrasting archaeal and bacterial community assembly processes and the importance of rare taxa along a depth gradient in shallow coastal sediments. *Sci. Total Environ.* **2022**, *852*, 158411. [CrossRef] [PubMed]
23. Jiao, S.; Wang, J.M.; Wei, G.H.; Chen, W.M.; Lu, Y.H. Dominant role of abundant rather than rare bacterial taxa in maintaining agro-soil microbiomes under environmental disturbances. *Chemosphere* **2019**, *235*, 248–259. [CrossRef] [PubMed]
24. Zhao, H.; Brearley, F.Q.; Huang, L.; Tang, J.; Xu, Q.; Li, X.; Huang, Y.; Zou, S.; Chen, X.; Hou, W. Abundant and rare taxa of planktonic fungal community exhibit distinct assembly patterns along coastal eutrophication gradient. *Microb. Ecol.* **2023**, *85*, 495–507. [CrossRef] [PubMed]
25. Geng, M.; Zhang, W.; Hu, T.; Wang, R.; Cheng, X.; Wang, J. Eutrophication causes microbial community homogenization via modulating generalist species. *Water Res.* **2022**, *210*, 118003. [CrossRef] [PubMed]
26. Zhang, G.Q.; Duan, S.Q. Lakes as sentinels of climate change on the Tibetan Plateau. *All Earth* **2021**, *33*, 161–165. [CrossRef]
27. Ni, Z.K.; Wang, S.R. Historical accumulation and environmental risk of nitrogen and phosphorus in sediments of Erhai Lake, Southwest China. *Ecol. Eng.* **2015**, *79*, 42–53. [CrossRef]
28. Pan, X.; Lin, L.; Huang, Z.; Liu, M.; Dong, L.; Chen, J.; Crittenden, J. Distribution characteristics and pollution risk evaluation of the nitrogen and phosphorus species in the sediments of Lake Erhai, Southwest China. *Environ. Sci. Pollut. Res.* **2019**, *26*, 22295–22304. [CrossRef] [PubMed]
29. Callahan, B.J.; McMurdie, P.J.; Rosen, M.J.; Han, A.W.; Johnson, A.J.A.; Holmes, S.P. DADA2: High-resolution sample inference from Illumina amplicon data. *Nat. Methods* **2016**, *13*, 581–583. [CrossRef] [PubMed]
30. Dai, Y.; Yang, Y.Y.; Wu, Z.; Feng, Q.Y.; Xie, S.G.; Liu, Y. Spatiotemporal variation of planktonic and sediment bacterial assemblages in two plateau freshwater lakes at different trophic status. *Appl. Microbiol. Biotechnol.* **2016**, *100*, 4161–4175. [CrossRef]
31. Xue, Y.Y.; Chen, H.H.; Yang, J.R.; Liu, M.; Huang, B.Q.; Yang, J. Distinct patterns and processes of abundant and rare eukaryotic plankton communities following a reservoir cyanobacterial bloom. *ISME J.* **2018**, *12*, 2263–2277. [CrossRef]
32. Bastian, M.; Heymann, S.; Jacomy, M. Gephi: An open source software for exploring and manipulating networks. In Proceedings of the International AAAI Conference on Web and Social Media, San Jose, CA, USA, 17–20 May 2009; pp. 361–362.
33. Parks, D.H.; Tyson, G.W.; Hugenholtz, P.; Beiko, R.G. STAMP: Statistical analysis of taxonomic and functional profiles. *Bioinformatics* **2014**, *30*, 3123–3124. [CrossRef] [PubMed]
34. Lynch, M.D.J.; Neufeld, J.D. Ecology and exploration of the rare biosphere. *Nat. Rev. Microbiol.* **2015**, *13*, 217–229. [CrossRef] [PubMed]
35. Zhang, Y.; Yao, P.; Sun, C.; Li, S.; Shi, X.; Zhang, X.H.; Liu, J. Vertical diversity and association pattern of total, abundant and rare microbial communities in deep-sea sediments. *Mol. Ecol.* **2021**, *30*, 2800–2816. [CrossRef] [PubMed]
36. Kuang, B.; Xiao, R.; Wang, C.; Zhang, L.; Wei, Z.; Bai, J.; Zhang, K.; Campos, M.; Jorquera, M.A. Bacterial community assembly in surface sediments of a eutrophic shallow lake in northern China. *Ecohydrol. Hydrobiol.* **2022**. [CrossRef]
37. Jung, J.; Park, W. species as model microorganisms in environmental microbiology: Current state and perspectives. *Appl. Microbiol. Biotechnol.* **2015**, *99*, 2533–2548. [CrossRef] [PubMed]
38. Zhang, J.X.; Yang, Y.Y.; Zhao, L.; Li, Y.Z.; Xie, S.G.; Liu, Y. Distribution of sediment bacterial and archaeal communities in plateau freshwater lakes. *Appl. Microbiol. Biotechnol.* **2015**, *99*, 3291–3302. [CrossRef] [PubMed]

39. Fuentes, S.; Méndez, V.; Aguila, P.; Seeger, M. Bioremediation of petroleum hydrocarbons: Catabolic genes, microbial communities, and applications. *Appl. Microbiol. Biotechnol.* **2014**, *98*, 4781–4794. [CrossRef] [PubMed]
40. Gomes, B.C.; Adorno, M.A.T.; Okada, D.Y.; Delforno, T.P.; Gomes, P.C.F.L.; Sakamoto, I.K.; Varesche, M.B.A. Analysis of a microbial community associated with polychlorinated biphenyl degradation in anaerobic batch reactors. *Biodegradation* **2014**, *25*, 797–810. [CrossRef] [PubMed]
41. Cupples, A.M. RDX degrading microbial communities and the prediction of microorganisms responsible for RDX bioremediation. *Int. Biodeterior. Biodegrad.* **2013**, *85*, 260–270. [CrossRef]
42. Li, J.F.; Zhang, J.Y.; Liu, L.Y.; Fan, Y.C.; Li, L.S.; Yang, Y.F.; Lu, Z.H.; Zhang, X.G. Annual periodicity in planktonic bacterial and archaeal community composition of eutrophic Lake Taihu. *Sci. Rep.* **2015**, *5*, 15488. [CrossRef]
43. Yang, Y.Y.; Chen, J.F.; Chen, X.L.; Jiang, Q.S.; Liu, Y.; Xie, S.G. Cyanobacterial bloom induces structural and functional succession of microbial communities in eutrophic lake sediments. *Environ. Pollut.* **2021**, *284*, 117157. [CrossRef] [PubMed]
44. Haller, L.; Tonolla, M.; Zopfi, J.; Peduzzi, R.; Wildi, W.; Poté, J. Composition of bacterial and archaeal communities in freshwater sediments with different contamination levels (Lake Geneva, Switzerland). *Water Res.* **2011**, *45*, 1213–1228. [CrossRef] [PubMed]
45. Alegría-Gómez, J.; Castañón-González, J.H.; Hernández-García, J.A.; González-Terreros, E.; Velázquez-Ríos, I.O.; Ruíz-Valdiviezo, V.M. Changes in the abundance and diversity of bacterial and archaeal communities at different depths in a eutrophic freshwater lake in southwestern Mexico. *Environ. Sci. Pollut. Res.* **2023**, *30*, 98362–98376. [CrossRef] [PubMed]
46. Xing, P.; Tao, Y.; Luo, J.; Wang, L.; Li, B.; Li, H.; Wu, Q.L. Stratification of microbiomes during the holomictic period of Lake Fuxian, an alpine monomictic lake. *Limnol. Oceanogr.* **2019**, *65*, S134–S148. [CrossRef]
47. Offre, P.; Spang, A.; Schleper, C. Archaea in biogeochemical cycles. *Annu. Rev. Microbiol.* **2013**, *67*, 437–457. [CrossRef] [PubMed]
48. Sorokin, D.Y.; Makarova, K.S.; Abbas, B.; Ferrer, M.; Golyshin, P.N.; Galinski, E.A.; Ciordia, S.; Mena, M.C.; Merkel, A.Y.; Wolf, Y.I.; et al. Discovery of extremely halophilic, methyl-reducing euryarchaea provides insights into the evolutionary origin of methanogenesis. *Nat. Microbiol.* **2017**, *2*, 17081. [CrossRef] [PubMed]
49. Conrad, R.; Klose, M.; Enrich-Prast, A. Acetate turnover and methanogenic pathways in Amazonian lake sediments. *Biogeosciences* **2020**, *17*, 1063–1069. [CrossRef]
50. Coyte, K.Z.; Schluter, J.; Foster, K.R. The ecology of the microbiome: Networks, competition, and stability. *Science* **2015**, *350*, 663–666. [CrossRef] [PubMed]
51. Dini-Andreote, F.; Stegen, J.C.; van Elsas, J.D.; Salles, J.F. Disentangling mechanisms that mediate the balance between stochastic and deterministic processes in microbial succession. *Proc. Natl. Acad. Sci. USA* **2015**, *112*, E1326–E1332. [CrossRef]
52. Wang, J.L.; Wang, C.J.; Hu, M.; Bian, L.H.; Qu, L.N.; Sun, H.M.; Wu, X.F.; Ren, G.L. Bacterial co-occurrence patterns are more complex but less stable than archaea in enhanced oil recovery applied oil reservoirs. *Process Biochem.* **2023**, *130*, 40–49. [CrossRef]
53. Hou, D.W.; Zhou, R.J.; Wei, D.D.; Zeng, S.Z.; Weng, S.P.; Yan, Q.Y.; He, J.G.; Huang, Z.J. Abundant and rare microbial communities respectively contribute to an aquaculture pond ecosystem. *Front. Mar. Sci.* **2022**, *9*, 856126. [CrossRef]
54. Delgado-Baquerizo, M.; Oliverio, A.M.; Brewer, T.E.; Benavent-González, A.; Eldridge, D.J.; Bardgett, R.D.; Maestre, F.T.; Singh, B.K.; Fierer, N. A global atlas of the dominant bacteria found in soil. *Science* **2018**, *359*, 320–325. [CrossRef] [PubMed]
55. Zheng, B.H.; Dong, P.C.; Zhao, T.; Deng, Y.T.; Li, J.; Song, L.R.; Wang, J.N.; Zhou, L.; Shi, J.Q.; Wu, Z.X. Strategies for regulating the intensity of different cyanobacterial blooms: Insights from the dynamics and stability of bacterioplankton communities. *Sci. Total Environ.* **2024**, *918*, 170707. [CrossRef] [PubMed]
56. Banerjee, S.; Schlaeppi, K.; van der Heijden, M.G.A. Keystone taxa as drivers of microbiome structure and functioning. *Nat. Rev. Microbiol.* **2018**, *16*, 567–576. [CrossRef] [PubMed]
57. Zhang, D.D.; Li, M.Y.; Yang, Y.C.; Yu, H.; Xiao, F.S.; Mao, C.Z.; Huang, J.; Yu, Y.H.; Wang, Y.F.; Wu, B.; et al. Nitrite and nitrate reduction drive sediment microbial nitrogen cycling in a eutrophic lake. *Water Res.* **2022**, *220*, 118637. [CrossRef] [PubMed]
58. Zhang, L.; Zhao, T.T.; Wang, Q.; Li, L.; Shen, T.T.; Gao, G. Bacterial community composition in aquatic and sediment samples with spatiotemporal dynamics in large, shallow, eutrophic Lake Chaohu, China. *J. Freshw. Ecol.* **2019**, *34*, 575–589. [CrossRef]
59. Zhang, L.; Cheng, Y.; Gao, G.; Jiang, J.H. Spatial-temporal variation of bacterial communities in sediments in Lake Chaohu, a large, shallow eutrophic lake in China. *Int. J. Environ. Res. Public Health* **2019**, *16*, 3966. [CrossRef]
60. Te, S.H.; Tan, B.F.; Thompson, J.R.; Gin, K.Y.H. Relationship of microbiota and cyanobacterial secondary metabolites in dominated bloom. *Environ. Sci. Technol.* **2017**, *51*, 4199–4209. [CrossRef]
61. He, S.M.; Stevens, S.L.R.; Chan, L.K.; Bertilsson, S.; del Rio, T.G.; Tringe, S.G.; Malmstrom, R.R.; McMahon, K.D. Ecophysiology of Freshwater Verrucomicrobia Inferred from Metagenome-Assembled Genomes. *Msphere* **2017**, *2*, e00277-17. [CrossRef]
62. Ayala-Muñoz, D.; Macalady, J.L.; Sánchez-España, J.; Falagán, C.; Couradeau, E.; Burgos, W.D. Microbial carbon, sulfur, iron, and nitrogen cycling linked to the potential remediation of a meromictic acidic pit lake. *ISME J.* **2022**, *16*, 2666–2679. [CrossRef]
63. Gao, P.F.; Wang, P.; Ding, M.J.; Zhang, H.; Huang, G.X.; Nie, M.H.; Wang, G.W. A meta-analysis reveals that geographical factors drive the bacterial community variation in Chinese lakes. *Environ. Res.* **2023**, *224*, 115561. [CrossRef]
64. Shang, Y.; Wu, X.; Wang, X.; Wei, Q.; Ma, S.; Sun, G.; Zhang, H.; Wang, L.; Dou, H.; Zhang, H. Factors affecting seasonal variation of microbial community structure in Hulun Lake, China. *Sci. Total Environ.* **2022**, *805*, 150294. [CrossRef]
65. McIlroy, S.J.; Albertsen, M.; Andresen, E.K.; Saunders, A.M.; Kristiansen, R.; Stokholm-Bjerregaard, M.; Nielsen, K.L.; Nielsen, P.H. 'Candidatus Competibacter'-lineage genomes retrieved from metagenomes reveal functional metabolic diversity. *ISME J.* **2014**, *8*, 613–624. [CrossRef] [PubMed]

66. Westerholm, M.; Calusinska, M.; Dolfing, J. Syntrophic propionate-oxidizing bacteria in methanogenic systems. *FEMS Microbiol. Rev.* **2022**, *46*, fuab057. [CrossRef] [PubMed]
67. Lovley, D.R.; Ueki, T.; Zhang, T.; Malvankar, N.S.; Shrestha, P.M.; Flanagan, K.A.; Aklujkar, M.; Butler, J.E.; Giloteaux, L.; Rotaru, A.E.; et al. Geobacter: The microbe electric's physiology, ecology, and practical applications. *Adv. Microb. Physiol.* **2011**, *59*, 1–100. [CrossRef] [PubMed]
68. Zhang, S.; Chang, J.L.; Lin, C.; Pan, Y.R.; Cui, K.P.; Zhang, X.Y.; Liang, P.; Huang, X. Enhancement of methanogenesis via direct interspecies electron transfer between and conducted by granular activated carbon. *Bioresour. Technol.* **2017**, *245*, 132–137. [CrossRef]
69. Li, T.; Zhou, Q.X. The key role of in regulating emissions and biogeochemical cycling of soil-derived greenhouse gases. *Environ. Pollut.* **2020**, *266*, 115135. [CrossRef]
70. Yan, X.C.; Xu, X.G.; Ji, M.; Zhang, Z.Q.; Wang, M.Y.; Wu, S.J.; Wang, G.X.; Zhang, C.; Liu, H.C. Cyanobacteria blooms: A neglected facilitator of CH4 production in eutrophic lakes. *Sci. Total Environ.* **2019**, *651*, 466–474. [CrossRef]

microorganisms

MDPI

Communication

Biogenic Phosphonate Utilization by Globally Distributed Diatom *Thalassiosira pseudonana*

Huilin Shu [1,2,4], Yuan Shen [1,3], Hongwei Wang [1,3], Xueqiong Sun [1,3], Jian Ma [1,2] and Xin Lin [1,3,*]

1 State Key Laboratory of Marine Environmental Science, Xiamen University, Xiamen 361005, China; shuhuilin@jxnu.edu.cn (H.S.); yuanshen@xmu.edu.cn (Y.S.); wanghongwei@stu.xmu.edu.cn (H.W.); 22320220156409@stu.xmu.edu.cn (X.S.); jma@xmu.edu.cn (J.M.)
2 College of the Environment and Ecology, Xiamen University, Xiamen 361005, China
3 College of Ocean and Earth Sciences, Xiamen University, Xiamen 361005, China
4 College of Life Sciences, Jiangxi Normal University, Nanchang 330022, China
* Correspondence: xinlin@xmu.edu.cn

Abstract: Phosphonates are a class of organic phosphorus (P) compounds that contribute ~25% of dissolved organic P. Recent studies reveal the important role of phosphonates mediated by prokaryotes in the marine P redox cycle. However, its bioavailability by eukaryotic phytoplankton is under debate. 2-Aminoethylphosphonic acid (2-AEP) and 2-amino-3-phosphonopropionic acid (2-AP3) are two biogenic phosphonates in the marine environment. Here, *Thalassiosira pseudonana*, a common diatom species in the ocean, is able to recover growth from P starvation when provided with 2-AEP and 2-AP3. Moreover, 2-AEP cultures exhibited a more similar growth rate at 12 °C than at 25 °C when compared with inorganic P cultures. The cellular stoichiometry of 2-AEP groups was further determined, the values of which are in-between the P-depleted and DIP-replete cultures. This study provides evidence that biogenic phosphonates could be adopted as alternative P sources to support diatom growth and may provide physiological adaptation.

Keywords: phosphonate; bioavailability; diatom; 2-AEP; 2-AP3

1. Introduction

Phosphorus (P) is an essential nutrient for the growth of living organisms. Dissolved inorganic P (DIP), the preferred form of P used by phytoplankton, is often a limiting nutrient for phytoplankton in marine environments [1–3]. Thus, dissolved organic P (DOP) has emerged as a prominent alternative P source [4,5].

Phosphonates are a class of organic P with a chemically stable C-P bond. Besides synthetic compounds (e.g., herbicide glyphosate), biogenic phosphonates produced by various organisms are present in the ocean [6,7]. Two biogenic phosphonate compounds, 2-aminoethylphosphonate (2-AEP) and its derivative, 2-amino-3-phosphonopropionic acid (2-AP3), are the composition of membrane phospholipids in many organisms, such as prokaryotes and mollusks [8,9]. A recent study shows that *Prochlorococcus* likely allocates over 40% of cellular P towards phosphonate production in the ocean [10]. Genome surveys suggest that de novo synthesis of 2-AEP is performed in corals [11]. Therefore, exploring the bioavailability of biogenic phosphonates by phytoplankton has considerable importance.

Compared with the well-elucidated metabolism of phosphonates in prokaryotes [12–14], discrepancies were found in a few studies of eukaryotic phytoplankton [15,16]. Picoprasinophyte *Micromonas commode* and coccolithophore *Emiliania huxleyi* are able to utilize 2-AEP, whereas diatom *Phaeodactylum tricornutum* failed [17]. However, our recent study showed that P-starvation-treated *P. tricornutum* can recover growth with a 2-AEP supplement, and the utilization is mediated by endocytosis and integration into membrane phospholipids (DAG-2-AEP, diacylglyceryl-2-AEP) [16]. Furthermore, an in silico analysis of the global meta-omic atlas suggests that the associated utilization functional genes are prevalent in diatom assemblages and actively expressed in the cold regions [16].

On these grounds, a hypothesis that the cosmopolitan diatom *Thalassiosira pseudonana* [18–20] is able to utilize biogenic phosphonates and the metabolic activity is temperature-sensitive is proposed. Here, the bioavailability of 2-AEP and 2-AP3, which share similar chemical structures and are components of membrane phospholipids, was investigated [8,9]. Then, the physiological responses of algae grown with phosphonate supplements were examined at different temperatures.

2. Materials and Methods

2.1. Cell Culture and Experiment Setup

T. pseudonana was provided by the Center for Collections of Marine Bacteria and Phytoplankton of Xiamen University, China. Two batch experiments were conducted to (batch 1) examine the bioavailability of 2-AEP and 2-AP3 (Sigma-Aldrich, St. Louis, MO, USA) and (batch 2) explore the physiological response under the conditions of significant temperature differences (Table 1). As documented in other studies, 12 °C and 25 °C (comparable to those of 20 °C) represent the temperatures with the lowest growth rate and the highest growth rate, respectively [21]. Before the experiments, seed cultures were subject to P starvation for 8~10 days until the ambient phosphate concentration was below ~0.3 μM and the cell growth ceased. Antibiotics were applied to inhibit the growth of bacteria (Table 1).

Table 1. Culture conditions (a, b: control groups in batch 1, b: control group in batch 2).

Culture	T (°C)	P Nutrient Concentration	Culture Condition
Seed		Starvation treated 8–10 days (<0.3 μM)	• f/2 medium (salinity = 30)
−P [a]		No addition	• 14:10 light: dark cycle
+P [b]		DIP (36 μM)	• photon flux: 180 μmol m^{-2} s^{-1}
Batch 1	20	2-AEP (36 μM)	• Antibiotics cocktail (final concentration in
		2-AP3 (36, 72 μM)	medium: 100 mg L^{-1} ampicillin, 50 mg L^{-1}
Batch 2	12/25	2-AEP (72 μM)	streptomycin, and 50 mg L^{-1} kanamycin)

2.2. Determination of Cell Density and Fv/Fm

Cell density was measured daily by using a CytoFLEX flow cytometer (Beckman Coulter, Brea, CA, USA) and estimated by gating areas in the chlorophyll A versus SSC-A dot plot generated from a 1 mL cell sample. Fv/Fm was determined using a FIRe Fluorometer System (Satlantic, Halifax, NS, Canada). Prior to the measurement, 1 mL of the cell sample was subject to dark adaption for 20 min and then processed following the manufacturer's protocol [22].

2.3. Cellular C and N Content

Cells were collected using pretreated GF/F membranes. The collected cells were dried at 60 °C for 8 h. Then, 500 μL of 1% HCl was dripped onto the filters, and the filters were dried at 60 °C for 12 h again. After the cells were pretreated [16], the cellular C and N contents were determined using a Vario EL cube analyzer (Elementar Analysensysteme GmbH, Langenselbold, Germany) in accordance with the reported method [23].

2.4. Determination of DIP and P Content

Cells were filtered onto GF/F membranes and resuspended in 25 mL of distilled water. The suspension was digested by adding 4 mL of 50 g/L potassium persulfate and autoclaving at 121 °C for 30 min [24]. Then, the cellular P content of the digested suspension and the DIP concentration of the filtrate were determined using the molybdenum method [24].

2.5. Statistical Analysis

Differences among different P and temperature treatment groups were measured with *t*-tests.

3. Results and Discussion

3.1. Differential Growth-Promoting Effects between 2-AEP and 2-AP3

Different growth-promoting effects were observed in the batch 1 culture. After P starvation, *T. pseudonana* was able to recover growth significantly in the medium supplied with 2-AEP (36 μM) and 2-AP3 (72 μM), respectively ($p < 0.05$), while failing to grow in the 2-AP3 supplement (36 μM, Figure 1a). The cells in the 2-AEP (36 μM) group continued growth and peaked at the maximum cell concentration of 8.24×10^5 cells mL^{-1} at Day 3, then declined gradually, and Fv/Fm exhibited a similar pattern accordingly (Figure 1a,b). In comparison, we observe mild growth promotion in the 2-AP3 group. After the supplement of 72 μM 2-AP3, instant cell growth was recovered in 24 h, showing a growth rate of 0.24 μ d^{-1}, which is about half of that in the 2-AEP (36 μM) group (Figure 1a). After then, cells ceased growth and declined towards the end of this experiment, accompanied by an abrupt decline in Fv/Fm (Figure 1b). Maximum cell density observed on D2 was 6.26×10^5 cells mL^{-1}, about half of that in the 2-AEP (36 μM) group.

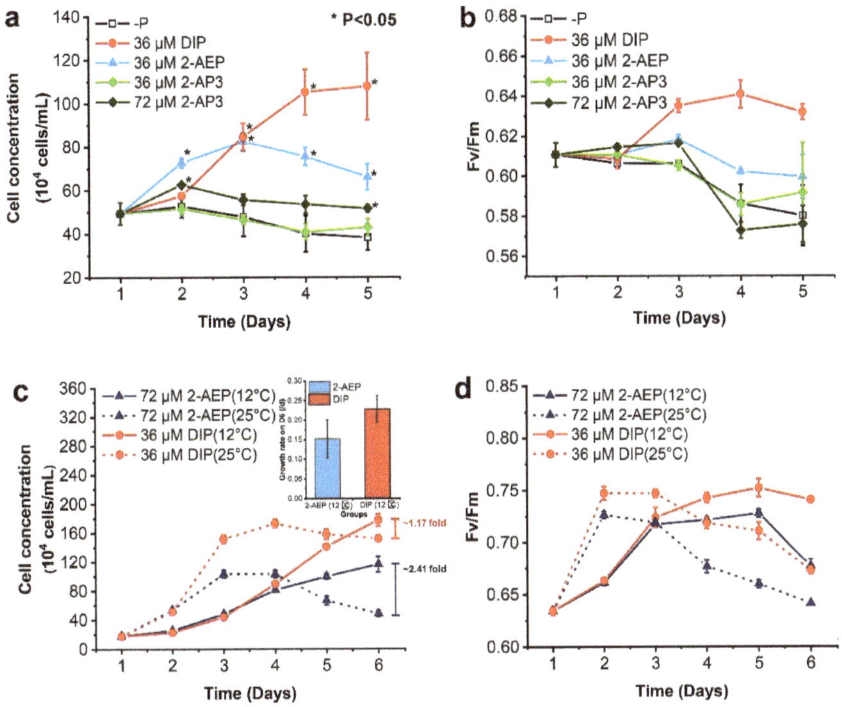

Figure 1. Physiological responses of *T. pseudonana* to different phosphorus levels ((**a**), growth curve; (**b**), Fv/Fm) and temperatures ((**c**), growth curve; (**d**), Fv/Fm). The culture temperature was 20 °C (**a**,**b**) and 12/25 °C (**c**,**d**). Each culture group was set up in biological triplicate. The error bar represents the standard deviation of the mean values. The inner panel of (**c**) represents the growth rate on D6 at 12 °C, and * represents a significant difference ($p < 0.05$) between P-depleted and other groups.

DIP was barely detected (lower than the detection limit; ~0.3 mM) in the 2-AEP and 2-AP3 groups, demonstrating that 2-AEP and 2-AP3 can be utilized as an alternative P source by *T. pseudonana*. Though provided in the same or higher concentration, lower cell density acquired in both phosphonate groups suggests limited utilization efficiency compared with DIP, which is common according to previous reports [15,17]. Furthermore, lower Fv/Fm indicates repressed photosynthesis, suggesting that *T. pseudonana* cells were under P stress in both phosphonate groups.

2-AP3 is the derivative of 2-AEP, known as a component of phospholipids in cell membranes. Studies have demonstrated that 2-AEP and its derivatives can be incorporated into phospholipids in cell membranes [25–27]. 2-AP3 can be decomposed via a transamination reaction and decarboxylation to 2-AEP in *Tetrahymena* [28]. Given the limited utilization of 2-AP3 observed in the present study, the other possible metabolic pathways cannot be completely excluded. The underlying mechanism of 2-AP3 utilization by *T. pseudonana* and its potential bioavailability need to be further investigated to address the knowledge gap in eukaryotic phytoplankton.

3.2. *Different Growth Strategies under Variable P Nutrients and Temperatures*

A recent study of the global ocean gene atlas shows an enriched distribution of representative genes of the proposed 2-AEP utilization mechanism by diatoms in low-temperature waters [16]. Therefore, the batch 2 experiment was conducted to further explore the cellular physiological response grown with different P nutrients and temperatures, in which 72 μM of 2-AEP was provided to obtain more significant differences for the comparative analysis.

Different growth patterns that are temperature-dependent were observed (Figure 1c,d). *T. pseudonana* cells exhibited sustained growth at 12 °C with higher cell density, and a short-time rapid growth at 25 °C with lower cell density regardless of P condition. At 12 °C, the cell density almost showed no change during the first 24 h and then increased steadily in the 2-AEP and DIP groups until D6, sharing no difference in the first 4 days (Figure 1c). When cultured at 25 °C, the cell density exhibited a rapid growth in the first 48 h and then entered the stationary phase after D4 in the DIP group. The cells ceased growth after D3 in the 2-AEP group and then decreased significantly. Consistently, the growth rate of *T. pseudonana* is higher under 8 °C~17 °C than that under 17 °C~25 °C with sufficient DIP supply [29].

A significant difference was found in the promotion effect under different P conditions (Figure 1c). In the DIP-replete groups, the final cell concentration at 12 °C was slightly higher than at 25 °C. Meanwhile, in the 2-AEP groups, the final cell concentration in the 12 °C culture was about three times higher than that in the 25 °C culture. When cultured at 12 °C, the highest growth rate on D6 in the 2-AEP group was 0.152 ± 0.049 $\mu\ d^{-1}$, which is 66.4% of that in the DIP-replete group (0.228 ± 0.034 $\mu\ d^{-1}$, Figure 1c).

Variations of Fv/Fm showed a comparable pattern in accordance with the growth curve (Figure 1d). The dramatic increase in Fv/Fm values in the first 24 h indicated that instant recovery of photosynthetic capacity accounts for the rapid growth in both cultures under 25 °C. After day 3, the Fv/Fm values declined continuously along with the cells entering the stationary phase in 25 °C cultures. In contrast, the Fv/Fm values increased steadily to meet sustained cell growth under 12 °C, and they were higher in the DIP-replete group than in the 2-AEP group.

These results showed that *T. pseudonana* cells cultured with 2-AEP exhibited better physiological adaptability under lower temperatures, resulting in significantly increased cell density. Meanwhile, the Fv/Fm value represents a good consensus of higher photosynthetic capacity and a higher growth rate.

3.3. *Changes in Cellular Elemental Stoichiometry of* T. pseudonana

Many studies have revealed that marine phytoplankton elemental stoichiometric ratios deviate from the empirical Redfield ratio of 106C:16N:1P [30], thus playing a major role in shaping the environmental stoichiometry ratio [31].

3.3.1. Stoichiometry Variation under Different P Conditions

In this study, the N:P ratio of the DIP-replete group (~6:1) (Figure 2b) was far below that of the Redfield ratio and the group-specific optimal value of 14:1 [32]. This finding can be explained by the luxury uptake of DIP and storage in the form of polyP after P starvation, which is typical in diatoms [33,34]. Regarding the initial P starvation state, the C:P and N:P ratios were 117.72 ± 23.14 and 24.25 ± 7.94, respectively, consistent with reported cellular stoichiometry in diatoms under insufficient nutrient conditions [35]. Afterwards, the C:P and N:P ratios decreased significantly to 36.93 ± 3.96 and 5.63 ± 0.56, respectively, at D5 in the DIP-replete group, whereas they barely changed or increased under other different P conditions (Figure S2a).

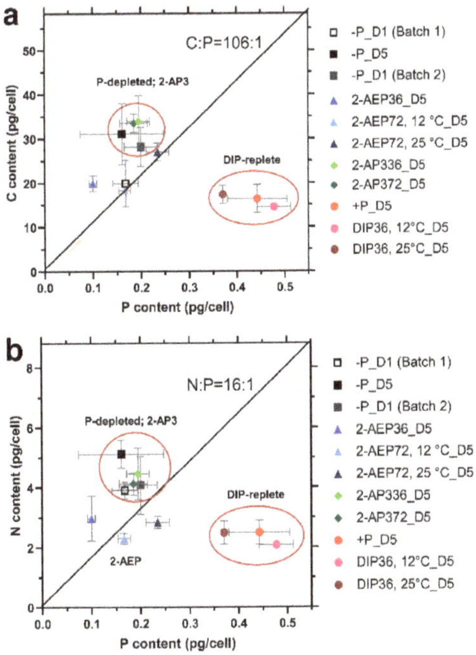

Figure 2. Cellular stoichiometry of *T. pseudonana* under different conditions. (-P, 2-AEP36, 2-AEP72, 2-AP336, and 2-AP372 represent P-depleted, 36 μM 2-AEP, 72 μM 2-AEP, 36 μM 2-AP3, and 72 μM 2-AP3 groups, respectively. D1 and D5 represent the first day and fifth day of the cultural period, respectively). (**a**) C:P ratio; (**b**) N:P ratio. Each culture group was set up in biological triplicate. The error bar represents the standard deviation of the mean values.

The stoichiometry of 2-AP3 groups (36 μM and 72 μM) was similar to that of the P-depleted group (Figure 2), in line with the growth pattern. The P-depleted cells and those treated with 2-AP3 were grouped together, showing higher C:P and N:P ratios, mainly because of higher cellular C and N contents and lower cell P content than that in the DIP-replete groups (Figure S3a).

3.3.2. Effect of Temperature on Cellular Stoichiometry

In the DIP-replete groups, no significant difference was identified between 12 °C and 25 °C (Figures 2 and S2b). In the 2-AEP group, the C:P and N:P ratios were lower than those in P-depleted *T. pseudonana* and higher than those in the DIP-replete group. In the 72 μM 2-AEP cultures (12 °C and 25 °C), the C:P and N:P values were 111.2 ± 11.1~115.8 ± 12.1 and 13.9 ± 1.8~12.1 ± 0.6, respectively, closer to the Redfield ratio with higher 2-AEP concentration and lower temperature.

The interactions between environmental conditions and cell growth are the key factors driving stoichiometric variation [36]. Temperature is the major factor due to its direct effect on cell growth [34,37]. Global research on phytoplankton stoichiometry has found that C:P and N:P ratios decrease with temperature [37–39], but laboratory evidence regarding species differences is insufficient. The findings of the present study show the synergetic effect of temperature and phosphonates on cellular stoichiometry in *T. pseudonana*, higher N:P ratios when cultured with 2-AEP under lower temperature. In the DIP-replete group, the lower temperature significantly decreased the C:P and N:P ratios of *T. pseudonana*, consistent with previous reports [37,40,41].

Cells increase ribosome concentration and cellular P content to compensate for low translation efficiency of ribosomes at low temperature [42]. Such a hypothesis is consistent with the observation of the strong temperature dependency of C:P and N:P in high-latitude ecosystems [43]. In this study, the cellular P content of the DIP-replete groups was significantly higher at 12 °C than at 25 °C (Figure S3b), indicating that low temperature may increase the P demand and promote the P absorption of *T. pseudonana*. In addition, temperature had a significant effect on the C:P and N:P ratios in the DIP group. In the 2-AEP group, temperature effects were barely observed on the stoichiometry of *T. pseudonana*. This finding may be attributed to the stable chemical properties of 2-AEP, which is hard to be hydrolyzed into phosphate during cellular P determination.

3.4. High Variability of Stoichiometry in Diatom

Nutrient availability is considered to be the major driver shaping phytoplankton stoichiometry; for example, P limitation accounts for the increase in C:P and N:P ratios of phytoplankton [44,45]. In this research, the C:P and N:P ratios of *T. pseudonana* declined rapidly after 36 µM DIP supplementation. The C:P and N:P in this study were much lower under the same or similar P conditions than those in other studies (Figure 3) [37,45–47]. In agreement with the previous results, the C:P and N:P ratios of DIP-depleted *T. pseudonana* were higher than the Redfield ratio [45,48,49]. According to the observation data from ALOHA and BATS stations, the C:P ratio of suspended particles varies mostly from 100 to 200 and below the P-stressed threshold [50]. Our results show that *T. pseudonana* grown with 2-AEP is close to the classical Redfield ratio.

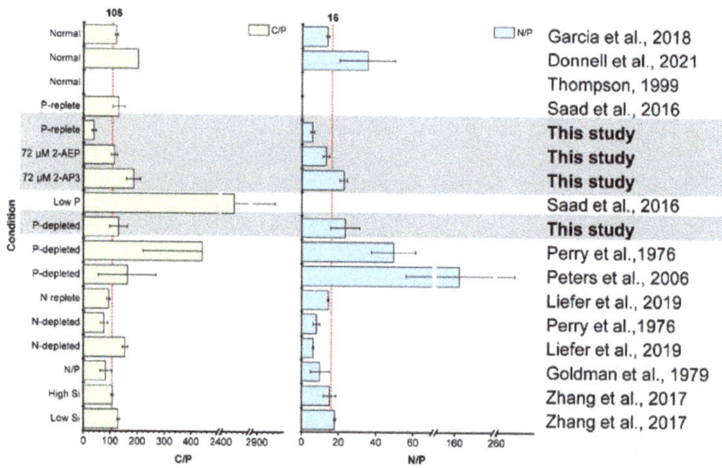

Figure 3. Comparison between the elemental stoichiometry of *T. pseudonana* in this study and that in previous research. Normal represents the nutrients in adequate culture conditions, P-replete represents DIP resupplied to the phosphorus-starvation cells, P-depleted represents the phosphorus-starvation condition, and low P represents a very low phosphorus concentration in the medium [37,45–49,51–53].

The effects of other nutrients, such as N or Si, have been studied. N deficiency leads to a decreased N:P ratio [48,51,52], and Si concentration has no significant effect on both C:P and N:P ratios (Figure 3) [53]. Through summarizing and comparing with previous reports, our study provides fundamental information for addressing temperature and P nutrient effects on stoichiometry variation in *T. pseudonana*.

4. Conclusions

Overall, this study has three major findings. (1) Biogenic phosphonates 2-AEP and 2-AP3 can be utilized by diatom *T. pseudonana* to support cell growth, and 2-AEP is more preferable than 2-AP3, as evidenced by higher cell density. (2) Disparate growth strategies are identified under different temperatures, and the significantly promoted cell growth of the 2-AEP culture under lower temperature than mild temperature indicates its adaptive function. (3) The mediate value of C:P and N:P ratios in the 2-AEP groups between that in P-depleted (2-AP3) and DIP-replete groups suggests the potential effect on environmental elemental stoichiometry.

The results of this study provide new insights into interpreting the alternative P nutrient strategy adopted by diatoms in different scenarios. In sub-polar regions, diatoms represent the major primary producers and may benefit from taking 2-AEP to support growth. Another bold hypothetical scenario is in inorganic nutrient-scarce coral reef ecosystems, where diatoms may obtain advantages by using biogenic phosphonates released by metazoans [54], which require further study. The findings of stoichiometry variation of *T. pseudonana* under different P and temperature conditions provide further understanding of the diatom ecophysiology in marine biogeochemical cycling.

Supplementary Materials: The following supporting information can be downloaded at: https://www.mdpi.com/article/110.3390/microorganisms12040761/s1, Figure S1: Growth curve of *P. tricornutum* under 2-AP3 treatment. Figure S2: Stoichiometry of *T. pseudonana* under different phosphorus and temperature conditions. Figure S3: Variations of cellular C, N, and P of *T. pseudonana* under different phosphorus and temperature conditions.

Author Contributions: H.S., J.M. and X.L. designed the experiments. H.S., H.W. and X.S. performed the experiments. H.S. and X.L. analyzed the data. H.S., Y.S., J.M. and X.L. wrote the manuscript. All authors have read and agreed to the published version of the manuscript.

Funding: This research was supported by the National Key R&D Program of China grant 2022YFC3105302, and National Natural Science Foundation of China (NSFC) 42322602, 42106040.

Data Availability Statement: Data are contained within the article and Supplementary Materials. They are available on request from the authors.

Acknowledgments: We wish to thank Junhui Chen for assisting in cellular C and N determination.

Conflicts of Interest: The authors declare no conflicts of interest.

References

1. Karl, D.M.; Tien, G. Temporal variability in dissolved phosphorus concentrations in the subtropical North Pacific Ocean. *Mar. Chem.* **1997**, *56*, 77–96. [CrossRef]
2. Wu, J.F.; Sunda, W.; Boyle, E.A.; Karl, D.M. Phosphate depletion in the western North Atlantic Ocean. *Science* **2000**, *289*, 759–762. [CrossRef] [PubMed]
3. Mills, M.M.; Ridame, C.; Davey, M.; La Roche, J.; Geider, R.J. Iron and phosphorus co-limit nitrogen fixation in the eastern tropical North Atlantic. *Nature* **2004**, *429*, 292–294. [CrossRef]
4. Lomas, M.W.; Burke, A.L.; Lomas, D.A.; Bell, D.W.; Shen, C.; Dyhrman, S.T.; Ammerman, J.W. Sargasso Sea phosphorus biogeochemistry: An important role for dissolved organic phosphorus (DOP). *Biogeosciences* **2010**, *7*, 695–710. [CrossRef]
5. Torres-Valdes, S.; Roussenov, V.M.; Sanders, R.; Reynolds, S.; Pan, X.; Mather, R.; Landolfi, A.; Wolff, G.A.; Achterberg, E.P.; Williams, R.G. Distribution of dissolved organic nutrients and their effect on export production over the Atlantic Ocean. *Glob. Biogeochem. Cycles* **2009**, *23*, GB4019. [CrossRef]
6. Dawson, R.; Kemp, P. The aminoethylphosphonate-containing lipids of rumen protozoa. *Biochem. J.* **1967**, *105*, 837–842. [CrossRef]
7. Taro, H.; Ikuko, A. Isolation and characterization of new sphingolipids containing N, N-acylmethylaminoethylphosphonic acid and N-acylaminoethylphosphonic acid from the mussel, Corbicula sandai. *BBA-Lipids Lipid Metab.* **1969**, *176*, 898–900. [CrossRef]

8. Quin, L.D.; Quin, G.S. Screening for carbon-bound phosphorus in marine animals by high-resolution ^{31}P-NMR spectroscopy: Coastal and hydrothermal vent invertebrates. *Comp. Biochem. Phys. B* **2001**, *128*, 173–185. [CrossRef] [PubMed]
9. Hilderbrand, R.L. *The Role of Phosphonates in Living Systems*; CRC Press Inc: Boca Raton, FL, USA, 1983.
10. Acker, M.; Hogle, S.L.; Berube, P.M.; Hackl, T.; Coe, A.; Stepanauskas, R.; Chisholm, S.W.; Repeta, D.J. Phosphonate production by marine microbes: Exploring new sources and potential function. *Proc. Natl. Acad. Sci. USA* **2022**, *119*, e2113386119. [CrossRef]
11. Shoguchi, E.; Tanaka, M.; Takeuchi, T.; Shinzato, C.; Satoh, N. Probing a Coral Genome for Components of the Photoprotective Scytonemin Biosynthetic Pathway and the 2-Aminoethylphosphonate Pathway. *Mar. Drugs* **2013**, *11*, 559–570. [CrossRef]
12. Rott, E.; Steinmetz, H.; Metzger, J.W. Organophosphonates: A review on environmental relevance, biodegradability and removal in wastewater treatment plants. *Sci. Total Environ.* **2018**, *615*, 1176–1191. [CrossRef] [PubMed]
13. Dyhrman, S.T.; Chappell, P.D.; Haley, S.T.; Moffett, J.W.; Orchard, E.D.; Waterbury, J.B.; Webb, E.A. Phosphonate utilization by the globally important marine diazotroph *Trichodesmium*. *Nature* **2006**, *439*, 68–71. [CrossRef] [PubMed]
14. Chin, J.P.; Quinn, J.P.; McGrath, J.W. Phosphate insensitive aminophosphonate mineralisation within oceanic nutrient cycles. *ISME J.* **2018**, *12*, 973–980. [CrossRef] [PubMed]
15. Wang, C.; Lin, X.; Li, L.; Lin, S. Differential growth responses of marine phytoplankton to herbicide glyphosate. *PLoS ONE* **2016**, *11*, e0151633. [CrossRef] [PubMed]
16. Shu, H.L.; You, Y.C.; Wang, H.W.; Wang, J.T.; Li, L.; Ma, J.; Lin, X. Transcriptomic-guided phosphonate utilization analysis unveils evidence of clathrin-mediated endocytosis and phospholipid synthesis in the model diatom, *Phaeodactylum tricornutum*. *Msystems* **2022**, *7*, e00563. [CrossRef] [PubMed]
17. Whitney, L.P.; Lomas, M.W. Phosphonate utilization by eukaryotic phytoplankton. *Limnol. Oceanogr. Lett.* **2019**, *4*, 18–24. [CrossRef]
18. Lim, A.S.; Jeong, H.J.; Jang, T.Y.; Jang, S.H.; Franks, P.J. Inhibition of growth rate and swimming speed of the harmful dinoflagellate *Cochlodinium polykrikoides* by diatoms: Implications for red tide formation. *Harmful Algae* **2014**, *37*, 53–61. [CrossRef]
19. Kang, Y.Y.; Liang, J.R.; Gao, Y.H.; Lin, R.C.; Gao, H.; Xing, X.L.; Ma, J.; Luo, Q.Q. Influence of the concentration ratio of nitrogen to phosphorus on the growth and interspecies competition of two red tide algae. *Acta Oceanol. Sin.* **2007**, *26*, 107–115.
20. Zheng, J.W.; Mao, X.T.; Ye, M.H.; Li, H.Y.; Liu, J.S.; Yang, W.D. Allelopathy and underlying mechanism of *Karenia mikimotoi* on the diatom *Thalassiosira pseudonana* under laboratory condition. *Algal Res.* **2021**, *54*, 102229. [CrossRef]
21. Sheehan, C.E.; Baker, K.G.; Nielsen, D.A.; Petrou, K. Temperatures above thermal optimum reduce cell growth and silica production while increasing cell volume and protein content in the diatom *Thalassiosira pseudonana*. *Hydrobiologia* **2020**, *847*, 4233–4248. [CrossRef]
22. Li, M.Z.; Shi, X.G.; Guo, C.T.; Lin, S.J. Phosphorus deficiency inhibits cell division but not growth in the dinoflagellate *Amphidinium carterae*. *Front. Microbiol.* **2016**, *7*, 826. [CrossRef] [PubMed]
23. Sharp, J.H. Improved analysis for "particulate" organic carbon and nitrogen from seawater 1. *Limnol. Oceanogr.* **1974**, *19*, 984–989. [CrossRef]
24. Jeffries, D.S.; Dieken, F.; Jones, D. Performance of the autoclave digestion method for total phosphorus analysis. *Water Res.* **1979**, *13*, 275–279. [CrossRef]
25. Tamari, M. Digestion, absorption and excretion of free-form-ciliatine in rats. *J. Agric. Chem. Soc. Jpn.* **1971**, *45*, 433–440.
26. Curley, J.M.; Henderson, T.O. The incorporation of 2-aminoethylphosphonic acid into rat liver diacylglyceroaminoethylphosphonate. *Lipids* **1972**, *7*, 676–679. [CrossRef] [PubMed]
27. Horigane, A.; Horiguchi, M.; Matsumoto, T. Metabolism of 2-amino-3-phosphonopropionic acid in rats. *BBA-Lipids Lipid Metab.* **1979**, *572*, 385–394. [CrossRef] [PubMed]
28. Akira, H.; Masaaki, H.; Tatsuro, M. Metabolism of 2-amino-3-phosphono [3-14C] propionic acid in cell-free preparations of *Tetrahymena*. *BBA-Lipids Lipid Metab.* **1980**, *618*, 383–388. [CrossRef]
29. Berges, J.A.; Varela, D.E.; Harrison, P.J. Effects of temperature on growth rate, cell composition and nitrogen metabolism in the marine diatom *Thalassiosira pseudonana* (Bacillariophyceae). *Mar. Ecol. Prog. Ser.* **2002**, *225*, 139–146. [CrossRef]
30. Redfield, A.C. The biological control of chemical factors in the environment. *Am. Sci.* **1958**, *46*, 205–221.
31. Chien, C.T.; Pahlow, M.; Schartau, M.; Li, N.; Oschlies, A. Effects of phytoplankton physiology on global ocean biogeochemistry and climate. *Sci. Adv.* **2023**, *9*, eadg1725. [CrossRef]
32. Sharoni, S.; Halevy, I. Nutrient ratios in marine particulate organic matter are predicted by the population structure of well-adapted phytoplankton. *Sci. Adv.* **2020**, *6*, eaaw9371. [CrossRef] [PubMed]
33. Dyhrman, S.T.; Jenkins, B.D.; Rynearson, T.A.; Saito, M.A.; Mercier, M.L.; Alexander, H.; Whitney, L.P.; Drzewianowski, A.; Bulygin, V.V.; Bertrand, E.M. The transcriptome and proteome of the diatom *Thalassiosira pseudonana* reveal a diverse phosphorus stress response. *PLoS ONE* **2012**, *7*, e33768. [CrossRef] [PubMed]
34. Cruz de Carvalho, M.H.; Sun, H.X.; Bowler, C.; Chua, N.H. Noncoding and coding transcriptome responses of a marine diatom to phosphate fluctuations. *New Phytol.* **2016**, *210*, 497–510. [CrossRef]
35. Quigg, A.; Finkel, Z.V.; Irwin, A.J.; Rosenthal, Y.; Ho, T.Y.; Reinfelder, J.R.; Schofield, O.; Morel, F.M.M.; Falkowski, P.G. The evolutionary inheritance of elemental stoichiometry in marine phytoplankton. *Nature* **2003**, *425*, 291–294. [CrossRef] [PubMed]
36. Moreno, A.R.; Martiny, A.C. Ecological stoichiometry of ocean plankton. *Ann. Rev. Mar. Sci.* **2018**, *10*, 43–69. [CrossRef] [PubMed]

37. Garcia, N.S.; Sexton, J.; Riggins, T.; Brown, J.; Lomas, M.W.; Martiny, A.C. High variability in cellular stoichiometry of carbon, nitrogen, and phosphorus within classes of marine eukaryotic phytoplankton under sufficient nutrient conditions. *Front. Microbiol.* **2018**, *9*, 241843. [CrossRef] [PubMed]
38. Martiny, A.C.; Pham, C.T.; Primeau, F.W.; Vrugt, J.A.; Moore, J.K.; Levin, S.A.; Lomas, M.W. Strong latitudinal patterns in the elemental ratios of marine plankton and organic matter. *Nat. Geosci.* **2013**, *6*, 279–283. [CrossRef]
39. Yvon-Durocher, G.; Dossena, M.; Trimmer, M.; Woodward, G.; Allen, A.P. Temperature and the biogeography of algal stoichiometry. *Glob. Ecol. Biogeogr.* **2015**, *24*, 562–570. [CrossRef]
40. Arrigo, K.R.; Robinson, D.H.; Worthen, D.L.; Dunbar, R.B.; DiTullio, G.R.; VanWoert, M.; Lizotte, M.P. Phytoplankton community structure and the drawdown of nutrients and CO_2 in the Southern Ocean. *Science* **1999**, *283*, 365–367. [CrossRef]
41. Arrigo, K.R.; Dunbar, R.B.; Lizotte, M.P.; Robinson, D.H. Taxon-specific differences in C/P and N/P drawdown for phytoplankton in the Ross Sea, Antarctica. *Geophys. Res. Lett.* **2002**, *29*, 44-1–44-4. [CrossRef]
42. Toseland, A.; Daines, S.J.; Clark, J.R.; Kirkham, A.; Strauss, J.; Uhlig, C.; Lenton, T.M.; Valentin, K.; Pearson, G.A.; Moulton, V. The impact of temperature on marine phytoplankton resource allocation and metabolism. *Nat. Clim. Chang.* **2013**, *3*, 979–984. [CrossRef]
43. Tanioka, T.; Garcia, C.A.; Larkin, A.A.; Garcia, N.S.; Fagan, A.J.; Martiny, A.C. Global patterns and predictors of C:N:P in marine ecosystems. *Commun. Earth Environ.* **2022**, *3*, 271. [CrossRef] [PubMed]
44. Sanudo-Wilhelmy, S.A.; Tovar-Sanchez, A.; Fu, F.X.; Capone, D.G.; Carpenter, E.J.; Hutchins, D.A. The impact of surface-adsorbed phosphorus on phytoplankton Redfield stoichiometry. *Nature* **2004**, *432*, 897–901. [CrossRef] [PubMed]
45. Saad, E.M.; Longo, A.F.; Chambers, L.R.; Huang, R.; Benitez-Nelson, C.; Dyhrman, S.T.; Diaz, J.M.; Tang, Y.; Ingall, E.D. Understanding marine dissolved organic matter production: Compositional insights from axenic cultures of *Thalassiosira pseudonana*. *Limnol. Oceanogr.* **2016**, *61*, 2222–2233. [CrossRef]
46. Thompson, P. The response of growth and biochemical composition to variations in daylength, temperature, and irradiance in the marine diatom *Thalassiosira pseudonana* (Bacillariophyceae). *J. Phycol.* **1999**, *35*, 1215–1223. [CrossRef]
47. O'Donnell, D.R.; Beery, S.M.; Litchman, E. Temperature-dependent evolution of cell morphology and carbon and nutrient content in a marine diatom. *Limnol. Oceanogr.* **2021**, *66*, 4334–4346. [CrossRef]
48. Perry, M. Phosphate utilization by an oceanic diatom in phosphorus-limited chemostat culture and in the oligotrophic waters of the central North Pacific 1. *Limnol. Oceanogr.* **1976**, *21*, 88–107. [CrossRef]
49. Peters, F.; Arin, L.; Marrasé, C.; Berdalet, E.; Sala, M.M. Effects of small-scale turbulence on the growth of two diatoms of different size in a phosphorus-limited medium. *J. Mar. Syst.* **2006**, *61*, 134–148. [CrossRef]
50. Kwon, E.Y.; Sreeush, M.G.; Timmermann, A.; Karl, D.M.; Church, M.J.; Lee, S.S.; Yamaguchi, R. Nutrient uptake plasticity in phytoplankton sustains future ocean net primary production. *Sci. Adv.* **2022**, *8*, eadd2475. [CrossRef]
51. Goldman, J.C.; McCarthy, J.J.; Peavey, D.G. Growth rate influence on the chemical composition of phytoplankton in oceanic waters. *Nature* **1979**, *279*, 210–215. [CrossRef]
52. Liefer, J.D.; Garg, A.; Fyfe, M.H.; Irwin, A.J.; Benner, I.; Brown, C.M.; Follows, M.J.; Omta, A.W.; Finkel, Z.V. The macromolecular basis of phytoplankton C:N:P under nitrogen starvation. *Front. Microbiol.* **2019**, *10*, 763. [CrossRef] [PubMed]
53. Zhang, S.W.; Liu, H.B.; Ke, Y.; Li, B. Effect of the silica content of diatoms on protozoan grazing. *Front. Mar. Sci.* **2017**, *4*, 202. [CrossRef]
54. Thomas, S.; Burdett, H.; Temperton, B.; Wick, R.; Snelling, D.; McGrath, J.W.; Quinn, J.P.; Munn, C.; Gilbert, J.A. Evidence for phosphonate usage in the coral holobiont. *ISME J.* **2010**, *4*, 459–461. [CrossRef] [PubMed]

microorganisms

MDPI

Review

Phosphate-Solubilizing Bacteria: Advances in Their Physiology, Molecular Mechanisms and Microbial Community Effects

Lin Pan [1] and Baiyan Cai [1,2,*]

[1] Engineering Research Center of Agricultural Microbiology Technology, Ministry of Education & Heilongjiang Provincial Key Laboratory of Ecological Restoration and Resource Utilization for Cold Region, Key Laboratory of Molecular Biology, College of Heilongjiang Province, School of Life Sciences, Heilongjiang University, Harbin 150080, China; panlin0610@126.com

[2] Hebei Key Laboratory of Agroecological Safety, Hebei University of Environmental Engineering, Qinhuangdao 066102, China

* Correspondence: caibaiyan@126.com

Abstract: Phosphorus is an essential nutrient for all life on earth and has a major impact on plant growth and crop yield. The forms of phosphorus that can be directly absorbed and utilized by plants are mainly HPO_4^{2-} and $H_2PO_4^-$, which are known as usable phosphorus. At present, the total phosphorus content of soils worldwide is 400–1000 mg/kg, of which only 1.00–2.50% is plant-available, which seriously affects the growth of plants and the development of agriculture, resulting in a high level of total phosphorus in soils and a scarcity of available phosphorus. Traditional methods of applying phosphorus fertilizer cannot address phosphorus deficiency problems; they harm the environment and the ore material is a nonrenewable natural resource. Therefore, it is imperative to find alternative environmentally compatible and economically viable strategies to address phosphorus scarcity. Phosphorus-solubilizing bacteria (PSB) can convert insoluble phosphorus in the soil into usable phosphorus that can be directly absorbed by plants, thus improving the uptake and utilization of phosphorus by plants. However, there is no clear and systematic report on the mechanism of action of PSB. Therefore, this paper summarizes the discovery process, species, and distribution of PSB, focusing on the physiological mechanisms outlining the processes of acidolysis, enzymolysis, chelation and complexation reactions of PSB. The related genes regulating PSB acidolysis and enzymatic action as well as genes related to phosphate transport and the molecular direction mechanism of its pathway are examined. The effects of PSB on the structure and abundance of microbial communities in soil are also described, illustrating the mechanism of how PSB interact with microorganisms in soil and indirectly increase the amount of available phosphorus in soil. And three perspectives are considered in further exploring the PSB mechanism in utilizing a synergistic multi-omics approach, exploring PSB-related regulatory genes in different phosphorus levels and investigating the application of PSB as a microbial fungicide. This paper aims to provide theoretical support for improving the utilization of soil insoluble phosphorus and providing optimal management of elemental phosphorus in the future.

Keywords: phosphorus; phosphorus-solubilizing bacteria; mechanisms of phosphorus solubilization; microbial community effects; molecular mechanism

1. Introduction

Phosphorus is an essential plant macronutrient second only to nitrogen and is directly involved in nucleic acid synthesis, cell division and the growth of new tissues and is also necessary for a variety of cellular processes such as photosynthesis, carbohydrate metabolism, energy production, redox homeostasis and signaling in plants [1–3]. Phosphorus is also essential for plant growth and development, with levels reaching 0.2% of the dry weight of plants [4]. However, arable land worldwide contains more total phosphorus and less available phosphorus, and the concentration of plant-available phosphorus seldom exceeds 10 μM, which seriously affects plant growth and agricultural development [5].

Soil organic phosphorus accounts for 30–65%, while the remaining 35–70% is in inorganic forms. Organic phosphorus generally exists in an inert form in the soil and is fixed to form insoluble phosphorus forms, which are difficult for plants to directly absorb and use, and inorganic phosphorus easily reacts with ions in the soil, such as Fe^{3+}, Al^{3+} and Ca^{2+}, generating insoluble phosphate; thus, the long-term application of phosphorus fertilizer overwhelmingly results in phosphate that is difficult for plants to directly absorb and cannot completely address phosphorus deficiencies in soils [6,7]. The overapplication of phosphate fertilizers can also lead to environmental problems such as negative impacts on the food chain, eutrophication, soil nutrient imbalances and destruction of the soil microbial environment. Global phosphate reserves are limited, and depletion will probably occur in 50–100 years (approximately 2059–2109) [8]. However, these data could be changed and updated as new phosphate ores have been recently discovered in Scandinavian countries such as Norway. Due to the associated deleterious effects of the application of phosphorus fertilizers and the decline in global phosphorus stocks, efforts must be made to find alternative environmentally compatible and economically viable strategies to improve phosphorus levels in low-phosphorus or phosphorus-deficient soils for plant growth [1].

The discovery of PSB provides a new way to solve the problem of effective phosphorus deficiency in soils; PSB play a key role in the soil phosphorus cycle by mineralizing organic phosphorus through the secretion of acid and hydrolyzing inorganic phosphorus minerals through enzyme activity, thus solubilizing insoluble phosphorus and increasing the amount of available phosphorus in soils [9,10]. Several studies in recent years have confirmed that PSB can convert the insoluble phosphorus forms present in soils into available phosphorus that can be directly absorbed and utilized by plants through different mechanisms. Examples include secretions of enzymes (phytase and phosphatase), acids (organic, inorganic) and chelation (siderophores, extracellular polysaccharides) to produce dissolved phosphate [10,11]. In addition to their own ability to alter phosphorus speciation, PSB can act synergistically with phosphate fertilizers, and their joint application has been shown to improve phosphorus utilization [12]. PSB are also gradually becoming raw materials for biofertilizers and inoculants; the effect of their application in agricultural fields is also noticeable and can significantly increase the productivity of agronomic crops in agroecological niches [13,14]. Currently, research on PSB mechanisms is limited in scope and lacks comprehensive studies from different perspectives. Due to the interference of PSB species and external factors, PSB metabolites have different levels of capacity, and many functional genes have different effects at different levels, while the ecological environment in soil is so complex that both the exogenous addition of PSB and fertilization will trigger changes in microbial communities, which indirectly change the phosphorus content in soil. Based on these findings, this paper summarizes the phosphorus solubilization mechanism of PSB in soil and the problems encountered from different perspectives, including physiological, molecular and microbial community effects, to provide a theoretical basis for reducing the use of chemical fertilizers and improving the utilization of soil phosphorus in the future (Figure 1).

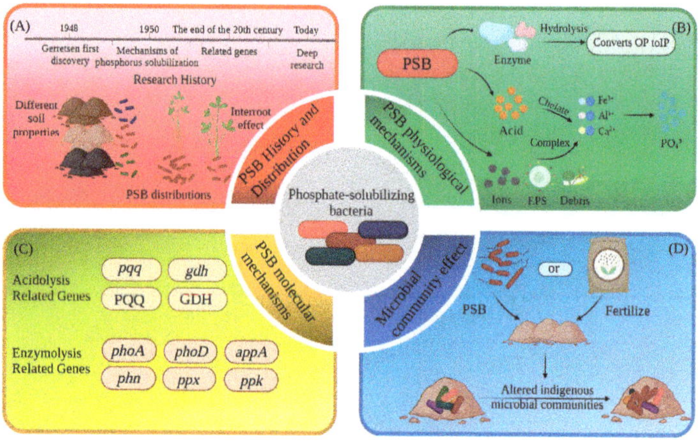

Figure 1. Research process of PSB phosphorus solubilization mechanism. Note: PSB = phosphorus-solubilizing bacteria; PQQ = pyrroloquinoline quinone; GDH = glucose dehydrogenase; PSB have been explored for their phosphorus-solubilizing mechanism from the time they were discovered and have been progressively refined from physiological to molecular to microbial community effects. (**A–D**) describe the developmental history and species distribution of PSBs, the physiological mechanisms of PSBs, the molecular mechanisms of PSBs, and the effects of PSBs on microbial communities, respectively.

2. Overview of PSB

A relationship between microorganisms and soil phosphorus was first noticed at the beginning of the 20th century, when organisms capable of mineralizing phosphate, called phosphorus-solubilizing microorganisms, were discovered for the first time in nature [15]. In 1948, Gerrestsen reported the effect of microorganisms on phosphorus uptake by plants and found that the dry weight of plants grown in normal soil increased by 72% to 188% compared with plants grown in sterilized soil and phosphorus uptake increased by 79% to 342%. In the same year, the first experiments on phosphorus-solubilizing microorganisms for plant growth were carried out, and it was found that phosphorus-solubilizing microorganisms could promote plant growth. Since the greatest proportion of microorganisms in the soil is bacteria, scientists began to study PSB [16]. Initial progress was made in the 1950s in the study of the mechanism of phosphorus solubilization, and it was noted that PSB could secrete different organic acids and their dephosphorylating ability was related to the pH of the surrounding environment [17].

Distribution and Species of PSB

PSB in soil can convert insoluble phosphorus forms into plant-available phosphorus; these bacteria are mostly Gram-negative (Table 1) [18]. There are many types of PSB, which can be classified into two main groups according to the different substrates they act on: organic PSB, which can mineralize organic phosphorus, and inorganic PSB, which can convert insoluble inorganic phosphorus into soluble organic phosphorus; many PSB are also able to dissolve inorganic phosphorus and mineralize organic phosphorus at the same time [19,20]. Phosphorus-solubilizing microorganisms are composed of phosphorus-solubilizing bacteria, phosphorus-solubilizing actinomycetes and phosphorus-solubilizing fungi, among which PSB are widely distributed, accounting for 1–50% of phosphorus-solubilizing microorganisms in the soil, phosphorus-solubilizing actinomycetes for 15–45% and phosphorus-solubilizing fungi for only 0.1% to 0.5%. [21,22]. Djuuna et al. found the highest proportion of PSB among the phosphorus-solubilizing microorganisms screened in the soil, with a total population ranging from 25 × 103 to 550 × 103 CFU g^{-1} [23]. The distribution of PSB is also strongly influenced by many factors. For example, soil properties, habitat types and rhizosphere effects all contribute to differences in the distribution of PSB, and these distributional differences can directly affect PSB species and abundance [17] (Figure 2). Currently, high-throughput sequencing technology is used to determine the abundance of PSB communities, to understand the state of their distribution and to analyze the diversity of PSB communities in the soil and their composition. The effects of different factors on the distribution of PSB are also different. Currently, *Bacillus* spp., *Burkholderia* spp. and *Pseudomonas* spp. show high distribution and abundance in all soil types. Numerous studies have identified *Bacillus* spp., *Pseudocystis* spp. and *Burkholderia* spp. in different types of soils (tea gardens, saline soil, soils with heavy metals, forest soil) and in crop rhizosphere soils, where they have a high relative abundance in the bacterial community as well as a strong phosphorus-solubilizing capacity [24–31]. Currently, other PSB such as *Enterobacter, Salmonella, Flavobacterium, Micrococcus, Thiobacillus, Azotobacter, Burkholderia, Enterobacteriaceae, Serratia marcescens* and *Baeyerlingia*, are also present in the soil (Table 1) [32,33]. In addition, many different PSB were also isolated using genome sequencing (high-throughput sequencing) and raw plate isolation at different distribution sites. Tables 2 and 3 describe the distribution of PSB by type of crop and by location, respectively.

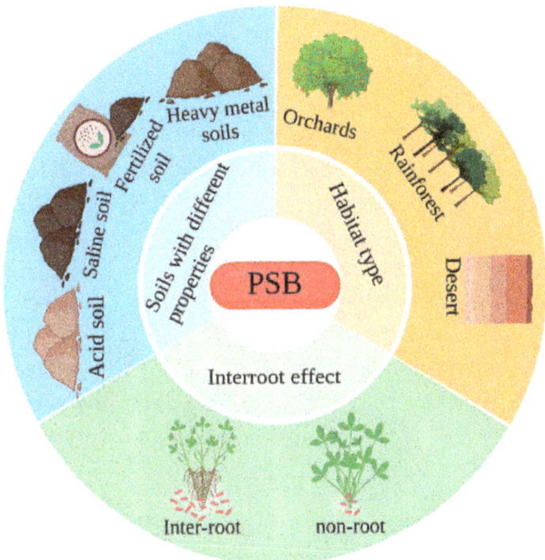

Figure 2. PSB distribution characteristics. Note: PSB = Phosphorus-solubilizing bacteria; the distribution of PSB is mainly influenced by three aspects: the nature of the soil, habitat type and rhizosphere effects.

The species and capacity of PSB isolated from different types of habitats vary. Tang et al. screened eight species of PSB in forest soils, of which more than one-third were from the genus *Burkholderia* and all of which had good phosphate-dissolving capacity [31]. Most of the PSB isolated from oil tea plantations by Liu et al. were *Bacillus*, with *Bacillus ayabatensis* (NC285) having the greatest ability to solubilize inorganic phosphorus [24]. Gao et al. screened the rhizosphere soils of sand dunes for PSB, which contained a high proportion of *Bacteroidetes* spp.; *Bacteroidetes* spp. had a higher proportion in soils with better nutrient supplies [34]. Xie et al. screened fruit tree rhizosphere soils in karst rock desertification (KRD) and non-karst rock desertification (NKRD) areas for PSB and found that the number of PSB screened in the KRD area was much higher than that in the NKRD area; *Pseudomonas* spp. were present in the soils of both the KRD and NKRD areas, whereas *Fusobacterium* spp. and *Bakkholderia* spp. appeared only in some of the soil samples from the KRD area [35]. Therefore, there is a need for targeted strain screening based on differences in geography and plant species.

Soil properties also affect the distribution of PSB, with different soil properties causing differences in the distribution of PSB with different functions. Krystal et al. isolated

44 strains of PSB from calcareous soil, of which 15 strains were identified as *Bacillus flexus*, *Beijerinckia fluminensis*, *Enterobacter ludwigii*, *Enterobacter* sp., and *Pantoea cypripedii*. Among these, *Enterobacter ludwigii* was the best PSB; most of the PSB isolated from soils with a high cadmium content belonged to *Bacillus* and *Enterobacteriaceae*, and the isolated PSB were highly tolerant of cadmium [25]. Rahul et al. isolated two PSB, *Pseudomonas carboxylans* and *Pseudomonas malodorans*, from soils heavily contaminated with heavy metals, and these bacteria were extremely tolerant of lead [26]. Zhang et al. found that bacteria isolated from saline soils belonged to *Bacillus*, such as *Bacillus amyloliquefaciens*, and these PSB were better for the amelioration of saline soils [36]. As a result of these distributional characteristics, PSB can be used to remediate specific soils.

The distribution of PSB also showed an obvious rhizosphere effect. The number and activity of PSB in rhizosphere soils were significantly higher than those in non-rhizosphere soils; this effect played a key role in the construction of PSB communities in rhizosphere soils and influenced the transformation of substances, nutrient cycling and energy flow in soils [37]. The rhizosphere effects of PSB have also been demonstrated in several experiments. Li et al. demonstrated in their experiments that the number of PSB isolated from rhizosphere soils was significantly greater than the number of PSB isolated from non-rhizosphere soils [38]. Zheng et al. analyzed the gene abundance of phosphatase and the *phoC* and *phoD* functional genes of PSB in maize rhizosphere and nonrhizosphere areas; *phoC* and *phoD* encode acid phosphomonolipase and alkaline phosphomonolipase, respectively [39]. The researchers found that phosphatase activity and the gene copy numbers of *phoC* and *phoD* were higher in rhizosphere soils than in non-rhizosphere soils, which indicated that rhizosphere soils possessed higher PSB activity [40]. The strains screened from different plant roots varied; for example, *Bacillus* spp. were the dominant strains in rice, *Fusobacterium* spp., *Bacillus* spp., *Enterobacteriaceae* and *Ochrobacterium* spp. were the dominant strains in wheat, and most of the PSB screened from tea roots were strains of *Burkholderia* spp. [41–43]. Similar results were found by Ponmurugan and Gopi, who determined the population densities of PSB in the rhizospheres of peanut, duckweed, sorghum and maize, with results of 14.9 × 105, 8.6 × 105, 7.3 × 105 and 7.7 × 105 Cfu/g, respectively, indicating that PSB had the highest population densities in the peanut rhizosphere, while PSB had relatively small population densities in the rhizosphere of the remaining three crops. The population densities were relatively low in the other three crops [23,44].

At present, people have a certain understanding of the distribution characteristics and patterns of PSB, but this understanding is still incomplete. PSB, as the largest proportion of phosphorus-solubilizing microorganisms, are distributed all over the world. In some off-the-beaten-track or extreme environments of unexplored areas, PSB and other characteristics of the distribution of the bacteria still need to be explored.

Table 1. Physiological and biochemical testing of different PSB.

PSB	Gram Stain	Glucose Hydrolysis	Starch Hydrolysis	Gelatin Liquefaction	Citrate Utilization	Hydrogen Sulfide Generation	V-P	Methyl Red	Reference
Pantoea roadsii	-	+		+	+	-	-	+	[45]
Pseudomonas donghuensis	-	+		+	+	-	-	+	[45]
Ochrobactrum pseudo-grignonense	-	+		-	-	-	-	+	[45]
Pseudomonas moraviensis	-	+	+	+	+		+		[46]
Bacillus safensis	+	+	+	+	+		+		[46]
Falsibacillus pallidus	-	+	-	-	-		-		[46]
Pseudomonas sp.	-		-	-				-	[47]
Acinetobacter calcoaceticus	-		-	-				-	[47]
Antagonistic Bacillus	+		+	+	+	+	+	-	[48]

Note: In the table, "+" is a positive reaction result and "-" is a negative reaction result.

Table 2. Distribution of PSB in rhizosphere soils of different crop types.

Crop Species	Distribution	PSB	Impact on Crop Performance	Reference
Cereals	Maize rhizosphere soil	*Pseudomonas fluorescens*, *Pseudomonas poae*, *Bacillus subtilis*	Increase in dry weight	[49]
	Wheat rhizosphere soil	*Bacillus safensis*, *Falsibacillus pallidus*	Root length, root surface area, root volume and number of root tips increased significantly	[46]
	Maize rhizosphere soil	*S. marcensens*, *P. brenneri*	Substantial increase in maize production	[50]
Pulses	Legumes in Fes-Meknes region rhizosphere soil	*PSB WJEF38*	Agronomic traits of peas and broad beans were improved	[51]
	Phaseolus vulgaris L. rhizosphere soil	*Pseudomonas kribbensis*	Biomass increased dramatically	[13]
	Peanut rhizosphere soil	*Bacillus amyloliquefaciens*	Significant increase in aboveground dry and fresh weights	[48]
Horticultural crops	Walnut, feijoa, jujube, apple rhizosphere soil	*Pantoea gavini*, *Acinetobacter* sp.	Increase in root length	[52]
	Chenopodium quinoa rhizosphere soil	*Licheniformis*, *Enterobacter, Asburiae*,	Increase in fresh weight and root length	[53]
	Tomato rhizosphere soil	*Acinetobacter*, *Stenotrophomonas maltophilia*	Tomato plant height and leaf area were increased	[54]
Cash crops (economics)	Cotton rhizosphere soil	*Bacillus halotolerans*	Increased cotton yield	[55]
	Tobacco rhizosphere soil	*Burkholderia cenocepacia*	Significant increase in plant height	[56]

Table 3. Distribution of PSB in different types of sites and screening tools.

Source	Source Location	PSB General Category	Screening Method	Reference
Soil	Arid land	*Pseudomonas azotoformans*, *Acinetobacter baumannii*, *Bacillus paramycoides*	Flatbed screening	[57]
	Cinnamomum camphora soil	*Bacteroidetes*, *Proteobacteria*, *Chloroflexi*, and *Gemmatimonadetes*	High-throughput sequencing	[45]
	Saline soil	*Bacillus amyloliquefaciens*	Flatbed screening	[36]
	Brazilian cerrado soil	*Pseudomonas aeruginosa*, *Bacillus cereus*	Flatbed screening	[58]
Rhizosphere	Rhizosphere of Taxus chinensis var. mairei	*Pseudomonas fluorescens*, *Bacillus cereus*, *Sinorhizobium meliloti*, *Bacillus licheniformis*	Flatbed screening	[59]
	Rice rhizosphere	*Pseudomonas aeruginosa*, *Bacillus subtilis strain*	Flatbed screening	[41]
	Blueberry plant rhizosphere	*Buttiauxella* sp.	High-throughput sequencing	[60]
	Characteristics of rhizosphere	*Ascomycetes*, *Acidobacteria*,	High-throughput sequencing	[61]
Sediment	Reservoir sediment	*Micromonospora* sp., *Aminobacter* sp.	High-throughput sequencing	[62]
	Surface sediment in the Changjiang or Yangtze River estuary	*Firmicutes*, *Proteobacteria*, *Actinobacteria*	Flatbed screening	[63]
	Lake Taihu sediment	*Burkholderia* sp.	Flatbed screening	[64]

Table 3. *Cont.*

Source	Source Location	PSB General Category	Screening Method	Reference
Plant parts	Roots, stems and leaves of moso bamboo	*Alkaloid-producing bacilli of the genus Bacillus, Enterobacter* spp., *Bacillus* spp.	Flatbed screening	[65]
	Stems of Oryza officinalis	*Acinetobacter, Cutibacterium, Dechloromonas*	Flatbed screening	[66]
	Corn roots	*Burkholderia* spp.	Flatbed screening	[67]

3. Mechanisms of Phosphorus Solubilization by PSB

Phosphorus solubilization mechanisms are mainly concerned with important components of the soil phosphorus cycle (dissolution–precipitation, mineralization–fixation and adsorption–desorption), which, in addition to being related to bacterial species, are also regulated by phospholysis-related genes. Most of the phosphate-solubilizing bacteria can mineralize or hydrolyze the insoluble phosphate in the soil by secreting acids and enzymes and also chelate or complex with metal cations (Ca^{2+}, Fe^{2+}, Al^{3+}) in the soil to release phosphate ions. A small number of phosphate-solubilizing bacteria can indirectly change the pH of the environment by releasing gas molecules (CO_2 through respiration and hydrogen sulfide from PSB) to achieve the effect of solubilizing insoluble phosphate. PSB can also increase the phosphorus content of soils by altering the microbial community in the soil. PSB in the soil change the diversity of the microbial community and the associated enzymes increase, thus changing the phosphorus content of the soil and enhancing crop growth and yield [68] (Figure 3). Due to the complexity and diversity of phosphate solubilization mechanisms, this paper focuses on three systematic aspects: the physiological mechanisms of PSB, molecular mechanisms and indirect mechanisms shaped by changes in the structure and abundance of microbial communities in the soil (Figure 4).

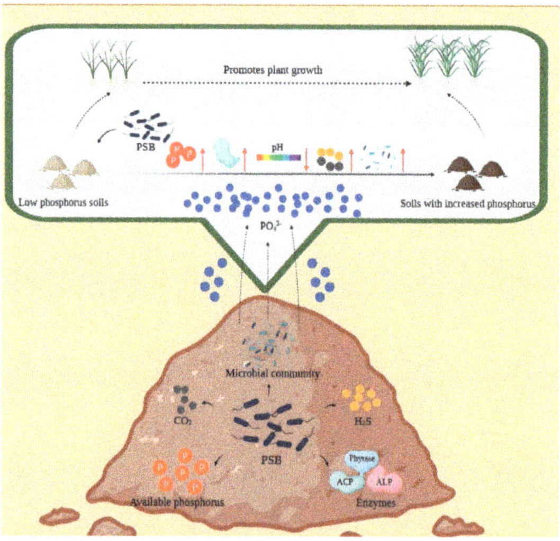

Figure 3. Substances involved in phosphorus solubilization by PSB. Note: PSB = phosphorus-solubilizing bacteria; CO_2 = carbon dioxide; H_2S = hydrogen sulfide. PSB are mainly affected by organic acids, related enzymes, gas molecules such as CO_2 and H_2S and microbial communities in the process of phosphorus solubilization, which generally leads to a decrease in pH and achieves a certain effect of phosphorus solubilization. Enhanced soil phosphorus levels increase crop growth and fields.

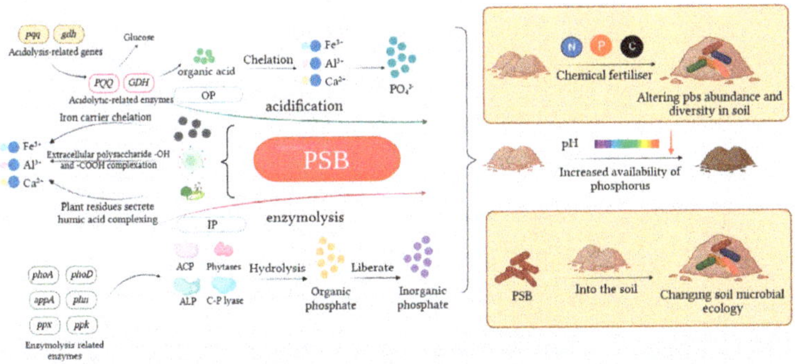

Figure 4. Mechanism of phosphorus solubilization by PSB. Note: PSB = phosphorus-solubilizing bacteria; OP = organic phosphate; IP = inorganic phosphate; ACP = acid phosphatase; ALP = alkaline phosphatase; PQQ = pyrroloquinoline quinone; GDH = glucose dehydrogenase; PSB dissolve/mineralize insoluble phosphorus forms in soil through different mechanisms. Acidolysis occurs in organic phosphorus soils, in which the pqq and gdh genes promote the secretion of organic acids under the regulation of PQQ and GDH, respectively, thereby chelating metal ions (Ca^{2+}, Al^{3+}, Fe^{3+}) in the soil to release PO_4^{3-}. Enzymatic degradation occurs in inorganic phosphorus soils, where phoA, appA, ppx, phoD, phn, ppk and other related genes promote the synthesis of related enzymes to hydrolyze organic phosphorus in the soil, thereby releasing inorganic phosphorus. In addition, PSB will cause iron ions, extracellular polysaccharides, plant and animal residues and metal ions to undergo chelating and complexation reactions to reduce soil pH and increase the available phosphorus in the soil. External fertilizers and access to PSB can change the abundance of the soil microbial community, which can increase the available phosphorus in the soil.

3.1. *Physiological Mechanisms*

3.1.1. Solubilizing Action of Acids

Acidolysis involves the action of both organic and inorganic acids; most PSB secrete organic acids and very few secrete inorganic acids. PSB bacteria can secrete low-molecular-weight organic acids such as malic acid, succinic acid, fumaric acid, acetic acid, tartaric acid, malonic acid, glutamic acid, propionic acid, butyric acid, lactic acid, 2-ketogluconic acid, saccharinic acid and oxalic acid during growth [69,70]. These low-molecular-weight organic acids can chelate under low pH conditions via hydroxyl and carboxyl groups with metal ions in the soil (Fe^{3+}, Al^{3+}, Ca^{2+}) (Figure 5) [71]. Their Ca^{2+} chelation ability is most significant, and they compete with phosphates for phosphorus adsorption sites in the soil, enhancing the soil uptake of phosphate and increasing the solubilizing capacity of inorganic phosphorus, resulting in increased solubility and availability of mineral phosphates [72,73]. Solubilization efficiency is also related to the nature of the organic acids; tricarboxylic and dicarboxylic acids are usually more effective than mono- and aromatic acids, and aliphatic acids are more effective than phenolic, citric and fumaric acids in solubilizing phosphates [74]. However, different types of organic acids are produced by different PSB, and different organic acids have very different adsorption capacities for phosphate (Table 4) [75,76]. A few PSB also secrete inorganic acids (e.g., hydrochloric, sulfuric, nitric and carbonic acids), which lower the soil pH and dissolve inorganic phosphorus but are not as effective as organic acids at the same pH [33].

PSB secretes organic acids that interact with plant roots [77]. This rhizosphere effect, which increases the rhizosphere's bacterial diversity and the relative abundance of dominant bacteria in the rhizosphere zone, indirectly alters the available phosphorus content of the soil [78,79]. The acidic environment created by organic acids is more suitable for the enrichment of PSB, but it is also somewhat targeted. For example, malic and citric acids can both attract *Pseudomonas fluorescens* WCS365, which can colonize tomato roots [80]. Watermelon roots can secrete citric and malic acids, which induce plants to promote the colonization of roots by a rhizosphere bacterial strain, *Bacillus polymyxa* SQR-21 [81]. Exploring the production of organic acids in the plant root system–PSB–plant root system relationship is of great value in refining the mechanism of phosphate solubilization.

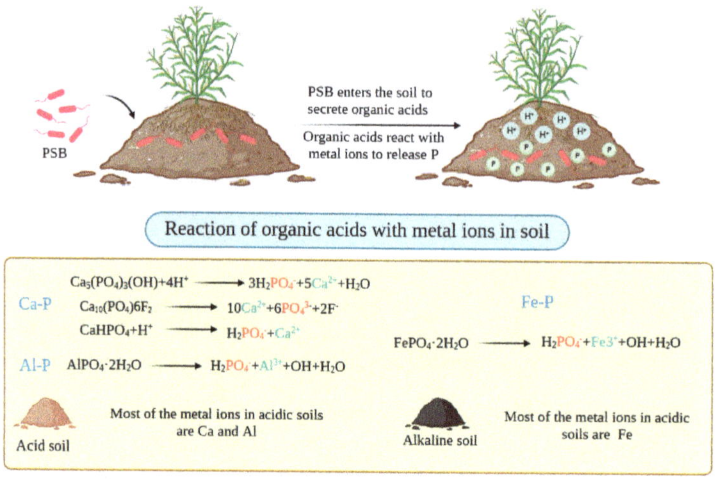

Figure 5. Phosphorus dissolution process in soil.

Although most studies have concluded that the secretion of organic acids is one of the main mechanisms of PSB, a few still have different views on this. Zhao et al. found that the organic acid content of the medium did not correlate with the phosphorus-solubilizing capacity of PSB in their experiments [82]. A similar conclusion was reached by Qin et al., who found that the application of PSB to over-acidified soils increased the pH and that there was no positive correlation between organic acids and the ability to dissolve phosphorus [83]. These results showed that the secretion of organic acids did not necessarily have a strong correlation with phosphate-solubilizing ability, and the important mechanism between organic acids and phosphate-solubilizing bacteria was not fully explored and still needs to be further investigated.

Table 4. Species of organic acids secreted by different PSB.

Phylum	PSB Species	PSB Source	Secreted Organic Acids	References
Proteobacteria	*Enterobacter aerogenes*	Mangrove rhizosphere soil	Lactic, succinic, isovaleric, isobutyric and acetic acids	[84]
	Pantoea dispersa	Corn rhizosphere soil	Citric, malic, succinic and acetic acids	[85]
	Pantoea sp.	Nectarine rhizosphere soil	Oxalic, formic, acetic and citric acids	[86]
Firmicutes	*Bacillus*	Rice paddy soil	Gluco-oxalic acid, citric acid, tartaric acid, succinic acid, formic acid and acetic acid	[87]
	Bacillus safensis	Turmeric rhizosphere soil	Gluconic acid, alpha-ketogluconic acid, succinic acid, oxalic acid and tartaric acid	[69]
	Bacillus siamensis	Wheat rhizosphere soil	Glycolic acid	[88]
	Bacillus amyloliquefaciens	Saline soil	Lactic acid, maleic acid and oxalic acid	[36]
Actinobacteria	*Tsukamurellatrosinosolvens*	Tea tree rhizosphere soil	Lactic acid, maleic acid and oxalic acid	[89]

3.1.2. Mineralization Action of Enzymes

The mineralization of organophosphorus by enzymes secreted by PSB is one of the main dephosphorylation mechanisms (Table 5). The main enzymes found to have a dephosphorylating effect are phosphatases, phytases and C-P-cleaving enzymes. It has also been well demonstrated in many experiments that PSB increase enzyme activity in the soil, resulting in an increase in the available phosphorus content of the soil, but different hydrolytic enzymes are involved in the mineralization of organic phosphorus in different ways [88,90]. Phosphatase, also known as phosphomonolipase, is mainly encoded by *olp*A and is classified into three types: acidic, alkaline and neutral phosphatases. ACP is more effective in mineralizing organic phosphorus in acidic soils with pH values less than 7. Alkaline phosphatases (ALPs) mainly catalyze the hydrolysis of phospholipids (i.e., phosphoglucose-6 and ATP) and release inorganic phosphorus in soils with pH values higher than 7. In contrast, neutral phosphatases, unlike acid and alkaline phosphatases, do not have a very pronounced effect on phosphorus mineralization and hydrolysis. [2,87,91]. Phytase is another type of phosphatase (encoded by *appA*) that mainly mineralizes organic phosphorus in phytate. As the ester bond in phytate is quite stable, phosphatases are not able to completely hydrolyze it, and it must be converted by phytase into a form that can be broken down by phosphatases, finally releasing inorganic phosphorus [92]. C-P lyase (encoded by *phnJ*) is involved in the mineralization of organophosphorus, as are several of the enzymes mentioned above, but differs in that the hydrolysis of organophosphorus by C-P lyase ultimately releases phosphorus in its free form [2].

Table 5. Secretion of related enzymes by different PSB.

PSB Species	PSB Source	Secreted Enzymes	Reference
Pantoea sp.	Forest soil	Phytase	[5]
Serratia liquiefaciencs	Laboratory storage	ALP	[93]
Acinetobacter pittii gp-1	Laboratory storage	ACP, phytase	[94]
aryabhattai	Laboratory storage	ACP	[95]
Pseudomonas asiatica	Ant hill soil	ALP	[96]
Burkholderia sp.	Zea mays rhizosphere soil	ACP	[97]
Pseudomonas sp.	Yellow sandalwood rhizosphere soil	ACP, ALP	[98]
Enterobacter sp.	rhizosphere soil of moso bamboo	ACP	[99]

3.1.3. Chelation and Complexation

Chelation and complexation are two important mechanisms of phosphorus solubilization that are based on the principle of the functional group binding to metal cations in the soil, releasing phosphate for the purpose of phosphorus solubilization. Currently, chelation and complexation are produced mainly through siderophores produced by phosphorus-solubilizing bacteria, extracellular polysaccharides and the decomposition of plant and animal residues. Siderophores are small molecules with low molecular weights, and under low iron stress, phosphate-dissolving bacteria produce siderophores that chelate Fe^{3+}, Al^{3+} and Ca^{2+} in the soil, releasing phosphate ions for phosphate dissolution [71,100,101]. The formed metal-iron carrier complexes can bind to iron carrier receptor proteins on cell membranes to enter cells, increasing iron uptake by plants and promoting plant growth [25]. Extracellular polysaccharides secreted by PSB are high-molecular-weight sugar polymers attached to bacterial surfaces that have special anionic functional groups (phosphate, carboxyl and succinate groups) and abundant -OH and -COOH acid groups on the surface that can complex with cations of heavy metal elements in the soil [102]. During the degradation of plant and animal residues by PSB, humic acid-like substances are formed that can also chelate Fe^{3+}, Al^{3+} and Ca^{2+} in the soil, releasing phosphate and increasing the available phosphorus content in the soil [17].

3.2. Molecular Mechanisms

In addition to the physiological mechanism, the molecular mechanism of PSB is also important, but at present the complete molecular mechanism of the PSB pathway is not clear in the research; it is mainly studied through the genome sequencing of PSB's metabolism of acid and enzyme-related functional genes, such as *GDH, pqq, ALP, ACP,* and so on. Thus, the mechanism of PSB metabolism is gradually formed through the combination of different functional genes and the metabolites they regulate.

3.2.1. Functional Genes Related to the Regulation of Acidolysis

The production and secretion of organic acids are the main mechanisms of phosphorus solubilization by PSB, but the molecular mechanism of acidolysis has not been studied in depth, and there are only a few genes that can play a role in regulating phosphate solubilization (Table 6) such as *gad*γ (Gluconate 2-dehydrogenase), *gdh* (Glucose dehydrogenase), *gcd* (Glucose dehydrogenase) and *ylil* (Aldose sugar dehydrogenase) [103–105]. In addition to the proven *gcd* genes, pyruvate dehydrogenase (*poxB*), l-lactate dehydrogenase (*ldh*), glyoxylate reductase (*gyaR*) and glyoxylate/hydroxypyruvate/6-ketoglucuronate reductase (*ghrB*) are several genes associated with organic acid metabolism that may be relevant to phosphate solubilization by PSB [104]. However, among the many genes, gluconic acid-related genes have been more intensively studied, and many Gram-negative bacteria use periplasmic glucose oxidation, mediated by pyrroloquinoline quinone (*PQQ*) and glucose dehydrogenase (*GDH*), to produce gluconic acid that encodes *GCD*; pyrroloquinoline quinone acts as an oxidative cofactor in glucose dehydrogenase and is mediated by the *pqq* manipulator genes (A, B, C, D, E, F) [2,105,106]. Most *PQQ* genes are present in the genus Pseudomonas [107]. Among the pqq A-E genes, *pqqA*, *pqqC*, *pqqD* and *pqqE* are prerequisite genes for phosphate catabolism by PSB [108]. Moreover, *pqqA* and *pqqC* are the two key genes; the gene *pqqA* consists of 22 amino acids, the peptides of glutamic and tyrosine that provide carbon and nitrogen to the *PQQ* biological system, and the gene *pqqC* that encodes the pyrroloquinoline synthase C (*PQQC*) gene, which catalyzes the conversion of 3a-(2-amino-2-carboxyethyl)-4,5-dioxo-4,5,6,7,8,9-hexahydroquinoline-7,9-dicarboxylic acid to pyrroloquinoline quinone (*PQQ*) [109]. It has been shown that the *pqqC* gene is highest in many PSB, such as *Burkholderia* (8.98–16.69%), *Azotobacter* (5.27–11.05%), *Pseudomonas aeruginosa* (3.31–9.28%) and *Mycobacterium* (1.25–9.8%), and the *pqqC* gene has also been identified to track the inorganic phosphate-solubilizing ability marker genes of microorganisms [99]. Joshi et al. found 16 bacterial isolates with 82 bp *pqqC* gene amplification positivity and 6 bacterial isolates with 72 bp *pqqA* gene amplification positivity. Bacterial isolates exhibiting *pqqA* gene amplicons were also positive for the *pqqC* gene [110]. The results of these experiments demonstrated that there are two key genes in the *PQQ* biosynthetic pathway. The production of many organic acids, such as citric acid, α-ketoglutaric acid, succinic acid and malic acid, requires the TCA cycle, and genes related to the dephosphorylation mechanism may be relevantly linked to the TCA cycle. Ding et al. showed that the inoculation of PSB in insoluble phosphorus form into a medium resulted in a significant enhancement of TCA-related genes such as citrate synthase *(CS)*, isocitrate dehydrogenase (*IDH*), α-ketoglutarate dehydrogenase (*OGDH*), succinate dehydrogenase (*SDH*) and fumarate hydratase *(FH)* compared to the CK (not inoculated with phosphorus-solubilizing bacteria) [111]. Studies on the TCA cycle and the acidolysis mechanism of PSB are scarce, and these topics still need to be explored in depth.

3.2.2. Functional Genes Related to the Regulation of Enzymolysis

Many of the carrier genes in PSB encode different enzymes (Table 7), such as alkaline phosphatases (*phoD* and *phoA*), acid phosphatases (*phoC*), phytase (*appA*), C-P cleavage enzyme (*phn*), extracellular polyphosphatase (*ppx*) and polyphosphate kinase (*ppk*), with alkaline phosphatases (*phoD* and *phoA*) being the most studied [39,112]. Alkaline phosphatase production in PSB is encoded by one of three homologous gene families (*phoA*, *phoD*, *phoX*) as part of the pho regulator. Macrogenomics studies have also shown that

32% of sequenced prokaryotic genomes contain at least one *phoA*, *phoD* or *phoX* gene [113]. Under inorganic phosphorus deficiency conditions, especially, *phoA* and *phoD* are the most common genes, with *phoD* also being the most strongly linked to alkaline phosphatase [20]. Wan et al. also showed that inoculation with PSB increased the abundance of genes related to organic and inorganic phosphorus cycling, including *phoD*, and that the *phoD* copy number was positively correlated with soil alkaline phosphatase activity [11,114]. There are also numerous genes that can participate in the regulation of the phosphorus deprivation response (*phoU*, *phoR* and *phoB*), which are closely linked to genes involved in phosphorus uptake and transport (*pst*) and control the expression of genes encoding alkaline phosphatases, especially under low phosphorus conditions [112]. In their experiments, Dai et al. found that under low phosphorus conditions, the release of phosphatase could be enhanced by regulating a gene involved in the regulation of the phosphorus starvation response (*phoR*) to mineralize soil organophosphorus. In contrast, a consistently higher abundance of the *phoR* gene was observed only in low-phosphorus soils compared to high-phosphorus soils [112]. Although alkaline phosphatase is more likely to be secreted by soil PSB, acid phosphatase is still secreted in small amounts. Additionally, bacterial genera carrying pho genes may contain acid phosphatase genes, such as *Pseudomonas*, *Stemonella* and *Lysobacillus* [109,114].

Phytase is mainly encoded by the *bpp* gene, which is more widely distributed than other phytate degradation-related genes (*ptp* and *hap*), and there is also a strong positive correlation between *bpp* gene abundance and the number of phytate-degrading bacteria [94]. The *bpp* gene has a six-bladed propeller-folded structure and contains a cleavage, an affinity phosphate binding site and six calcium binding sites, three each for enzyme stability and catalysis. To date, only a small number of *bpp* in PSB have been isolated and investigated, e.g., *Bacillus subtilis* and *Bacillus licheniformis* [5]. In addition, the *phn* gene in C-P lyase and the genes for extracellular polyphosphatase (*ppx*) and polyphosphate kinase (*ppk*) biosynthesis can all solubilize inorganic phosphate (*polyP*) [115].

The study of phospholysis-related genes is of transgenerational significance, revealing the metabolic mechanism of phosphate-solubilizing microorganisms in solubilizing insoluble phosphorus at the genetic level, and we have witnessed the excavation of many phospholysis genes. Nevertheless, information about some unknown phosphate solubilization mechanisms and their related genes is still limited, and some pathways are still waiting to be discovered. Most of the studies at the present stage mainly focus on the genes coding for the synthesis of organic acids, such as gluconic acid and phosphatases. Studies on individual genes are restricted to the level of functional validation. In-depth research on the metabolic mechanisms of the upstream and downstream processes is still relatively limited, and the interactions and links between the related genes are not yet clear [60].

Table 6. Genes related to the acidolytic action of different PSB.

PSB Species	PSB Source	Related Genes	Functions	Reference
Pseudomonas putida	Laboratory storage	*gcd*	Encode glucose dehydrogenase, which promotes the solubilization of inorganic P	[116]
Pseudomonas sp.	Wheat rhizosphere soil	*gcd*	Encode glucose dehydrogenase, which promotes the solubilization of inorganic P	[105]
Acinetobacter	Soils in rocky desertification areas	*gcd*	Mediate the production of gluconic acid	[35]
Acinetobacter pittii gp-1	Laboratory storage	*gcd*	Promotion of the solubilization of inorganic and organic phosphorus	[94]
Ochrobactrum haematophilum	Sweet potato rhizosphere soil	*CS, ACO, ODGH, SFD, FH, MDA*	Tricarboxylic acid cycle-related genes	[111]

Table 6. *Cont.*

PSB Species	PSB Source	Related Genes	Functions	Reference
Ochrobactrum haematophilum	Sweet potato rhizosphere soil	*POX, LDH*	Acetic acid and lactic acid regulatory genes	[111]
Acinetobacter spp., *Pseudomonas* spp.	Rice rhizosphere soil	*pqq*C, *pqq*E	Regulation of gluconic acid production	[107]

Table 7. Genes related to the enzymolysis action of different PSB.

PSB Species	PSB Source	Related Genes	Functions	Reference
Ochrobactrum sp	Wheat rhizosphere soil	*pho*	Promotion of the solubilization of inorganic and organic phosphorus	[117]
Pantoea agglomerans	Wheat rhizosphere soil	*phy*	Participate in the dissolution of phytic acid	[117]
Acinetobacter pittii gp-1	Laboratory storage	*phoD, bpp,*	Promotion of the solubilization of inorganic and organic phosphorus	[94]
Arthrobacter	Corn rhizosphere soil	*Ppx, ppk*	Promotes the synthesis of exonuclease polyphosphatase and polyphosphate kinase	[94]
aryabhattai	Laboratory storage	*Phn, pho,*	Promotion of phosphorus metabolic pathway activity	[95]
Pseudomonas	Corn rhizosphere soil	*bpp*	Phytase-encoding genes	[118]
Pantoea brenneri	Soil samples of the Republic of Tatarstan	phnK	C-P lyase regulatory genes	[119]

3.3. *Mechanisms of Microbial Community Effects*

Current studies on PSB mechanisms mainly focus on bacteria. In addition to bacterial mechanisms, changes in soil nutrients also have a certain impact on the bacterial phosphorus-solubilizing effect. On the other hand, the relationship between PSB and indigenous microorganisms in the soil can also directly affect the structure and function of soil microorganisms, changing phosphorus transformation in the soil [120].

3.3.1. Effects of Soil Nutrient Changes on the Abundance of PSB Communities

The phosphorus-solubilizing capacity of PSB is related to the abundance of their community, which is closely related to nutrient changes in the soil. Exogenously applied fertilizers can improve soil nutrients and influence the abundance of phosphate-dissolving bacterial communities through pH and C:N:P stoichiometry [121]. Changes in the composition and activity of PSB communities are driven by the C:P and N:P ratios in the soil [116]. Several experiments have confirmed that the sustained, exogenous addition of inorganic phosphorus, biochar and organic carbon increased pH and that the abundance of PSB communities increased with pH, which was positively correlated with biomass, phosphorus content and phosphorus uptake [122–125]. By contrast, the effect of nitrogen fertilizer application on community abundance is not clear; moderate nitrogen fertilizer application can improve the soil nitrogen–phosphorus balance, improve the growth conditions of PSB, and increase the abundance of PSB communities [126]. However, high concentrations of nitrogen fertilizer can inhibit the growth and proliferation of PSB because nitrogen fertilizer application increases the nitrogen content of the soil, changes the nutrient environment and relative nitrogen–phosphorus ratios of soil microorganisms and decreases soil pH. PSB may be inhibited, which may lead to a decrease in their abundance [51,127,128]. Conversely, appropriate amounts of nitrogen reduction may be positive for the abundance of PSB communities [129]. The interventions of C, N and P all resulted in different changes in pH and corresponding changes in the structure of the PSB community. It has been shown that soil pH affects the abundance of PSB communities. In acidic environments, which contain more protons, it is easy to release insoluble phosphorus, so the abundance of PSB communities does not change significantly under normal circumstances, while in alkaline environments, because a large amount of residual phosphorus is immobilized by the binding of metal ions, the abundance of PSB communities rises significantly to assist in the release of insoluble phosphorus; therefore, the higher the pH value of the soil, the greater the abundance of the PSB community [130].

3.3.2. Effects of PSB on Soil Microcosm Systems

Current agricultural practices negatively impact soil microbial composition and biodiversity and are unsustainable [131]. In contrast, the effect of PSB on soil microorganisms is more positive. The introduction of PSB into soils can have an effect on both the diversity and abundance of the indigenous microbial community; the abundance and diversity of indigenous microorganisms may be enhanced, but some experiments have shown that the addition of PSB resulted in a significant decrease in the diversity of the indigenous bacterial community, an increase in the abundance of the inoculated PSB and an enhancement of the soil phosphorus content [132,133]. Bacterial communities with higher abundance are also positively correlated with the growth of some substances associated with the phospholysis mechanism (phosphatase, phytase, etc.) [134]. Thus, the contribution of the bacterial community to phosphorus solubilization can be indirectly improved by the interaction of PSB and indigenous microorganisms, suggesting that there is also a link between elevated soil phosphorus levels and changes in microecosystems, which indicates a potential mechanism for phosphorus enhancement. Therefore, further attention should be given to the effects of exogenous microbial inoculants, including PSB applied to calcareous soils, on the diversity and composition of soil bacterial communities.

In summary, the mechanism of phosphorus solubilization by PSB is complex and varied, and the physiological mechanism includes acidolysis, enzymatic action for mineralization and hydrolysis of insoluble phosphate and chelation and complexation for binding of cations in soil. The molecular direction covers the genes and pathways involved in the regulation of acidolysis and enzymolysis and with phosphate transport. In addition to the above internal regulatory mechanisms, PSB also have external regulatory mechanisms that alter the structure and abundance of microbial communities in the soil to indirectly achieve the dissolution of insoluble phosphorus. [134].

3.4. *PSB Regulates Plant Root Transporter Proteins*

PSB in the soil allow the conversion of insoluble phosphorus in the soil into phosphorus that can be directly absorbed and utilized by the plant, resulting in an increase in phosphorus content in the plant. This is not only related to the genes of PSB themselves but also to the interaction between PSB and the root system. PSB can regulate an inorganic phosphate transporter protein in plant roots to promote phosphorus uptake in plants in the presence of phosphorus deficiency. PSB can regulate an inorganic phosphate transporter protein in plant roots to promote phosphorus uptake in plants in the presence of phosphorus deficiency [20]. Srivastava et al. showed that *Pseudomonas* putida regulates *PHT1* and *PHT2* and relieves phosphorus deficiency under low phosphorus stress [135]. Similar conclusions were reached in the trial by Saia et al. where *PHT1* gene expression was increased in wheat due to inoculation with Bacillus, promoting phosphorus uptake in wheat [136]. Moreover, the elevated expression of inorganic phosphate transporter proteins due to PSB regulation will cause plants to develop smaller root systems with more lateral branches and root hairs, which stimulate phosphorus transporter proteins that produce phosphatases to solubilize phosphorus that is not available to plants [137]. Thus, PSB have significant implications for improving plant uptake of available phosphorus by inducing inorganic phosphate transport proteins in the plant root system.

4. Problems and Future Outlook

4.1. Multi-omic Synergy in the In-Depth Mining of Functional Genes of PSB

The use of multi-omic synergy can increase the rate of research progress on the mechanism of dephosphorylation. Macrogenomic technology can be used to analyze genes that are significantly up- or downregulated by PSB in different environments, GO (gene ontology) can be used to functionally annotate them and QS (group sensing) analysis can reveal the regulatory mechanism of PSB group effects. Joint transcriptome and metabolome analyses to uncover differential genes can be used to quickly determine the core regulatory networks and key candidate genes, clarify the functions of transcription factors and ultimately screen for functional genes in mechanistic studies. Ding et al. used multi-omic synergy to investigate the mechanism of PSB, and the macrogenomic results identified the genes of the main enzymes and their coenzymes involved in the synthesis of gluconic acid, 2-ketogluconic acid and indole acetic acid (IAA), and the transcriptomics analyses of differential genes in PSB with GO enrichment and KEGG enrichment revealed the relevant gene expression levels and their regulatory proteins related to the synthesis pathway of organic acids. Transcriptomics analysis of differential gene expression levels and regulatory proteins in PSB revealed the types and amounts of organic acids secreted by PSB, and metabolomics results also showed the types and amounts of organic acids secreted by phosphorus-solubilizing bacteria [111]. Currently, there are some similarities and differences in the related genes and pathways obtained by whole genome sequencing and enrichment analyses of different PSB. However, this can also prove that the study of the whole genome sequence of PSB may be an important and new research direction to gain insights into its mechanism (Table 8). Therefore, multi-omics crossover will also open up new avenues for the study of phospholysis mechanisms.

Table 8. Information on different PSB whole-genome sequencing and enrichment analyses.

PSB	Genome Size (bp)	G+C (%)	Predicted Number of Genes	Acidolysis of Associated Genes	Enzymatic Hydrolysis of Associated Genes	Genes Associated with Phosphorus Cycling	Related Pathways	Reference
Pseudomonas sp.	5,617,746	62.86	5097	*GDH, ppq, gcd, gdh*	*ppa, PPX*	*PstS, PstC, PstA, PstB, PhoU*	Entner–Doudoroff (ED)	[138]
Serratia marcescens	5,061,510	59.75	4742	*gcd, PQQ, ipdC*	*ppx, ppa, phoA, phnX, chi*	-	Secreted iron carrier-complete pathway	[139]
Bacillus subtilis	4,173,570	43.25	4604	*pyruvate, carboxylase*	*phoD, phoR*	*pit, pstS, pstB*	Carbon metabolism, biosynthesis of amino acids	[140]
Enterobacter cloacae	4,608,301	54.78	4450	*ppq, gcd, gdh*	-	*phnG, PhnH, PhnJ, PhnM, PhnP, PhnI*	-	[141]
Pseudomonas putida	5,957,620	61.55	5535	*GDH, pqq, gcd, aceE, aceF, lpd, IDH3, idhA, dld, maeB*	-	*PstS, PstC, PstA, PstB, PhoU*	Tricarboxylic acid cycle and pyruvate pathway	[142]
Pseudomonas fildesensis	6,789,479	60.82	6028	*GDH, pqq, gcd, aceE, aceF, lpd, IDH3, idhA, dld, maeB*	*phoD*	*PstS, PstC, PstA, PstB, PhoU*	Tricarboxylic acid cycle and pyruvate pathway	[142]
Pantoea	7,989,160	51.3	4548	*pqq, GCD*	-	*phnN, phnM*	Indole-3-pyruvate pathway	[143]

4.2. PSB Regulatory Genes in Soils with Different Phosphorus Levels

As the molecular mechanisms of PSB continue to be explored, the number of PSB regulatory genes has gradually increased, but soil phosphorus content and the types of refractory phosphorus will make the abundance and types of related genes different. Sun et al. succeeded in experimentally predicting five related genes that can solubilize inorganic phosphorus and mineralize organic phosphorus: *ppa* (encoding inorganic pyrophosphatase), *phoD* (encoding alkaline phosphatase D), *phnW* (encoding 2-aminoethylphosphonate pyruvate aminotransferase), *phnX* (encoding phosphorylglycolide hydrolase), *phnA* (encoding phosphonoacetic acid hydrolase) and *phnA* (encoding phosphonoacetic acid hydrolase). These five genes were significantly enriched in phosphorus-deficient soils, and their total abundance was higher than that in soils with normal phosphorus levels. However, the relative abundance of two other genes, *gcd* (encoding pentaprotein glucose dehydrogenase) and *appA* (encoding 4-phosphatase/acid phosphatase), was reduced in phosphorus-deficient soils compared to normal-phosphorus soils [134]. Lv et al. found the highest abundance of inorganic phosphorus-solubilizing genes (*gcd*, *pqqC* and *ppa*), the second-highest abundance of phosphorus deprivation-regulated genes (*phoR*) and the lowest abundance of genes involved in the mineralization of organic phosphorus (*appA* and 3-phytase) in the soil [144]. Therefore, continuous summarization and understanding of the types and abundance of genes regulated by PSB under different phosphorus levels are also among the components of the continuous deepening of understanding of the molecular mechanism of PSB.

4.3. Use of PSB for Making Microbial Preparations in Agriculture

PSB have the function of improving soil fertility, remediating land pollution, reducing heavy metals on land and restoring the ecological environment. Microbial agents made from dephosphorylating bacteria have great potential. For example, microbial preparations using *Pantoea* sp. and *Pseudomonas* sp. as formulas have been proven to be effective in promoting the growth of lucerne, and *Bacillus licheniformis* and *Pseudomonas aeruginosa*, alone or in combination, can also be effective in promoting the growth of Brassica campestri alone or in combination [134]. At present, about 80 countries in the world are carrying out the promotion of microbial preparations, but there are still some unsolved problems. For example, the backward production process leads to insufficient inoculum quantity, which is not enough to achieve the effect of a stable yield increase. There are fewer new varieties of microbial preparations that are applicable to a single environment [145]. In addition, in terms of the safety of dephosphorylating bacteria, sometimes there is a certain threat of dephosphorylating microorganisms promoting conditions for pathogenic bacteria, such as the strong pathogen *Klebsiella pneumoniae*, so the application should be screened to prevent the negative impacts of the spread of the application and to make safety the first priority [146]. In addition to the problems of microbial agents in production applications, there are also concerns about how much higher crop yields and economic benefits can be achieved by applying microbial agents to the soil. The emergence of machine learning models has been applied to assess the effectiveness of phosphorus-solubilizing microorganisms in agricultural remediation, and it is believed that the application of machine learning models to predict the effectiveness of microbial agents in agricultural production also has a promising future [147].

Author Contributions: B.C. was responsible for research concept and design, supervision, and review. L.P. was responsible for finding the literature and writing and editing. All authors have read and agreed to the published version of the manuscript.

Funding: This work was supported by a grant from the National Natural Science Foundation of China (NO. 31972502).

Conflicts of Interest: The authors declare no conflict of interest.

References

1. Siedliska, A.; Piotr, B.; Joanna, P.; Monika, Z.; Jaromir, K. Identification of plant leaf phosphorus content at different growth stages based on hyperspectral reflectance. *BMC Plant Biol.* **2021**, *21*, 28. [CrossRef]
2. Elhaissoufi, W.; Ghoulam, C.; Barakat, A.; Zeroual, Y.; Bargaz, A. Phosphate bacterial solubilization: A key rhizosphere driving force enabling higher P use efficiency and crop productivity. *J. Adv. Res.* **2022**, *38*, 13–28. [CrossRef] [PubMed]
3. Chouyia, F.; Ventorino, V.; Pepe, O. Diversity, mechanisms and beneficial features of phosphate-solubilizing Streptomyces in sustainable agriculture: A review. *Front. Plant Sci.* **2022**, *13*, 1035358. [CrossRef]
4. Alori, E.; Glick, B.; Babalola, O. Microbial phosphorus solubilization and its potential for use in sustainable agriculture. *Front. Microbiol.* **2017**, *8*, 971. [CrossRef] [PubMed]
5. Suleimanova, A.D.; Beinhauer, A.; Valeeva, L.R.; Chastukhina, I.B.; Balaban, N.P.; Shakiro, E.V.; Greiner, R.; Sharipova, M.R. Novel glucose-1-phosphatase with high phytase activity and unusual metal ion activation from soil *bacterium pantoea* sp. strain 3.5.1. Appl. *Environ. Microbiol.* **2015**, *81*, 6790–6799. [CrossRef] [PubMed]

6. Santos, H.; Silva, G.; Carnietto, M.; Oliveira, L.; Nogueira, C.; Silva, M. *Bacillus velezensis* associated with organomineral fertilizer and reduced phosphate doses improves soil microbial—Chemical properties and biomass of sugarcane. *Agronomy* **2022**, *12*, 2701. [CrossRef]
7. Singh, A.; Singh, J.; Singh, R.; Kantwa, S.; Jha, P.; Ahamad, S.; Singh, A.G.; Prasad, M.; Singh, S.; Singh, S.; et al. Understanding soil carbon and phosphorus dynamics under grass-legume intercropping in a semi-arid region. *Agronomy* **2023**, *13*, 1692. [CrossRef]
8. Cordell, D.; Schmid-Neset, T.; White, S.; Drangert, J.O. Preferred future phosphorus scenarios: A framework for meeting long-term phosphorus needs for global food demand. In Proceedings of the 2009 International Conference on Nutrient Recovery from Wastewater Streams, Vancouver, BC, Canada, 10–13 May 2009.
9. Gross, A.; Lin, Y.; Weber, P.K.; Pett-Ridge, J.; Silver, W.L. The role of soil redox conditions in microbial phosphorus cycling in humid tropical forests. *J. Ecol.* **2020**, *101*, e02928. [CrossRef]
10. Liang, J.; Liu, J.; Jia, P.; Yang, T.; Zeng, Q.; Zhang, S.; Liao, B.; Shu, W.; Li, T. Novel phosphate-solubilizing bacteria enhance soil phosphorus cycling following ecological restoration of land degraded by mining. *ISME J.* **2020**, *14*, 1600–1613. [CrossRef]
11. Neal, A.; Blackwell, M.; Akkari, E.; Guyomar, C.; Clark, I.; Hirsch, P. Phylogenetic distribution, biogeography and the effects of land management upon bacterial non-specific acid phosphatase gene diversity and abundance. *Plant Soil.* **2018**, *427*, 175–189. [CrossRef]
12. Udaondo, Z.; Duque, E.; Daddaoua, A.; Caselles, C.; Roca, A.; Pizarro-Tobias, P.; Ramos, J. Developing robust protein analysis profiles to identify bacterial acid phosphatases in genomes and metagenomic libraries. *Environ. Microbiol.* **2020**, *22*, 3561–3571. [CrossRef]
13. Kiprotich, K.; Muoma, J.; Omayio, D.; Ndombi, T.; Wekesa, C. Molecular characterization and mineralizing potential of phosphorus solubilizing bacteria colonizing common bean (*Phaseolus vulgaris* L.) rhizosphere in Western Kenya. *Int. J. Microbiol.* **2023**, *2023*, 6668097. [CrossRef] [PubMed]
14. Massucato, L.; Almeida, S.; Silva, M.; Mosela, M.; Zeffa, D.; Nogueira, A.; Filho, R.; Mian, S.; Higashi, A.; Teixeira, G.; et al. Efficiency of combining strains Ag87 (*Bacillus megaterium*) and Ag94 (*Lysinibacillus* sp.) as phosphate solubilizers and growth promoters in maize. *Microorganisms* **2020**, *10*, 1401. [CrossRef] [PubMed]
15. Stalstrom, V.A. Boitrag zur kennturs, and Ein-wisking. Sterier use in garung bofindlicher oranischer strofe auf dil losichket der phosphorson des tricalclum phosphate. *Zel Bakt.* **1903**, 724–732.
16. Gerretsen, F.C. The influence of micro-organism on the phosphate intaken by the plant. *Plant Soil* **1948**, *1*, 51–60. [CrossRef]
17. Chi, J.L.; Hao, M.; Wang, Z.X. Advances in research and application of phosphorus-solubilizing microorganism. *J. Microbiol.* **2021**, *41*, 1–7. (In Chinese)
18. Liu, J.; Liu, X.; Zhang, Q.; Li, S.; Sun, Y.; Lu, W.; Ma, C. Response of alfalfa growth to arbuscular mycorrhizal fungi and phosphate-solubilizing bacteria under different phosphorus application levels. *AMB Express.* **2020**, *10*, 200. [CrossRef] [PubMed]
19. Cao, Y.; Fu, D.; Liu, T.; Guo, G.; Hu, Z. Phosphorus solubilizing and releasing bacteria screening from the rhizosphere in a natural wetland. *Water* **2018**, *10*, 195. [CrossRef]
20. Timofeeva, A.; Galyamova, M.; Sedykh, S. Prospects for using phosphate-solubilizing microorganisms as natural fertilizers in agriculture. *Plants* **2022**, *11*, 2119. [CrossRef]
21. Rong, G.Q.; Qin, X.K.; Yao, Q.J.; Zhang, Z.Q.; Hu, Y.B. Research on application status of soil phosphate solubilizing bacteria in modern agriculture. *Farm Prod. Process.* **2021**, *13*, 86–89.
22. Li, M.; Teng, Z.D.; Zhu, J.; Song, M.Y. Research advances in heavy metal contaminated soil remediation by phosphate solubilizing microorganisms. *Acta Ecol. Sin.* **2018**, *38*, 3393–3402.
23. Djuuna, I.A.F.; Prabawardani, S.; Massora, M. Population distribution of phosphate-solubilizing microorganisms in agricultural soil. *Microbes Environ.* **2022**, *37*, ME21041. [CrossRef] [PubMed]
24. Liu, C.; Chen, L.; He, Z.; Zhang, Z.; Xu, Y.; Li, Z.; Peng, Y.; Deng, N.; Chen, Y. Integration and potential application ability of culturable functional microorganism in oil tea camellia. *Indian J. Microbiol.* **2021**, *61*, 1–9. [CrossRef]
25. Li, K.; Zeghbroeck, J.V.; Liu, Q.; Zhang, S. Isolating and characterizing phosphorus solubilizing bacteria from rhizospheres of native plants grown in calcareous soils. *Front. Environ. Sci.* **2021**, *9*, 802563. [CrossRef]
26. Sahu, S.; Rajbonshi, M.; Gujre, N.; Gupta, M.; Shelke, R.; Ghose, A.; Rangan, L.; Pakshirajan, K.; Mitra, S. Bacterial strains found in the soils of a municipal solid waste dumping site facilitated phosphate solubilization along with cadmium remediation. *Chemosphere* **2020**, *287*, 132320. [CrossRef]
27. Bahadur, I.; Maurya, B.; Meena, V.; Saha, M.; Kumar, A.; Aeron, A. Mineral release dynamics of tricalcium phosphate and waste muscovite by mineral-solubilizing rhizobacteria isolated from Indo-Gangetic plain of India. *Geomicrobiol. J.* **2017**, *34*, 454–466. [CrossRef]
28. Kashyap, A.; Manzar, N.; Rajawat, M.; Kesharwani, A.; Singh, R.; Dubey, S.; Abhijeet, S.K.; Pattanayak, D.; Dhar, S.; Lai, S.K.; et al. Screening and biocontrol potential of rhizobacteria native to gangetic plains and hilly regions to induce systemic resistance and promote plant growth in Chilli against Bacterial Wilt Disease. *Plants* **2021**, *10*, 2125. [CrossRef]
29. Zhu, Y.; Ku, Y.; Liu, J.; Le, T.; Zhao, Z. Community characteristics and functions of phosphate-solubilizing bacteria in rhizosphere soil of natural and planted *Pinus tabuliformis* forests on the Loess Plateau, Northwest China. *Ying Yong Sheng Tai Xue Bao* **2021**, *32*, 3097–3106. (In Chinese)

30. Ghosh, R.; Barman, S.; Mukherjee, R.; Mandal, N. Role of phosphate solubilizing *Burkholderia* spp. for successful colonization and growth promotion of *Lycopodium cernuum* L. (Lycopodiaceae) in lateritic belt of Birbhum district of West Bengal, India. *Microbiol. Res* **2016**, *183*, 80–91. [CrossRef]
31. Tang, A.; Haruna, A.; Majid, N.; Jalloh, M. Potential PGPR properties of cellulolytic, nitrogen-fixing, phosphate-solubilizing bacteria in rehabilitated tropical forest soil. *Microorganisms* **2020**, *8*, 442. [CrossRef]
32. Saeed, Q.; Xiukang, W.; Haider, F.; Kučerik, J.; Mumtaz, M.; Holatko, J.; Munaza, N.; Antonin, K.; Mukkaram, E.; Muhammad, N.; et al. Rhizosphere bacteria in plant growth promotion, biocontrol, and bioremediation of contaminated sites: A comprehensive review of effects and mechanisms. *Int. J. Mol. Sci.* **2021**, *22*, 10529. [CrossRef]
33. Gómez-Godínez, L.J.; Aguirre-Noyola, J.L.; Martínez-Romero, E.; Arteaga-Garibay, R.I.; Ireta-Moreno, J.; Ruvalcaba-Gómez, J.M. A Look at plant-growth-promoting bacteria. *Plants* **2023**, *12*, 1668. [CrossRef] [PubMed]
34. Gao, J.; Luo, Y.; Wei, Y.; Huang, Y.; Zhang, H.; He, W.; Sheng, H.; An, L. Screening of plant growth promoting bacteria (PGPB) from rhizosphere and bulk soil of *Caragana microphylla* in different habitats and their effects on the growth of Arabidopsis seedlings. *Biotechnol. Biotechnol. Equip.* **2019**, *33*, 921–930. [CrossRef]
35. Xie, J.; Yan, Z.; Wang, G.; Xue, W.; Li, C.; Chen, X.; Chen, D. A Bacterium isolated from soil in a karst rocky desertification region has efficient phosphate-solubilizing and plant growth-promoting ability. *Front. Microbiol.* **2021**, *11*, 625450. [CrossRef]
36. Zhang, C.; Chen, H.; Dai, Y.; Chen, Y.; Tian, Y.; Huo, Z. Isolation and screening of phosphorus solubilizing bacteria from saline alkali soil and their potential for Pb pollution remediation. *Front. Bioeng. Biotechnol.* **2023**, *11*, 1134310. [CrossRef]
37. Li, H.; Su, J.; Yang, X.; Zhu, Y. Distinct rhizosphere effect on active and total bacterial communities in paddy soils. *Sci. Total Environ.* **2019**, *649*, 422–430. [CrossRef] [PubMed]
38. Mirza, M.; Ahmad, W.; Latif, F.; Haurat, J.; Bally, R.; Normand, P.; Malik, K. Isolation, partial characterization, and the effect of plant growth-promoting bacteria (PGPB) on micro-propagated sugarcane in vitro. *Plant Soil* **2001**, *237*, 47–54. [CrossRef]
39. Guo, L.; Wang, C.; Shen, R. Stronger effects of maize rhizosphere than phosphorus fertilization on phosphatase activity and phosphorus-mineralizing-related bacteria in acidic soils. *Rhizosphere* **2022**, *23*, 100555. [CrossRef]
40. Zheng, M.M.; Wang, C.; Shen, R.F. Efects of calcium carbonate and rhizosphere on abundance of phosphate-solubilizing microorqanisms in acidic red soil. *Soils* **2022**, *52*, 704–709. (In Chinese)
41. Gupta, R.; Kumari, A.; Sharma, S.; Alzahrani, O.; Noureldeen, A.; Darwish, H. Identification, characterization and optimization of phosphate solubilizing rhizobacteria (PSRB) from rice rhizosphere. *Saudi J. Biol. Sci.* **2022**, *29*, 35–42. [CrossRef]
42. Yahya, M.; Islam, E.; Rasul, M.; Farooq, I.; Mahreen, N.; Tawab, A.; Muhammad, I.; Lubna, R.; Imran, A.; Sumera, Y. Differential Root Exudation and Architecture for Improved Growth of Wheat Mediated by Phosphate Solubilizing Bacteria. *Front. Microbiol.* **2021**, *12*, 744094. [CrossRef] [PubMed]
43. Panda, P.; Ray, P.; Mahato, B.; Paramanik, B.; Choudhury, A.; Karforma, J. Performance of phosphate solubilizing bacteria in tea (*Camellia sinensis* L.) Rhizosphere. *Natl. Acad. Sci. Lett.* **2021**, *44*, 561–564. [CrossRef]
44. Ponmurugan, P.; Gopi, C. Distribution pattern and screening of phosphate solubilizing bacteria isolated from different food and forage crops. *Agron. J.* **2006**, *5*, 600–604.
45. Chen, D.; Sun, W.; Xiang, S.; Zou, S. High-throughput sequencing analysis of the composition and diversity of the bacterial community in *Cinnamomum camphora* soil. *Microorganisms* **2021**, *10*, 72. [CrossRef] [PubMed]
46. Wang, Z.; Zhang, H.; Liu, L.; Li, S.; Xie, J.; Xue, X.; Jiang, Y. Screening of phosphate-solubilizing bacteria and their abilities of phosphorus solubilization and wheat growth promotion. *BMC Microbiol.* **2022**, *22*, 296. [CrossRef]
47. Huang, C.; Yang, K.Y.; Gao, P.; Liang, Y.P.; Han, L.J.; Zhao, X. Screening, ldentification and characteristics of phosphate-solubilizing microorganisms in *Lespedeza daurica*. *Acta Agrestia Sinica.* **2022**, *30*, 2345–2355.
48. Li, H.; Li, C.; Song, X.; Li, J.; Zhang, P.; Sun, F.; Geng, Z.; Liu, X. Isolation and identification of antagonistic *Bacillus amyloliquefaciens* HSE-12 and its effects on peanut growth and rhizosphere microbial community. *Front. Microbiol.* **2023**, *14*, 1274346. [CrossRef]
49. Mei, X.L.; Shan, A.Q.; Jiang, Y.; Wei, Z.; Wang, Y.Y.; Wang, S.M. Screening of phosphorus solubng bacteria adapted to maize and their effect on maize growth. *Acta Pedologica Sinica.* **2016**, *53*, 502–509.
50. Ateş, Ç.; Yalçin, G.R.; Taşpinar, K.; Alveroğlu, V. Isolation and characterization of phosphate solubilizing bacteria and effect of growth and nutrient uptake of maize under pot and field conditions. *Commun. Soil Sci. Plant Anal.* **2022**, *53*, 2114–2124. [CrossRef]
51. Janati, W.; Mikou, K.; Ghadraoui, L.; Errachidi, F. Growth stimulation of two legumes (*Vicia faba* and *Pisum sativum*) using phosphate-solubilizing bacteria inoculation. *Front. Microbiol.* **2023**, *14*, 1212702. [CrossRef]
52. Xiao, K.; Cui, Y.; Gao, D.Y.; Wang, Z.G.; Yan, A.H. Screening of rhizosphere phosphorus-dissolving bacteria of walnut and studies on their role in walnut promotion. *J. Hebei Agric. Univ.* **2018**, *41*, 49–54.
53. Mahdi, I.; Fahsi, N.; Hafidi, M.; Allaoui, A.; Biskri, L. Plant Growth Enhancement using Rhizospheric halotolerant phosphate solubilizing *Bacterium Bacillus* licheniformis QA1 and *Enterobacter asburiae* QF11 isolated from *Chenopodium quinoa* Willd. *Microorganisms* **2020**, *80*, 948. [CrossRef] [PubMed]
54. He, Z.D.; Gao, Y.F.; Wang, Y.; Li, C.X.; Gao, X.Y.; Zhang, Z.H. Study on phosphate-solubilizing strain selection of plant growth promoting rhizobacteria and its effect on tomato growth promotion. *Southwest China J. Agric. Sci.* **2020**, *33*, 2891–2896.
55. Shah, S.H.; Hussain, M.B.; Zahir, Z.A.; Haq, T.U.; Matloob, A. Thermal plasticity and cotton production enhancing attributes of phosphate-solubilizing bacteria from cotton rhizosphere. *J. Soil. Sci. Plant Nutr.* **2022**, *22*, 3885–3900. [CrossRef]
56. Liu, C.J.; Du, C.Y.; Liang, Z.J.; Xia, Z.J.; Zhang, D.Z.; Gao, Y.P. Identification of high-efficiency phosphate—Dissolving strain CT45-1 and its promoting effect on tobacco. *Shandong Agric. Sci.* **2019**, *51*, 74–78. (In Chinese)

57. Susilowati, L.; Kusumo, B.; Arifin, Z. Screening of the drought tolerant phosphate solubilizing bacteria in dissolving P-inorganic. *J. Phys. Conf. Ser.* **2019**, *1402*, 55082. [CrossRef]
58. Soares, A.; Nascimento, V.; De Oliveira, E.; Jumbo, L.; Dos Santos, G.; Queiroz, L.; Silva, R.; Filho, R.; Romero, M.; Aguiar, R. *Pseudomonas aeruginosa* and *Bacillus cereus* isolated from Brazilian cerrado soil act as phosphate-solubilizing bacteria. *Curr. Microbiol.* **2023**, *80*, 146. [CrossRef]
59. Ren, J.; Liu, H.; Wu, X.; Wang, Q.; Ren, Y.; Liu, Y. Screening, identification, and promoting effect of phosphate-solubilizing bacteria in rhizosphere of Taxus chinensis var. mairei. *Wei Sheng Wu Hsüeh Pao* **2012**, *52*, 295–303.
60. Wang, M.; Sun, H.; Xu, Z. Analysis of blueberry plant rhizosphere bacterial diversity and selection of plant growth promoting rhizobacteria. *Curr. Microbiol.* **2022**, *79*, 331. [CrossRef]
61. Wei, X.; Fu, T.; He, G.; Zhong, Z.; Yang, M.; Lou, F.; He, T. Characteristics of rhizosphere and bulk soil microbial community of Chinese cabbage (*Brassica campestris*) grown in Karst area. *Front. Microbiol.* **2023**, *14*, 1241436. [CrossRef]
62. Liu, Y.; Cao, X.; Li, H.; Zhou, Z.; Wang, S.; Wang, Z.; Song, C.; Zhou, Y. Distribution of phosphorus-solubilizing bacteria in relation to fractionation and sorption behaviors of phosphorus in sediment of the Three Gorges Reservoir. *Environ. Sci. Pollut. Res.* **2017**, *24*, 17679–17687. [CrossRef] [PubMed]
63. Lu, X.M.; Shang, K.; Liu, X.; Wei, Q.S.; Ran, X.B.; Du, G.X.; Zhao, S.J.; Qu, L.Y. Diversity characteristics of organic phosphate-solubilizing bacteria in surface sediments of the Yangtze River Estuary. *Adv. Mar. Biol.* **2019**, *37*, 495–507.
64. Qu, J.H.; Zhang, L.J.; Fu, Y.H.; Li, H.F.; Tian, H.L. Isolation Identification and phosphorus-dissolving capacity of an efficient phosphate-solubilizing bacterium. *J. Henan Agric. Sci.* **2018**, *47*, 55–91. (In Chinese)
65. Yuan, Z.; Liu, F.; Zhang, G. Characteristics and biodiversity of endophytic phosphorus- and potassium-solubilizing bacteria in moso bamboo (*Phyllostachys edulis*). *Acta. Biol. Hung.* **2015**, *66*, 449–459. [CrossRef]
66. Tian, Q.; Gong, Y.; Liu, S.; Ji, M.; Tang, R.; Kong, D.; Qin, S. Endophytic bacterial communities in wild rice (*Oryza officinalis*) and their plant growth-promoting effects on perennial rice. *Front. Plant Sci.* **2023**, *14*, 1184489. [CrossRef] [PubMed]
67. Baghel, V.; Thakur, J.K.; Yadav, S.S.; Manna, M.C.; Mandal, A.; Shirale, A.O.; Sharma, P.; Sinha, N.K.; Mohanty, M.; Singh, A.B.; et al. Phosphorus and potassium solubilization from rock minerals by *Endophytic burkholderia* sp. Strain FDN2-1 in soil and shift in diversity of bacterial endophytes of corn root tissue with crop growth stage. *Geomicrobiol. J.* **2020**, *37*, 550–563. [CrossRef]
68. Tang, D.X.; Gao, Y.Z. Advances on the strategies of soil phosphate solubilizing micioorganisms to promote plan. *Sheng Tai Xue Bao* **2023**, *43*, 4390–4399. (In Chinese)
69. Dinesh, R.; Srinivasan, V.; Praveena, R.; Subila, K.; George, P.; Das, A.; Shajina, O.; Anees, N.K.; Leela, P. Exploring the potential of P solubilizing rhizobacteria for enhanced yield and quality in turmeric (*Curcuma longa* L.). *Ind. Crops Prod.* **2022**, *189*, 115826. [CrossRef]
70. Patel, D.; Murawala, P.; Archana, G.; Naresh Kumar, G. Repression of mineral phosphate solubilizing phenotype in the presence of weak organic acids in plant growth promoting fluorescent pseudomonads. *Bioresour. Technol.* **2021**, *102*, 3055–3061. [CrossRef]
71. Etesami, H.; Jeong, B.; Glick, B. Contribution of Arbuscular Mycorrhizal Fungi, Phosphate–Solubilizing Bacteria, and Silicon to P Uptake by Plant. *Front. Plant Sci.* **2021**, *12*, 699618. [CrossRef]
72. Billah, M.; Khan, M.; Bano, A.; Hassan, T.; Munir, A.; Gurmani, A. Phosphorus and phosphate solubilizing bacteria: Keys for sustainable agriculture. *Geomicrobiol. J.* **2019**, *36*, 904–916. [CrossRef]
73. Dash, S.; Borah, S.S.; Kalamdhad, A.S. Study of the limnology of wetlands through a one-dimensional model for assessing the eutrophication levels induced by various pollution sources. *Ecol. Model.* **2019**, *416*, 108907. [CrossRef]
74. Kalayu, G. Phosphate solubilizing microorganisms: Promising approach as biofertilizers. *Agron. J.* **2019**, *2019*, 4917256. [CrossRef]
75. Wei, Y.; Zhao, Y.; Shi, M.; Cao, Z.; Lu, Q.; Yang, T.; Fang, Y.; Wei, Z. Effect of organic acids production and bacterial community on the possible mechanism of phosphorus solubilization during composting with enriched phosphate-solubilizing bacteria inoculation. *Bioresour. Technol.* **2018**, *247*, 190–199. [CrossRef] [PubMed]
76. Hu, H.; He, J.; Li, X.; Liu, F. Effect of several organic acids on phosphate adsorption by variable charge soils of central China. *Environ Int.* **2001**, *26*, 353–358. [CrossRef]
77. Li, G.; Wu, X.; Ye, J.; Yang, H. Characteristics of Organic Acid Secretion Associated with the Interaction between *Burkholderia multivorans* WS-FJ9 and Poplar Root System. *Biomed Res. Int.* **2018**, *2018*, 9619724. [CrossRef]
78. Yue, L.; Jiao, L.; Tao, M.; Xu, L.; Cao, X.; Chen, F.; Wang, C.; Cheng, B.; Wang, Z. Dynamics of organic acid exudation and rhizobacteria in maize rhizosphere respond to N-CDs. *Sci. Total Environ.* **2023**, *901*, 166500. [CrossRef]
79. Wang, Y.; Luo, D.; Xiong, Z.; Wang, Z.; Gao, M. Changes in rhizosphere phosphorus fractions and phosphate-mineralizing microbial populations in acid soil as influenced by organic acid exudation. *Soil Tillage Res.* **2023**, *225*, 105543. [CrossRef]
80. De Weert, S.; Vermeiren, H.; Mulders, I.; Kuiper, I.; Hendrickx, N.; BloeMberg, G.; Vanderleyden, J.; Mot, R.; Lugtenberg, B. Flagella-driven chemotaxis towards exudate components is an important trait for tomato root colonization by *Pseudomonas fluorescens*. *Mol. Plant Microbe Interact.* **2002**, *15*, 1173–1180. [CrossRef]
81. Rudrappa, T.; Czymmek, K.; Paré, P.; Bais, H. Root-Secreted Malic Acid Recruits Beneficial Soil Bacteria. *Plant Physiol.* **2008**, *148*, 1547–1556. [CrossRef]
82. Zhao, X.R.; Lin, Q.M.; Li, B.G. The Relationship Between Rock Phosphate Solubilization and pH and Organic Acid Production of Microorganisms. *J. Microbiol.* **2003**, *23*, 5–7.
83. Qin, L.J.; Wang, Z.H.; Chen, Q.B.; Bai, C.J. Effect of inoculation with phosphate-solubilizing bacteria and soil acidification on the growth of *Stylosanthes guianensis* Reyan No. 2. *Cao Ye Xue Bao* **2008**, *17*, 20–28.

84. Vazquez, P.; Holguin, G.; Puente, M.; Lopez-Cortes, A.; Bashan, Y. Phosphate-solubilizing microorganisms associated with the rhizosphere of mangroves in a semiarid coastal lagoon. *Biol. Fertil. Soils.* **2000**, *30*, 460–468. [CrossRef]
85. Li, X.; Wang, C.; Cao, X.; Cao, A.; Han, Y.; Zhao, N. Isolation and identification of saline tolerance phosphate-solubilizing bacteria derived from salt-affected soils and their mechanisms of p-solubilizing. *ICAB* **2014**, *2*, 1259–1266.
86. Wang, J.; Yan, A.; Wang, W.; Li, J.; Li, Y. Screening, identification and phosphate-solubilizing characteristics of phosphate-solubilizing bacteria strain D2 (Pantoea sp.) in rhizosphere of *Pinus tabuliformis* in iron tailings yard. *Ying Yong Sheng Tai Xue Bao* **2016**, *27*, 3705–3711. [PubMed]
87. Chawngthu, L.; Hnamte, R.; Lalfakzuala, R. Isolation and characterization of rhizospheric phosphate solubilizing bacteria from wetland paddy field of Mizoram, India. *Geomicrobiol. J.* **2020**, *37*, 366–375. [CrossRef]
88. Khourchi, S.; Elhaissoufi, W.; Loum, M.; Ibnyasser, A.; Haddine, M.; Ghani, R.; Barakat, A.; Zeroual, Y.; Rchiad, Z.; Delaplace, P. Phosphate solubilizing bacteria can significantly contribute to enhance P availability from polyphosphates and their use efficiency in wheat. *Microbiol. Res.* **2022**, *262*, 127094. [CrossRef]
89. Zhang, H.; Han, L.; Jiang, B.; Long, C. Identification of a phosphorus-solubilizing *Tsukamurella tyrosinosolvens* strain and its effect on the bacterial diversity of the rhizosphere soil of peanuts growth-promoting. *World J. Microbiol. Biotechnol.* **2021**, *37*, 109. [CrossRef]
90. Parastesh, F.; Alikhani, H.; Etesami, H. Vermicompost enriched with phosphate–solubilizing bacteria provides plant with enough phosphorus in a sequential cropping under calcareous soil conditions. *J. Clean. Prod.* **2019**, *221*, 27–37. [CrossRef]
91. Trujillo-Narcía, A.; Rivera-Cruz, M.; Magaña-Aquino, M.; Trujillo-Rivera, E. The burning of sugarcane plantation in the tropics modifies the microbial and enzymatic processes in soil and rhizosphere. *J. Soil Sci. Plant Nutr.* **2019**, *19*, 906–919. [CrossRef]
92. Paul, R.; Singh, R.; Patra, A.; Biswas, D.; Bhattacharyya, R.; Arunkumar, K. Phosphorus dynamics and solubilizing microorganisms in acid soils under different land uses of Lesser Himalayas of India. *Agrofor. Syst.* **2018**, *92*, 449–461. [CrossRef]
93. Blanco-Vargas, A.; Rodríguez-Gacha, L.; Sánchez-Castro, N.; Garzón-Jaramillo, R.; Pedroza-Camacho, L.; Poutou-Piñales, R.; Rivera-Hoyos, C.; Díaz-Ariza, L.; Pedroza-Rodríguez, A. Phosphate-solubilizing *Pseudomonas* sp., and *Serratia* sp., co-culture for *Allium cepa* L. growth promotion. *Heliyon* **2020**, *6*, E05218. [CrossRef] [PubMed]
94. He, D.; Wan, W. Phosphate-solubilizing bacterium *Acinetobacter pittii* gp-1 affects rhizosphere bacterial community to alleviate soil phosphorus limitation for growth of soybean (*Glycine max*). *Front. Microbiol.* **2021**, *12*, 737116. [CrossRef] [PubMed]
95. Xing, P.; Zhao, Y.; Guan, D.; Li, L.; Zhao, B.; Ma, M. Effects of *Bradyrhizobium* co-inoculated with *Bacillus* and *Paenibacillus* on the structure and functional genes of soybean rhizobacteria community. *Genes* **2022**, *13*, 1922. [CrossRef]
96. Bagewadi, Z.; Yaraguppi, D.; Mulla, S.; Deshpande, S. Response Surface methodology based optimization, partial purification and characterization of alkaline phosphatase isolated from *Pseudomonas asiatica* Strain ZKB1 and its application in plant growth promotion. *Mol. Biotechnol.* **2022**, *64*, 984–1002. [CrossRef] [PubMed]
97. Rombola, T.H.; Pedrinho, E.A.; de Macedo Lemos, E.G.; Gonçalves, A.M.; dos Santos, L.F.; Pizauro, J.M. Identification and enzymatic characterization of acid phosphatase from *Burkholderia gladioli*. *BMC Res. Notes* **2014**, *9*, 221. [CrossRef] [PubMed]
98. Dasila, H.; Sah, V.; Jaggi, V.; Kumar, A.; Tewari, L.; Taj, G.; Chaturvedi, S.; Perveen, K.; Bukhari, N.; Siang, T.; et al. Cold-tolerant phosphate-solubilizing *Pseudomonas* strains promote wheat growth and yield by improving soil phosphorous (P) nutrition status. *Front. Microbiol.* **2023**, *14*, 1135693. [CrossRef] [PubMed]
99. Shi, W.; Xing, Y.; Zhu, Y.; Gao, N.; Ying, Y. Diverse responses of pqqC- and phoD-harbouring bacterial communities to variation in soil properties of Moso bamboo forests. *Microb. Biotechnol.* **2022**, *15*, 2097–2111. [CrossRef]
100. Feng, Q.; Guo, W.; Wang, T.; Cristina, M.A.L.; Luo, M.; Ge, R.; Zhou, C.; Zhang, Q.; Luo, J. Iron coupling with carbon fiber to stimulate biofilms formation in aerobic biological film systems for improved decentralized wastewater treatment: Performance, mechanisms and implications. *Bioresour. Technol.* **2021**, *319*, 124151. [CrossRef]
101. Mónica, M.; Collavino, P.A.; Sansberro, L.A.; Mroginski, O.; Mario, A. Comparison of in vitro solubllization activity of diverse phosphate solubllizing bacteria native to acid soil and their ability to promote *Phaseolus vulgaris* growth. *Biol. Fertil. Soils* **2010**, *46*, 727–738.
102. Yi, Y.; Huang, W.; Ge, Y. Exopolysaccharide: A novel important factor in the microbial dissolution of tricalcium phosphate. *World J. Microbiol. Biotechnol.* **2008**, *24*, 1059–1065. [CrossRef]
103. Kour, D.; Rana, K.; Kaur, T.; Yadav, N.; Yadav, A.; Kumar, M.; Kumar, V.; Dhaliwal, H.S.; Saxena, A.K. Biodiversity, current developments and potential biotechnological applications of phosphorus-solubilizing and -mobilizing microbes: A review. *Pedosphere* **2021**, *31*, 43–75. [CrossRef]
104. Garcia-Sanchez, M.; Bertrand, I.; Barakat, A.; Zeroual, Y.; Oukarroum, A.; Plassard, C. Improved rock phosphate dissolution from organic acids is driven by nitrate assimilation of bacteria isolated from nitrate and $CaCO_3$-rich soil. *PLoS ONE* **2023**, *18*, E0283437. [CrossRef] [PubMed]
105. Suleman, M.; Yasmin, S.; Rasul, M.; Yahya, M.; Atta, B.; Mirza, M. Phosphate solubilizing bacteria with glucose dehydrogenase gene for phosphorus uptake and beneficial effects on wheat. *PLoS ONE* **2018**, *13*, E0204408. [CrossRef] [PubMed]
106. Rawat, P.; Das, S.; Shankhdhar, D.; Shankhdhar, S. Phosphate-Solubilizing Microorganisms: Mechanism and Their Role in Phosphate Solubilization and Uptake. *J. Soil. Sci. Plant Nutr.* **2021**, *21*, 49–68. [CrossRef]
107. Rasul, M.; Yasmin, S.; Suleman, M.; Zaheer, A.; Reitz, T.; Tarkka, M.; Islam, E.; Mirza, M. Glucose dehydrogenase gene containing phosphobacteria for biofortification of Phosphorus with growth promotion of rice. *Microbiol. Res.* **2019**, *223–225*, 1–12. [CrossRef]

108. Sonnenburg, E.; Sonnenburg, J. The ancestral and industrialized gut microbiota and implications for human health. *Nat. Rev. Nat. Microbiol.* **2019**, *17*, 383–390. [CrossRef]
109. Eeshita, B.; Renuka, D.; Yasmin, B.; Yasmin, B.; Sunil, K.M. Study of *Pyrroloquinoline Quinine* from phosphate-solubilizing microbes responsible for plant growth: In silico approach. *Front. Agron.* **2021**, *3*, 667339.
110. Joshi, S.; Gangola, S.; Jaggi, V.; Sahgal, M. Functional characterization and molecular fingerprinting of potential phosphate solubilizing bacterial candidates from Shisham rhizosphere. *Sci. Rep.* **2023**, *13*, 7003. [CrossRef]
111. Ding, Y.; Yi, Z.; Fang, Y.; He, S.; Li, Y.; He, K.; Zhao, H.; Jin, Y. Multi-omics reveal the efficient phosphate-solubilizing mechanism of bacteria on rocky soil. *Front. Microbiol.* **2021**, *12*, 761972. [CrossRef]
112. Dai, Z.; Liu, G.; Chen, H.; Chen, C.; Wang, J.; Ai, S.; Wen, D.; Li, D.; Ma, B.; Tang, C.; et al. Long-term nutrient inputs shift soil microbial functional profiles of phosphorus cycling in diverse agroecosystems. *ISME J.* **2020**, *14*, 757–770. [CrossRef] [PubMed]
113. Hegyi, A.; Nguyen, T.; Posta, K. Metagenomic analysis of bacterial communities in agricultural soils from Vietnam with special attention to phosphate solubilizing bacteria. *Microorganisms* **2021**, *9*, 1796. [CrossRef] [PubMed]
114. Wan, W.; Qin, Y.; Wu, H.; Zuo, W.; He, H.; Tan, J.; Wang, Y.; He, D. Isolation and Characterization of Phosphorus Solubilizing Bacteria with Multiple Phosphorus Sources Utilizing Capability and Their Potential for Lead Immobilization in Soil. *Front. Microbiol.* **2020**, *11*, 752. [CrossRef] [PubMed]
115. Moe, L.; An, R. Regulation of pyrroloquinoline quinone-dependent glucose dehydrogenase activity in the model rhizosphere-dwelling bacterium *Pseudomonas putida* KT2440. *Appl. Environ. Microbiol.* **2016**, *82*, 4955–4964.
116. Luo, G.; Sun, B.; Li, L.; Li, M.; Liu, M.; Zhu, Y.; Guo, S.; Ling, N.; Shen, Q. Understanding how long-term organic amendments increase soil phosphatase activities: Insight into phoD- and phoC-harboring functional microbial populations. *Soil Biol. Biochem.* **2019**, *139*, 107632. [CrossRef]
117. Maria, R.; Sumera, Y.; Mahreen, Y.; Claudia, B.; Mika, T.; Reitz, T. The wheat growth-promoting traits of *Ochrobactrum* and *Pantoea* species, responsible for solubilization of different P sources, are ensured by genes encoding enzymes of multiple P-releasing pathways. *Microbiol. Res.* **2021**, *246*, 126703.
118. Cotta, S.R.; Cavalcante, F.D.; Seldin, L.A.; Andreote, F.D.; van Elsas, J.D. The diversity and abundance of phytase genes (β-propeller phytases) in bacterial communities of the maize rhizosphere. *Lett. Appl. Microbiol.* **2016**, *62*, 264–268. [CrossRef]
119. Suleimanova, A.; Bulmakova, D.; Sokolnikova, L.; Egorova, E.; Itkina, D.; Kuzminova, O.; Gizatullina, A.; Sharipova, M. Phosphate solubilization and plant growth promotion by *Pantoea brenneri* soil isolates. *Microorganisms* **2023**, *11*, 1136. [CrossRef]
120. Ducousso-Détrez, A.; Fontaine, J.; Lounès-Hadj Sahraoui, A.; Hijri, M. Diversity of phosphate chemical forms in soils and their contributions on soil microbial community structure changes. *Microorganisms* **2022**, *10*, 609. [CrossRef]
121. Zhang, Y.; Gao, W.; Ma, L.; Luan, H.; Tang, J.; Li, R.; Li, M.; Huang, S.; Wang, L. Long-term partial substitution of chemical fertilizer by organic amendments influences soil microbial functional diversity of phosphorus cycling and improves phosphorus availability in greenhouse vegetable production. *Agric. Ecosyst. Environ.* **2023**, *341*, 108193. [CrossRef]
122. Tian, J.; Kuang, X.; Tang, M.; Chen, X.; Huang, F.; Cai, Y. Biochar application under low phosphorus input promotes soil organic phosphorus mineralization by shifting bacterial phoD gene community composition. *Sci. Total Environ.* **2021**, *779*, 146556. [CrossRef] [PubMed]
123. Chen, S.; Gao, J.; Chen, H.; Zhang, Z.; Huang, J.; Lv, L.; Tan, J.; Jiang, X. The role of long-term mineral and manure fertilization on P species accumulation and phosphate-solubilizing microorganisms in paddy red soils. *Soil* **2023**, *9*, 101–116. [CrossRef]
124. Kumar, A.; Ral, L. Soil organic carbon and phosphorus availability regulate abundance of culturable phosphate-solubilizing bacteria in paddy fields. *Pedosphere* **2020**, *30*, 405–413. [CrossRef]
125. Kaur, G.; Sudhakara Reddy, M. Influence of P-solubilizing bacteria on crop yield and soil fertility at multilocational sites. *Eur. J. Soil Biol.* **2014**, *61*, 35–40. [CrossRef]
126. Li, C.L.; Xun, Y.L.; Wang, Y.; Dang, Y.H.; Song, Y. Effects of long-term fertilization on soil nitrogen-transforming bacteria and phosphate-solubilizingbacteria in rainfed cropland of Loess Plateau, China. *Chin. J. Plant Ecol.* **2020**, *39*, 3658–3667.
127. Zheng, B.; Hao, X.; Ding, K.; Zhou, G.; Chen, Q.; Zhang, J.; Zhu, Y. Long-term nitrogen fertilization decreased the abundance of inorganic phosphate solubilizing bacteria in an alkaline soil. *Sci. Rep.* **2017**, *7*, 42284. [CrossRef] [PubMed]
128. Widdig, M.; Schleuss, P.; Weig, A.; Guhr, A.; Biederman, L.; Borer, E.; Crawley, M.; Kirkman, K.; Seabloom, E.; Wragg, P.; et al. Nitrogen and Phosphorus Additions Alter the Abundance of Phosphorus-Solubilizing Bacteria and Phosphatase Activity in Grassland Soils. *Front. Environ. Sci.* **2019**, *7*, 185. [CrossRef]
129. Tang, J.; Su, L.; Fang, Y.; Wang, C.; Meng, L.; Wang, J.; Zhang, J.; Xu, W. Moderate Nitrogen Reduction Increases Nitrogen Use Efficiency and Positively Affects Microbial Communities in Agricultural Soils. *Agriculture* **2023**, *13*, 796. [CrossRef]
130. Zheng, B.; Zhang, D.; Wang, Y.; Hao, X.; Wadaan, M.; Hozzein, W.; Peñuelas, P.; Zhu, Y.; Yang, X. Responses to soil pH gradients of inorganic phosphate solubilizing bacteria community. *Sci. Rep.* **2019**, *9*, 25. [CrossRef]
131. Zhu, Y.; Su, J.; Cao, Z.; Xue, K.; Quensen, J.; Guo, G.; Yang, Y.; Zhou, J.; Chu, H.; Tiedje, M. A buried Neolithic paddy soil reveals loss of microbial functional diversity after modern rice cultivation. *Sci. Bull.* **2016**, *61*, 1052–1060. [CrossRef]
132. Xu, H.; Lv, J.; Yu, C. Combined phosphate-solubilizing microorganisms jointly promote *Pinus massoniana* growth by modulating rhizosphere environment and key biological pathways in seedlings. *Ind. Crops Prod.* **2023**, *191*, 116005. [CrossRef]
133. He, T.; Xu, Z.; Wang, J.; Zhang, K.; Wang, F.; Li, W.; Tian, P.; Li, Q. Inoculation of *Escherichia coli* enriched the key functional bacteria that intensified cadmium accumulation by halophyte *Suaeda salsa* in saline soils. *J. Hazard. Mater.* **2023**, *458*, 131922. [CrossRef] [PubMed]

134. Sun, R.; Zhang, W.; Liu, Y.; Yun, W.; Luo, B.; Chai, R.; Zhang, C.; Xiang, X.; Su, X. Changes in phosphorus mobilization and community assembly of bacterial and fungal communities in rice rhizosphere under phosphate deficiency. *Front. Microbiol* **2022**, *13*, 953340. [CrossRef]
135. Srivastava, S.; Srivastava, S. Prescience of endogenous regulation in *Arabidopsis thaliana* by *Pseudomonas putida* MTCC 5279 under phosphate starved salinity stress condition. *Sci. Rep.* **2020**, *10*, 5855. [CrossRef] [PubMed]
136. Saia, S.; Rappa, V.; Ruisi, P.; Abenavoli, M.R.; Sunseri, F.; Giambalvo, D.; Frenda, A.S.; Martinelli, F. Soil inoculation with symbiotic microorganisms promotes plant growth and nutrient transporter genes expression in durum wheat. *Front. Plant Sci.* **2015**, *6*, 815. [CrossRef] [PubMed]
137. Abbasi, S. Plant-microbe interactions ameliorate phosphate-mediated responses in the rhizosphere: A review. *Front. Plant Sci.* **2023**, *14*, 1074279. [CrossRef] [PubMed]
138. Wang, Z.; Chen, Z.; Fu, X. Integrated Effects of Co-Inoculation with Phosphate-Solubilizing Bacteria and N 2 -Fixing Bacteria on Microbial Population and Soil Amendment Under C Deficiency. *Int. J. Environ. Res. Public Health* **2019**, *16*, 2442. [CrossRef]
139. Wang, S.; LV, H.R.; Zhang, H.; Wu, Z.W.; Xiao, C.H.; Sun, D.M. Whole-Genome Sequencing Identification of Phosphate-solubilizing Bacteria PSB-R and Analysis of Its Phosphate-solubilizing Properties. *Biol. Bull.* **2023**, *39*, 274–283. (In Chinese)
140. Wang, K.X. *Genomic Analysis of the Grassland soil Bacterium Baillus sp. SP1 and its Growth Promoting Effect on Forage*; Inner Mongolia University: Neimenggu, Chian, 2022. (In Chinese)
141. Xi, H. *The Phosphate-Solubilizing Activity, Genome Analysis, Clone And Expression Of Genes Related to Phosphate Dissolution of a Strain A02*; Nanjing University: Nanjing, China, 2017. (In Chinese)
142. Li, S.S. *Study on Phosphorus-Soluble Characteristics of Two Pseudomonas strains and Their Effects on the Soil Phosphorus Solublizing*; Henan University of Technology: Luoyang, China, 2023. (In Chinese)
143. Ma, Q.; He, S.; Wang, X.; Rengel, Z.; Chen, L.; Wang, X.; Zhang, X. Isolation and characterization of phosphate-solubilizing bacterium *Pantoea rhizosphaerae* sp. nov. from *Acer truncatum* rhizosphere soil and its effect on *Acer truncatum* growth. *Front. Plant Sci.* **2023**, *14*, 1218445. [CrossRef]
144. Lv, H.; Yang, J.; Su, S.; Liu, Y.; Feng, J.; Sheng, Y.; Wang, T.; Pan, J.; Tang, L.; Chen, L.; et al. Distribution of Genes and Microbial Taxa Related to Soil Phosphorus Cycling across Soil Depths in Subtropical Forests. *Forests* **2023**, *14*, 1665. [CrossRef]
145. Wei, L.Q. *Isolation and Screening of Phosphate-solubilizing Bacteria and the Effect on Phosphorus Release from the Sediments in Rongcheng Swan Lake*; Yantai University: Yantai, China, 2021. (In Chinese)
146. Chen, D.Y.; Li, H.Q.; Zhang, B.H.; Dai, C.M.; Yang, J.Y. Phosphate solubilization activities and action mechanisms of two phosphate-solubilizing bacteria. *Chin. J. Eco-Agric.* **2017**, *25*, 410–418. (In Chinese)
147. Wu, J.; Zhao, F. Machine learning: An effective technical method for future use in assessing the effectiveness of phosphorus-dissolving microbial agroremediation. *Front. Bioeng. Biotechnol.* **2023**, *11*, 1189166. [CrossRef] [PubMed]

microorganisms

MDPI

Article

Seasonal Dynamics of Eukaryotic Microbial Communities in the Water-Receiving Reservoir of the Long-Distance Water Diversion Project, China

Yingying Yang [1], Fangfang Ci [2], Ailing Xu [1], Xijian Zhang [3], Ning Ding [1], Nianxin Wan [2], Yuanyuan Lv [1] and Zhiwen Song [1,*]

1 School of Environmental and Municipal Engineering, Qingdao University of Technology, Qingdao 266520, China; yangyingyingqd@163.com (Y.Y.); xalcsu@sina.com (A.X.)
2 Qingdao Branch of Shandong Water Transfer Project Operation and Maintenance Center, Qingdao 266525, China
3 Binzhou Branch of Shandong Water Transfer Project Operation and Maintenance Center, Binzhou 256600, China
* Correspondence: songzhiwen@qut.edu.cn

Abstract: Inter-basin water transfer projects, such as the Yellow River to Qingdao Water Diversion Project (YQWD), are essential for addressing water scarcity, but impact local aquatic ecosystems. This study investigates the seasonal characteristics of eukaryotic microbial communities in the Jihongtan Reservoir, the main water-receiving body of YQWD, over a one-year period using 18S rDNA amplicon sequencing. The results showed that the eukaryotic microbial diversity did not exhibit significant seasonal variation ($p > 0.05$), but there was a notable variance in the community structure ($p < 0.05$). Arthropoda and *Paracyclopina*, representing the most dominant phylum and the most dominant genus, respectively, both exhibited the lowest abundance during the winter. The Chlorophyta, as the second-dominant phylum, demonstrates its higher abundance in the spring and winter. The Mantel test and PLS-PM (Partial Least Squares Path Modeling) revealed that water temperature (WT), dissolved oxygen (DO), and pH influenced the seasonal dynamic of eukaryotic microbial communities significantly, of which WT was the primary driving factor. In addition to environmental factors, water diversion is likely to be an important influencing factor. The results of the co-occurrence network and robustness suggested that the spring network is the most complex and exhibits the highest stability. Moreover, keystone taxa within networks have been identified, revealing that these key groups encompass both abundant and rare species, with specificity to different seasons. These insights are vital for understanding the seasonal variation of microbial communities in the Jihongtan Reservoir during ongoing water diversions.

Keywords: eukaryotic microorganisms; storage reservoir; seasonal dynamics; environmental driving factors; co-occurrence network patterns

1. Introduction

Inter-basin water transfer projects represent a critical engineering intervention designed to address the discord between the escalating water demand and its availability [1]. The Yellow River to Qingdao Water Diversion Project (YQWD) was constructed to effectively mitigate the water scarcity in Qingdao, China. The Jihongtan Reservoir, serving as the only water-receiving reservoir of YQWD, also holds the distinction of being the largest man-made dammed plain reservoir in Asia. Since its completion in 1989, the reservoir has been in operation for 35 years, having transferred over 5 billion cubic meters of water from both the Yellow River and Yangtze River, making it responsible for more than 90% of the water used by the urban population of Qingdao [2]. Although the water diversion project has alleviated water shortages to a certain extent, it simultaneously changes the hydrological regime and the characteristics of receiving area water bodies and provides a path for biological invasions, subsequently affecting the local ecology [3,4].

Eukaryotic microbes are widely distributed in aquatic environments, serving as both producers of organic matter and major consumers of bacterial biomass, playing a significant ecological role in maintaining multifunctionality, facilitating material cycling, and contributing to energy transfers within food webs [5]. They have more complexly structured cellulars than prokaryotes, containing organelles with diverse functions, such as vacuolar constrictors for osmoregulation, peduncles for phagocytic feeding, and chloroplasts for photosynthesis. These structures may exhibit a wide range of responses to environmental heterogeneity, making eukaryotic microbes highly sensitive to local stresses [6–8]. Water reservoirs are highly dynamic environments that are influenced by local human activities, such as artificial water transfers, as well as regional climate changes like temperature and precipitation [9,10]. Therefore, driven by environmental heterogeneity, microorganisms in reservoirs are temporally variable. For example, it has been reported that the succession of eukaryotic microbial communities occurs along with the season [11–13]. Understanding the dynamic variations remained a considerable challenge due to the complexity of eukaryotic microbial communities, especially in water-receiving areas with obvious seasonality, facing environments under compounded long-term impacts [14]. It is crucial to monitor the dynamic changes in eukaryotic microorganisms in drinking water to provide early warning of microbial hazards and ensure the safe management of drinking water [15].

Both environmental factors and biological interactions can drive the dynamics of microorganisms in aquatic ecosystems [16]. Some studies have demonstrated that the community structure of eukaryotic microorganisms is associated with changes in environmental factors. Zhang et al. [17] reported that nutrients and temperature significantly influenced the eukaryotic microbial diversity in a drinking water reservoir in northwest China. Yang et al. [18]. investigated the relationship between eukaryotic microbial communities and environmental factors in extreme environments on the Tibetan Plateau, finding that the water temperature, pH, and total phosphorus were the primary driving factors. Co-occurrence network analysis is an effective method that has been extensively applied to explore the interconnectivity among microorganisms and elucidate the complexity and stability of these ecological associations based on topological characteristics [19,20]. A positive correlation can be interpreted as cooperative behaviors such as syntrophic interactions or shared environmental requirements. By contrast, negative relationships may be generally explained as competition as well as distinctive environmental niches and spatial isolation [21–23]. Beyond co-occurrence patterns, microbial networks can also be utilized to statistically recognize keystone species [24]. Microbial keystone taxa are highly connected taxa that individually or in a guild exert a considerable influence on the microbiome structure and functioning, having a unique and crucial role in microbial communities [25]. Recent studies have revealed that key taxa exert a positive influence on the stability of microbial symbiotic networks [26,27].

This study intends to investigate the seasonal characteristics of eukaryotic microorganisms in the Jihongtan reservoir under conditions of year-round water diversion, based on 18S rDNA amplicon sequencing technology. The specific objectives were (1) to investigate the seasonal variations in the community composition and diversity of eukaryotic microbial communities in the reservoir; (2) to identify environmental factors driving eukaryotic microbial diversity and explore how dominant environmental variables affect that diversity; and (3) to reveal the co-occurrence patterns of eukaryotic microbial communities in different seasons.

2. Materials and Methods

2.1. Study Area and Sample Collection

Jihongtan Reservoir, located in Qingdao (36°20′55″~36°23′31″ N, 120°12′27″~120°14′55″ E), Shandong province, China, is an eight-sided storage reservoir for drinking water (Figure 1). The reservoir has a volume of approx. 1.46×10^8 m^3, with a design water level of 14.2 m, a dam length of 14.2 km, and a surface area of 14.4 km^2. Water is transferred from the Eastern Route of the South-to-North Water Diversion Project (ESNWD), the Yellow River to Qingdao Water Diversion Project (YQWD), and Xiashan Reservoir (XR) through the aqueduct to the reservoir. ESNWD extracts water from the lower section of the Yangtze River, specifically from Sanjiangying in the Yangzhou area. It supplies water to Tianjin and the Jiaodong Peninsula of Shandong Province by utilizing the existing river channels of the Beijing–Hangzhou Grand Canal, connecting a chain of large lakes, including the Hongze, Luoma, Nansi, and Dongping Lakes. YQWD is located in Shandong province and it channels the water from the Yellow River in Binzhou City down to Jihongtan Reservoir in Qingdao, involving 6 cities of Binzhou, Dongying, Weifang, Yantai, Weihai, and Qingdao, with a total length of 290 km. XR is a drinking water reservoir in Weifang of Shandong Province, used to temporarily replenish the Jihongtan Reservoir during water shortages in summer and autumn. Table 1 shows the cumulative inflow amount of water diversion projects in four seasons. The local climate delineates the year into four distinct seasons.

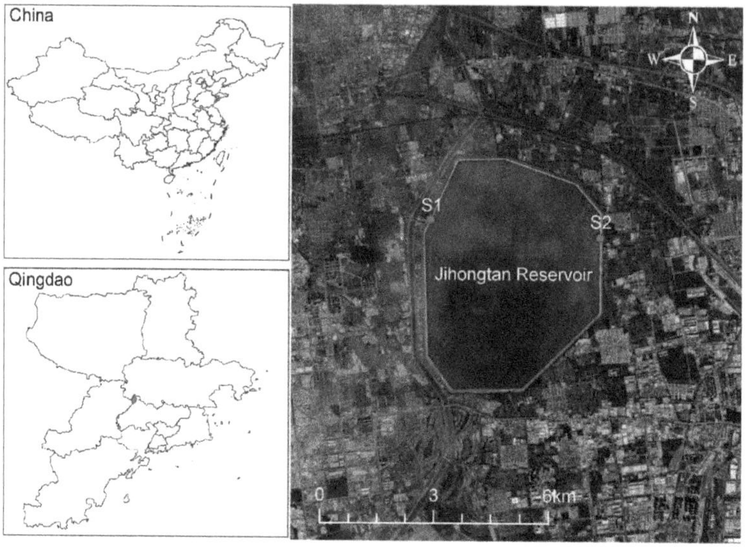

Figure 1. Geographical location of Jihongtan Reservoir and sampling sites.

Table 1. Cumulative inflow amount of water diversion project to Jihongtan Reservoir (10^8 m^3).

Sources	Spr	Sum	Aut	Win
ESNWD	0.7127	-	0.1607	1.2141
YQWD	0.5428	0.2630	-	0.2571
XR	-	0.2839	0.3845	-
Total	1.2555	0.5469	0.5452	1.4712

To explore the eukaryotic microorganisms in the Jihongtan Reservoir throughout the four seasons, we collected water samples from 2 sites every month, obtaining 6 samples for each season and a total of 24 samples for the entire year. Sampling was conducted from site S1 (120°12′28″ E; 36°20′15″ N) and site S2 (120°14′55″ E; 36°22′22″ N), respectively located at the inlets and outlets of the reservoir (Figure 1). Then, 10 L water samples were collected 0.5 m below the water surface. All samples were collected at approximately 10:00–12:00 and examined within 48 h.

2.2. Physicochemical Parameter Monitoring

Water temperature (WT), pH, and dissolved oxygen (DO) were measured by HQ40d (HACH, USA). Other physical and chemical parameters, including the total nitrogen (TN), total phosphorus (TP), chemical oxygen demand (COD), ammonium nitrogen (NH_4^+-N), nitrite nitrogen (NO_2^--N), nitrate nitrogen (NO_3^--N), and chlorophyll-a (Chl-a), were determined according to the environmental monitoring method standards of Ministry of Ecology and Environment of the People's Republic of China [28]. Chl-a was determined according to Li et al [29].

2.3. DNA Extraction and High-Throughput Sequencing

The 0.22 μm filter membranes were used to filter the water samples, then cut into fractions for DNA extraction on 24 samples using the E.Z.N.A.® soil DNA Kit (Omega Bio-tek, Norcross, GA, USA). The quality and concentration of DNA were determined by 1.0% agarose gel electrophoresis and a NanoDrop2000 spectrophotometer (Thermo Scientific, Waltham, MA, USA) and kept at −80 °C prior to further use. The amplification of the V4 region of the 18S rRNA was conducted using the primer set TAReuk454FWD1 (5′-CCAGCASCYGCGGTAATTCC-3′) and TAReukREV3R (5′-ACTTTCGTTCTTGATYRA-3′) [30], with amplicon length of 405 bp. PCR amplification cycling conditions were as follows: initial denaturation at 95 °C for 3 min, followed by 35 cycles of denaturing at 95 °C for 30 s, annealing at 55 °C for 30 s, extension at 72 °C for 45 s, single extension at 72 °C for 10 min, and end at 10 °C. The PCR product was extracted from 2% agarose gel and purified using the PCR Clean-Up Kit (YuHua, Shanghai, China) according to manufacturer's instructions and quantified using Qubit 4.0 (Thermo Fisher Scientific, Waltham, MA, USA). The sequencing was completed on the Illumina MiSeq platform (Illumina, San Diego, CA, USA) using a paired-end (2 × 250 bp) approach by Majorbio Bio-Pharm Technology Co., Ltd. (Shanghai, China).

2.4. Sequencing Data Processing

Raw FASTQ files were processed using an in-house Perl script for de-multiplexing, followed by quality filtering with fastp v0.19.6 [31]. Sequences were merged using FLASH v1.2.7 [32] under specific criteria: (i) truncation at sites with an average quality score below 20 within a 50 bp sliding window, discarding the truncated reads shorter than 50 bp and those with ambiguous characters; (ii) assembly of overlapping sequences longer than 10 bp with a maximum mismatch ratio of 0.2 in the overlap region, discarding non-assemblable reads; and (iii) sample identification based on barcode and primer matching, allowing for 2 nucleotide mismatches in primers. The resultant sequences were clustered into operational taxonomic units (OTUs) using UPARSE v7.1 [33] at a 97% similarity threshold, with the most abundant sequence per OTU selected as the representative. Each representative sequence was assigned to taxonomic categories against the 18S rRNA database (Silva v138) using the Ribosomal Database Project (RDP) classifier [34] at a confidence threshold of 0.7. To minimize the effects of sequencing depth on alpha and beta diversity measure, the samples were resampled randomly to the lowest number of retrieved sequences across all samples.

2.5. Statistical Analysis

Sequencing data in this study were processed using QIIME 2. A one-way analysis of variance (ANOVA) with a post hoc Tukey's Honestly Significant Difference (HSD) test was carried out to examine the seasonal variations of water physicochemical properties and alpha diversity indices. Before conducting the ANOVA, tests for homogeneity and normality were performed. If the data were found to be non-normal, the non-parametric Kruskal–Wallis test was used to assess significant differences. Statistical significance was accepted as $p < 0.05$. The Venn diagram constructed with shared and unique operational taxonomic units (OTUs) was used to depict the similarities and differences among communities. Heatmap of the top 50 genera were generated via the "ComplexHeatmap" package (v 2.20.0) in R [35]. Ordination analysis of eukaryotic microbial communities was carried out using non-metric multidimensional scaling (NMDS) based on the distance algorithm of Aitchison [36]. Analysis of similarity (ANOSIM) was performed to investigate the variability of eukaryotic microbial communities between groups [37]. To determine the characteristics and variations of eukaryotic microorganisms in different seasons, the Linear discriminant analysis effect size (LEfSe) was utilized via the R package "microbiomeMarker" (v 1.10.0) [38]. We implemented a Mantel test to assess the correlations between environmental factors and eukaryotic microbial communities based on Spearman rank correlation, and the analysis was performed using the "linkET" package (v 0.0.7.4) [28]. To further reveal the effects of environmental variables on microbial community structure, a Partial Least Squares Path Model (PLS-PM) was conducted by R package "plspm" (v 0.5.1) [39]. Co-occurrence network analysis was performed to investigate the interactions between microbial composition. The computation of network properties and the identification of keystone taxa were carried out using the R package "ggClusterNet" (v 0.1.0) [40]. The network was visualized using the Gephi v0.9.2.

3. Results

3.1. Seasonal Variations in Water Physicochemical Properties

The seasonal variations of water physicochemical properties, namely WT, pH, DO, COD, TN, TP, NH_4^+-N, NO_3^--N, NO_2^--N, and Chl-a, are shown in Figure 2. All the above-mentioned environmental variables, except for NH_4^+-N, exhibited significant seasonal differences ($p < 0.05$). A maximum average WT of 26.03 °C and minimum average WT of 5.17 °C were observed in the summer and winter, respectively. On the contrary, the highest average DO of 11.55 mg/L emerged in winter, and the lowest average DO of 5.97 mg/L was recorded in summer. pH mean values differed significantly between spring and summer, with a maximum value of 8.51 in the summer and a minimum value of 8.11 in the spring. The mean COD concentrations varied significantly between the summer and autumn, with the highest value of 12.8 mg/L in the autumn and the lowest value of 9.8 mg/L in the summer. TN, TP, NO_3^--N, NO_2^--N, and Chl-a did not show significant differences in the spring, summer, and winter, while the highest or lowest average concentration of these variables was exhibited in the autumn. Specifically, the minimum average NO_3^--N and TN were observed in the autumn, with their values of 0.64 mg/L and 1.03 mg/L, respectively; the maximum average NO_2^--N, TP, and Chl-a concentrations were detected in the autumn, with their values of 0.026 mg/L, 0.033 mg/L, and 19.00 mg/m^3, respectively.

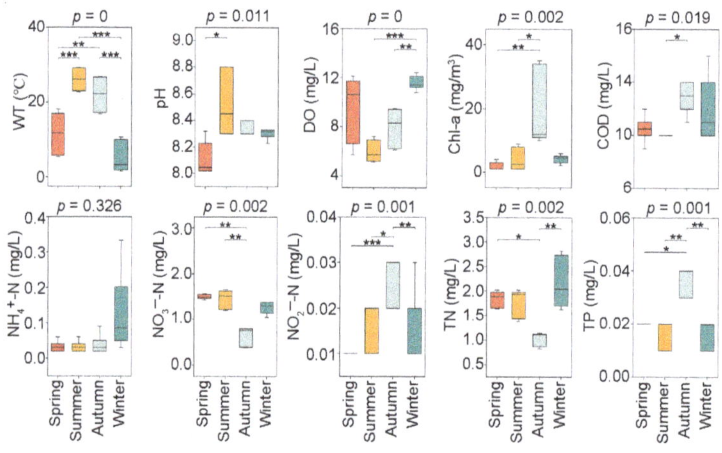

Figure 2. Seasonal variation of water physicochemical properties in Jihongtan Reservoir. (Significance: * $p < 0.05$, ** $p < 0.01$, *** $p < 0.001$). (WT: water temperature; DO: dissolved oxygen; Chl-a: chlorophyll-a; COD: chemical oxygen demand; NH_4^+-N: ammonia nitrogen; NO_3^--N: nitrate nitrogen; NO_2^--N: nitrite nitrogen; TN: total nitrogen; TP: total phosphorus).

3.2. Seasonal Changes in OUTs and Alpha Diversity

The Venn diagram in Figure 3a indicates that 1341, 1334, 1178, and 842 OTUs were identified from spring to winter. Among these, 260 OTUs were shared across all seasons, while 309, 406, 228, and 98 OTUs were unique to the spring, summer, autumn, and winter, respectively.

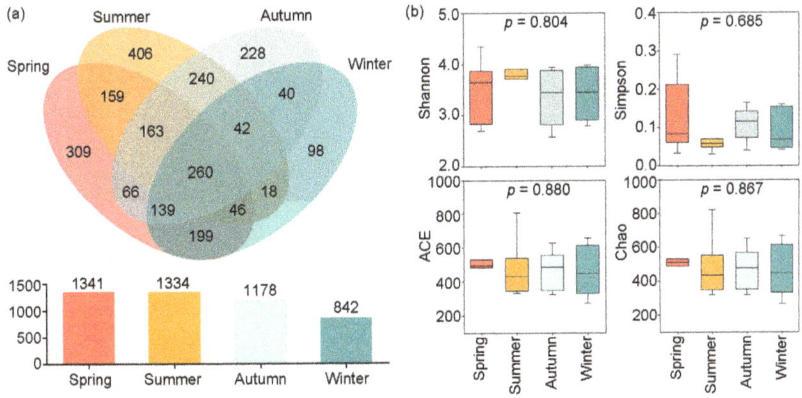

Figure 3. Seasonal variations in OTUs and Alpha diversity in Jihongtan Reservoir. (**a**) Venn diagram of the OTUs among the four seasons; (**b**) Alpha diversity indices of eukaryotic microbial communities in the four seasons.

The Alpha diversity indices (Shannon, Simpson, ACE, and Chao indices) exhibited seasonal variations, but showed no significance ($p > 0.05$) (Figure 3b). The mean values of the four alpha diversity indices were higher in the spring and summer than in the autumn and winter. Specifically, the ACE and Chao indices showed the highest average values in the spring, but the lowest in the winter. The maximum mean values of the Shannon and Simpson indices occurred in the spring and summer, respectively; the minimum mean values of the Shannon and Simpson indices were observed in the autumn and winter, respectively.

3.3. Eukaryotic Microbial Community Composition and Biomarkers Analysis

Figure 4a reveals an overview of the eukaryotic microbial community composition of the top 1% phyla across different seasons. A total of 51 eukaryotic phyla were identified in the reservoir throughout the year, with the dominant phyla being Arthropoda (19.34%), Chlorophyta (13.36%), Ochrophyta (7.85%), Cryptophyta (7.64%), Ciliophora (3.19%), Cryptomycota (2.18%), Cercozoa (1.22%), Dinoflagellata (1.22%), and Haptophyta (1.21%). The relative abundances of the dominant taxa exhibited some seasonal dynamics. Arthropoda was the most dominant phylum in the spring (18.65%), summer (30.97%), and autumn (22.24%); however, its dominance decreased in winter, accounting for only 5.52%, making it the third most dominant phylum. Chlorophyta was the predominant phylum in the winter (34.02%) and the second-dominant phylum in the spring (14.97%), but its abundance was low in the summer (2.49%) and autumn (1.92%). Ochrophyta was the second-dominant phylum in the summer (9.32%) and the third-dominant phylum in both the spring (8.25%) and the autumn (9.32%), with a lower relative abundance in the winter (4.54%) compared to other seasons. As the second-dominant phylum in both the autumn (14.92%) and the winter (8.35%), Cryptophyta gradually decreased from the autumn to the summer, where it accounted for only 0.52%. Conversely, Cryptomycota gradually increased in relative abundance from the autumn to the summer, eventually becoming the third-dominant phylum in the summer (5.06%).

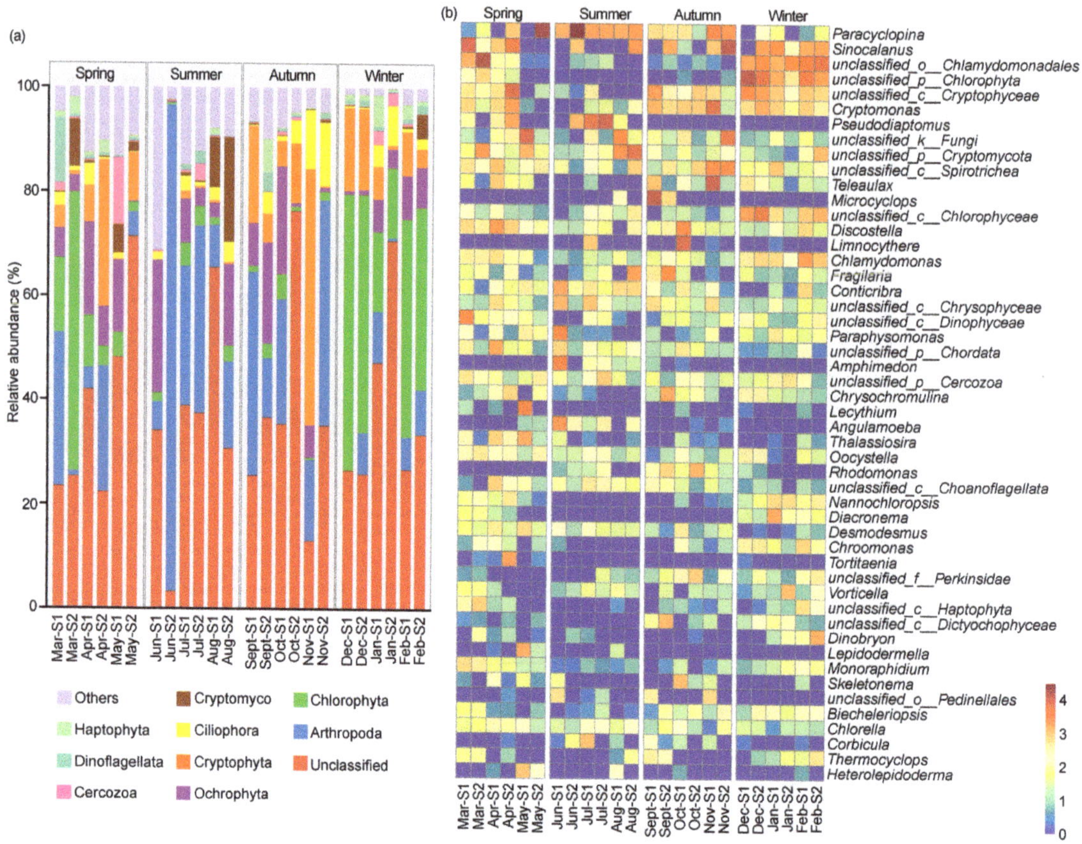

Figure 4. The compositions of eukaryotic microbial community on phyla level (**a**) and genus level (**b**) in Jihongtan Reservoir. Relative abundance less than 1% is defined as others.

Figure 4b shows the composition of the eukaryotic microbial communities based on the genus level. Excluding the unclassified genera, the dominant genera were identified including *Paracyclopina* (8.42%), *Sinocalanus* (5.46%), *Pseudodiaptomus* (2.41%), *Microcyclops* (1.47%), and *Limnocythere* (1.07%) belonging to Arthropoda; *Cryptomonas* (2.42%) and *Teleaulax* (1.64%) belonging to Cryptophyta; and *Discostella* (1.33%) belonging to Ochrophyta. *Paracyclopina* was the dominant genus in the spring (9.47%) and summer (19.59%), less abundant in the autumn (4.50%), and the least abundant in the winter (0.15%). *Sinocalanus* was the predominant genus in the autumn (7.27%) and the winter (4.93%), the second-dominant genus in the spring (7.72%), and had the lowest relative abundance in the summer (1.91%). Similarly, the lowest abundance of *Cryptomonas* occurred in the summer (0.10%), and higher proportions were in the spring (2.01%), autumn (5.19%), and winter (2.37%). *Teleaulax*, *Microcyclops*, and *Limnocythere* peaked in the autumn (5.55%, 5.74%, and 3.72%, respectively), while all of their relative abundance was less than 1% in the other seasons. *Discostella* was enriched in the spring (2.46%) and autumn (2.12%), but less frequent in the summer (0.27%) and winter (0.46%).

NMDS ordination and ANOSIM tests were used to analyze the seasonal differences of eukaryotic microbial communities across four seasons (Figure 5a,b). The NMDS results showed a distinct separation of seasonal changes in eukaryotic microbial communities (stress value = 0.15). In addition, the ANOSIM test of eukaryotic microbial communities showed that significant changes emerged in four seasonal groups (Global R = 0.626,

$p = 0.001$), indicating population shifts of eukaryotic microorganisms among different seasons in the Jihongtan Reservoir.

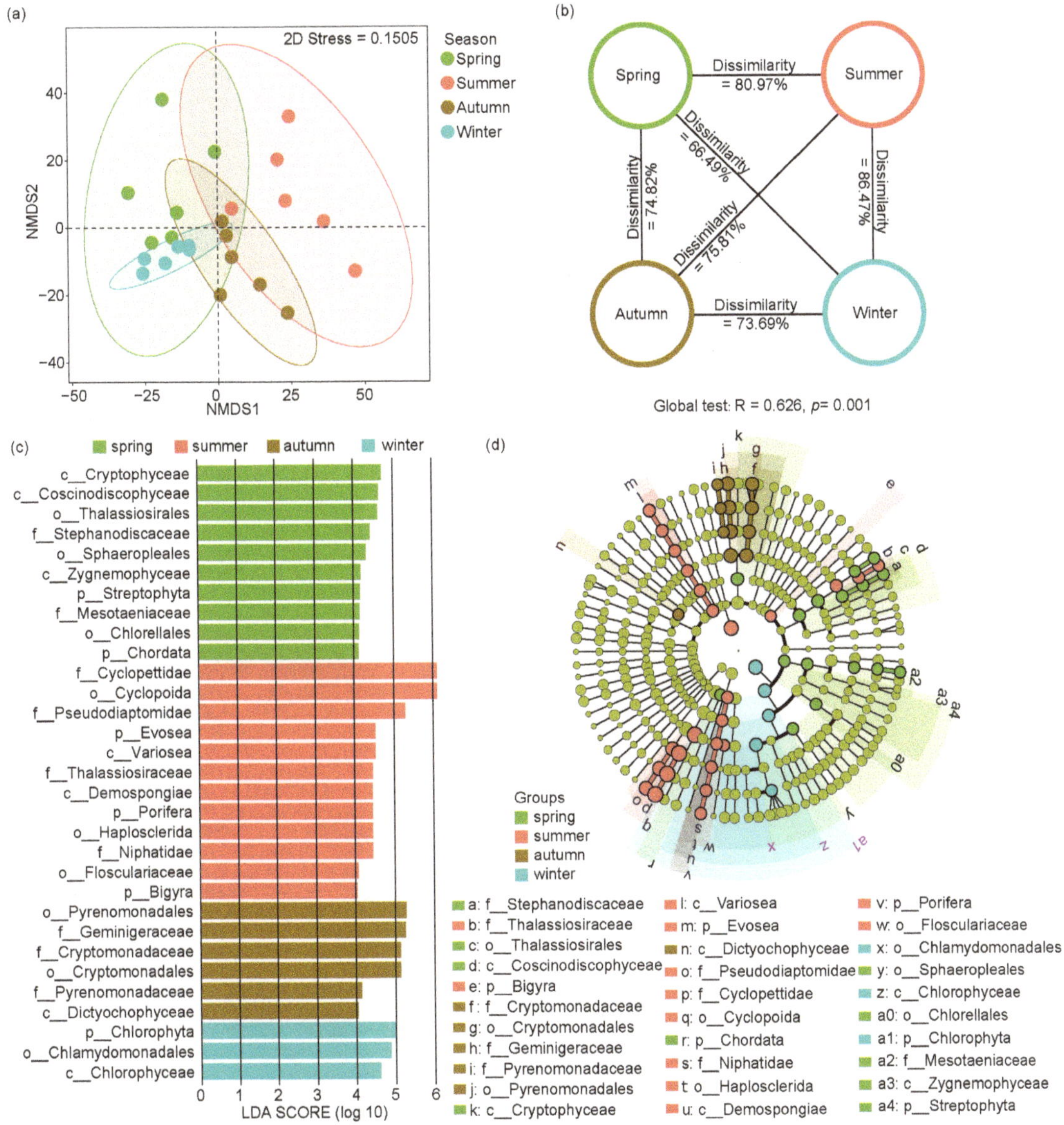

Figure 5. Non-metric multidimensional scaling analysis (NMDS), analysis of similarities (ANOSIM), and linear discriminant analysis effect size (LEfSe) analysis of eukaryotic microbial community within four seasons in Jihongtan Reservoir. (**a**) NMDS ordination plot produced based on Aitchison distance; (**b**) ANOSIM test; (**c**,**d**) LEfSe analysis; (**c**) Linear discriminant analysis (LDA) Score diagram shows differentially abundant taxa [LDA score = 4]; (**d**) Cladogram showing the phylogenetic structure of the eukaryotic microorganisms.

To identify the eukaryotic biomarkers with a significantly different relative abundance across the four seasons, we utilized LEfSe (Figure 5c,d), choosing linear discriminant analysis scores greater than four to distinguish significantly different groups. Consequently, we identified 31 biomarkers throughout the year, with 10 in spring, 12 in summer, 6 in autumn, and 3 in winter. In spring, the identified biomarkers belonged mainly to Cryptophyta, Bacillariophyta, Chlorophyta, Streptophyta, and Chordata. In summer, the main enriched members belonged to Arthropoda, Evosea, Ochrophyta, Porifera, and Bigyra. In autumn, the biomarkers mainly included Dictyochophyceae and members from Cryptophyta. Chlorophyta and its subgroups, Chlorophyceae and Chlamydomonadales, were more prevalent in winter.

3.4. Correlations between Eukaryotic Microbial Communities and Environmental Variables

The shifts regarding the abundance of eukaryotic microbes at the phylum and genus levels, along with various environmental factors, indicated a strong link between the eukaryotic community and environmental changes. Thus, we conducted further research to better understand the relationship between eukaryotic microbial communities and environmental variables. The Mantel test (Figure 6a) revealed several important environmental variables, including WT (r = 0.441, p = 0.001), DO (r = 0.498, p = 0.001), and pH (r = 0.192, p = 0.033), that played a major role in the variabilities of eukaryotic microorganisms in different seasons, which showed a positive correlation with eukaryotic microbial communities (Table S1). Compared to pH, WT and DO showed a higher correlation. Furthermore, we observed correlations among various water quality parameters and found that DO was significantly negatively correlated with WT.

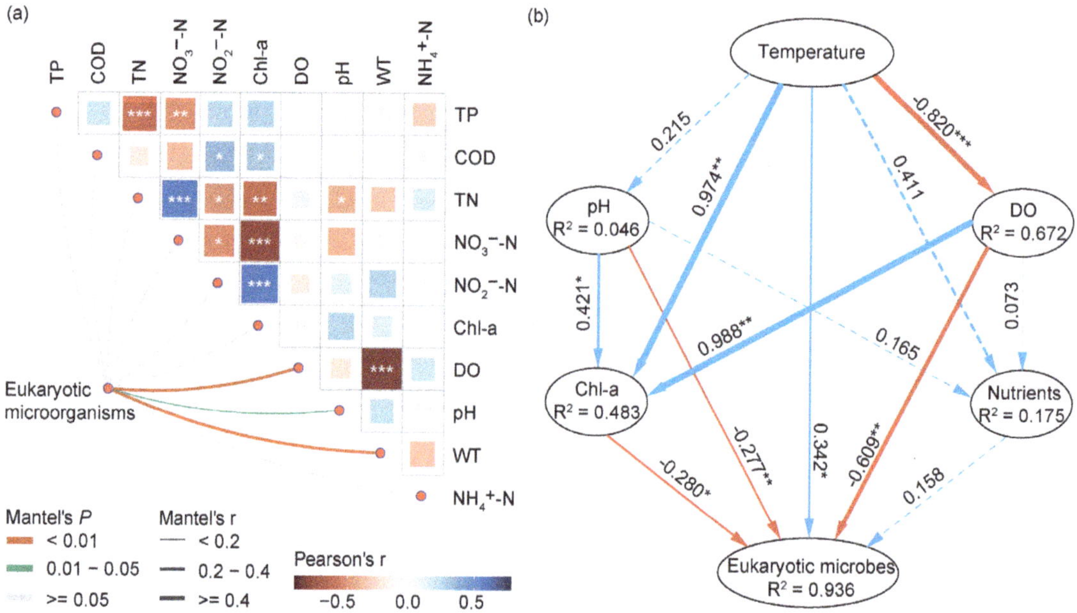

Figure 6. Environmental factors affecting eukaryotic microbial communities in Jihongtan Reservoir. (**a**) Pairwise comparisons of environmental factors are visually represented using a color gradient to indicate Spearman's correlation coefficients. The correlations between the eukaryotic microbial community and each environmental factor are evaluated using Mantel tests. (**b**) Partial least squares path modeling (PLS-PM) represents the direct and indirect effects of environmental variables on eukaryotic microbial communities. The blue line: a positive relationship; the red line: a negative relationship. Significance level: $p < 0.001$ ***; $p < 0.01$ **; $p < 0.05$ *.

How do the WT, DO, pH, and other environmental factors influence the eukaryotic microbial communities: through synergistic or direct ways? To address this issue, we constructed the partial least squares path model (PLS-PM), as shown in Figure 6b. The environmental drivers were categorized into six block variables: WT, pH, DO, nutrients (TN, TP, NH_4^+-N, NO_3^--N, and NO_2^--N), and algal properties (Chl-a). The result showed that the WT significantly influenced the eukaryotic microbial communities directly and positively, with the effects of 0.342 ($p < 0.05$), while DO, pH, and Chl-a showed a direct negative impact, with the effects of −0.609 ($p < 0.01$), −0.277 ($p < 0.01$), and −0.280 ($p < 0.05$), respectively. In addition, the WT exerted an indirect influence on the eukaryotic community by significantly affecting mainly DO and Chl-a, with the effects of −0.820 ($p < 0.001$) and 0.974 ($p < 0.01$), respectively. As a result, the WT could be the primary factor driving the seasonal succession of eukaryotic compositions in the Jihongtan Reservoir.

3.5. Co-Existence Network Analysis

Eukaryotic microbial communities are affected not only by abiotic factors, but also by biotic interactions. Co-occurrence networks of eukaryotic microbial communities under different seasons were constructed (Figure 7a) and the topological parameters are shown in Table S3. Across the four networks, the spring network exhibited a higher complexity, evident from its greater number of modes, linkage numbers, and highest average degree. Furthermore, the network in spring contained the highest graph density, shortest path length, and high clustering coefficient, indicating that spring eukaryotic microorganisms were more closely associated and correlated, with greater efficiency in the transferring of information, energy, and matter between taxa. Additionally, OTUs exhibited mainly positive correlations with each other, and this positive correlation is strongest in the summer and weakest in the autumn. In general, the positive interactions of the co-occurrence network represent cooperative or mutualistic relationships, while the negative correlations reflect competitive or antagonistic relationships [20]. The large proportion of positive links in the four seasons suggested that eukaryotic microbes are inclined to cooperate as a survival strategy in the Jihongtan Reservoir.

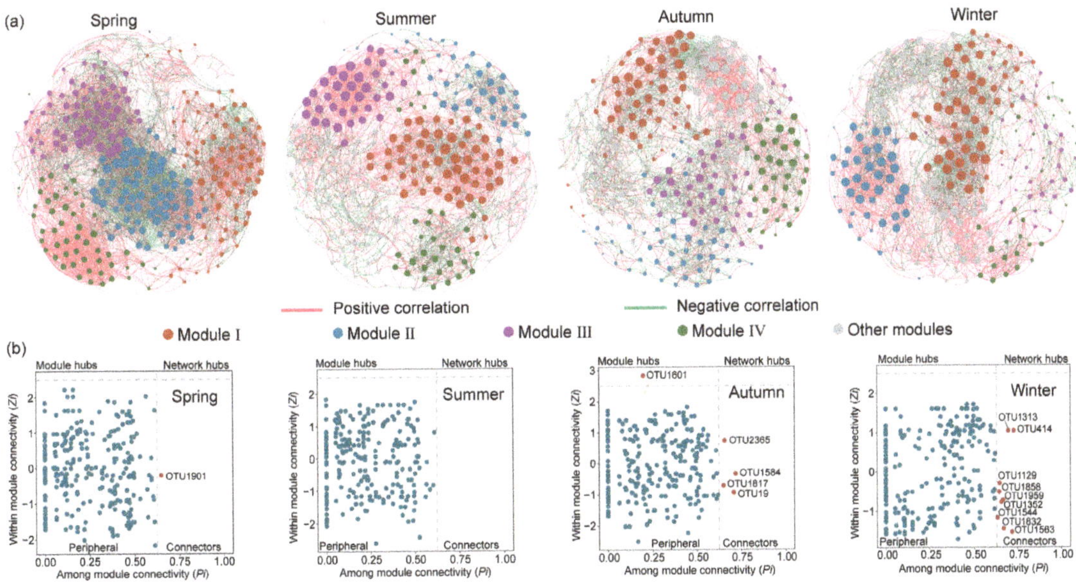

Figure 7. Seasonal co-occurrence network patterns of eukaryotic microbial communities in Jihongtan Reservoir. (**a**) Co-occurrence networks under different seasons; (**b**) keystone species analysis in different seasons.

Microbial networks can help identify keystone taxa, in addition to revealing co-occurrence patterns. According to the Zi (within-module connectivity) and Pi (among-module connectivity) values, nodes in a network can be categorized into four groups: network hubs (Zi > 2.5, Pi > 0.62), module hubs (Zi > 2.5, Pi < 0.62), connectors (Zi < 2.5, Pi > 0.62), and peripherals (Zi < 2.5, Pi < 0.62) [41]. Specifically, network hubs refer to nodes that can be highly connected within their own module as well as to other modules; module hubs are nodes that are only strongly connected within one module; connectors are nodes that connect different modules; and peripherals have limited connectivity within other nodes. In general, except for the peripherals, the other three categories are considered to contain keystone taxa due to their important roles in network topology, which may play a unique and crucial role in maintaining the microbiome structure and functioning [25]. According to Figure 7b, the majority of nodes in the four networks were peripheral, and only a minimal number of nodes were identified as key species. More detailed taxonomic information about keystones is presented in Table S2, indicating that the key species varied in both quantity and composition across the four seasonal communities. In the spring network, one connector was found, but as an unknown species. In the autumn network, one node assigned to the unknown species was identified as a module hub, and four nodes were identified as connectors belonging to Arthropoda, Porifera, and Cryptophyta. In the winter network, nine nodes were classified as connectors, of which four connectors were unassigned and the others mainly belonged to Ochrophyta, Ciliophora, and Chlorophyta.

4. Discussion

4.1. Seasonal Dynamics of Eukaryotic Microbial Diversity and Structure

The analyses of OTUs and alpha diversity revealed that the eukaryotic microbial biodiversity was higher in the spring and summer compared to the autumn and winter. Fang et al. studied the microbiome characteristics of urban rivers in different seasons and the seasonal changes in diversity were consistent with our results, suggesting that such changes could be partly explained by temperature variations [42]. Typically, severe conditions, especially the lower temperatures, lead to low eukaryotic microbial diversity [37]. For example, a low temperature narrows the niche for species with different ecological requirements, resulting in a decline in the species number and a consequent decrease in the diversity [43]. Other things being equal, there are more species in warm environments than in cold ones [44]. Notably, we found that the microbial diversity in the Jihongtan reservoir showed no significant differences across the four seasons, which is inconsistent with the results of previous studies of reservoirs [17,45]. A water diversion is a possible reason that might explain this phenomenon. Exotic microorganisms along with various kinds of nutrients can be introduced into the reservoir through the water transfer project, contributing to altering microbial community diversity directly and indirectly [46,47]. Previous studies indicated that an external water transfer can impact the microbial diversity in the water-receiving area [48,49]. For instance, Yang et al. [50] discovered that the Shannon–Wiener diversity index, Margalef richness index, Simpson dominance index, and Pielou evenness index of phytoplankton in receiving rivers were increased greatly after the two water transfers in the summer and winter.

This study demonstrated that the microbial community structure exhibited significant seasonal changes. Arthropoda was the most dominant phylum in all seasons except winter. Most of the dominant genera belong to the phylum Arthropoda, including *Paracyclopina*, *Sinocalanus*, *Pseudodiaptomus*, and *Microcyclops*, all of which are assigned to Copepoda. The abundance of Copepoda is highest in the summer and lowest in the winter. As primary and secondary consumers, Copepoda transfers organic carbon and energy from primary producers to higher trophic levels, playing a major role in aquatic food webs [51]. The abundance of *Paracyclopina* was much higher than that of *Sinocalanus* in the summer, whereas in the winter, the opposite trend was observed, reflecting changes in the water eutrophication level to some extent [52]. Among the identified dominant phyla and biomarkers, a significant portion belonged to algae, which play an essential role in maintaining primary productivity and matter-energy flows as autochthonous autotrophic producers in aquatic ecosystems [7,53].

It is notable that Chlorophyta dominates in the spring and winter, whereas its relative abundance is low in the summer and autumn, which is not consistent with previous studies showing that the higher water temperature and abundant nutrient conditions are more suitable for the mass reproduction of Chlorophyta [17,54]. We speculate that water diversion is the possible reason that might explain this phenomenon. Some studies have found dramatic differences in the phytoplankton composition before and after water transfers in freshwater ecosystems [55,56]. Dai et al. [57] stated that the main causes of variations in lacustrine phytoplankton communities were the direct input of exogenous species and changes in aquatic habitat conditions of the receiving lake, rather than nutrient perturbation. It has been found that most of the Chlorophyta species have faster growth rates and are more adaptive to the turbulent flowing environment [58]. Yang et al. [50] observed that Chlorophyta in the urban river increased by a maximum increment of 150% after the water diversion project in winter. As illustrated in Table 1, the primary sources of water replenishment in the spring and winter are ESNWD and YQWD, and the water recharge amount for each season is approximately equal to the capacity of the Jihongtan Reservoir, of which, ESNWD provided a water recharge of 57% in the spring and 83% in the winter, indicating ESNWD may have a greater impact on the Jihongtan Reservoir in the spring and winter. Chlorophyta has also been found to be one of the dominant algae phyla in ESNWD [59,60]. Dongping Lake, as part of ESNWD, experienced a rapid increase in the proportion of Chlorophyta during the water transfer period in March [55]. Moreover, according to Figure 3, there were no significant changes in the relative abundance of Chlorophyta in the inlets and outlets. Taken together, we prefer that mainly direct inputs of exogenous microorganisms caused the high abundance of Chlorophyta in the spring and winter.

4.2. Environmental Factors and Eukaryotic Microbial Communities

In this study, the mantel test and PLS-PM were conducted to analyze the correlation between environmental variables and the eukaryotic community structure. Among the environmental factors measured in the reservoir, WT, pH, and DO were found to be significantly associated with the eukaryotic microbial community structure, of which WT was the most critical factor driving seasonal succession among eukaryotic microorganisms. Many studies have reported the key role of WT in regulating variation in eukaryotic microbial communities [29,61]. According to the results of PLS-PM analysis, we can explain the important role of WT in shaping community diversity from two aspects (direct and indirect impacts). On the one hand, the water temperature can directly impact the cellular activity and metabolic rates, thereby influencing the life cycle of microorganisms through their growth and metabolic capacity [62]. For instance, the Copedoda species shows a positive correlation with the water temperature [63], and favorable warmer conditions support the growth of copepods. On the other hand, the water temperature, as a key indicator of seasonal characteristics, could indirectly affect the eukaryotic community structure by changing other environmental variables, such as DO and Chl-a. Chl-a is widely used as an indicator of water column phytoplankton biomass within aquatic ecosystems [64], and Zhang et al. [17] reported that there are interactions between phytoplankton and micro-eukaryotes. Being a key factor for phytoplankton reproduction and metabolism, DO was significantly negatively correlated with WT, indicating that a higher water temperature contributed to the lower DO concentration and vice versa. Moreover, pH was observed to have an obviously negative impact on the community of eukaryotic microorganisms, and an elevated pH may affect the nutrient uptake by affecting enzyme synthesis and cellular osmotic pressure, consequently affecting growth, metabolism, and the other life activities of aquatic organisms [65,66].

4.3. Co-Occurrence Patterns of Eukaryotic Microbial Communities and Keystone Taxa

Co-occurrence networks could offer insight into microbial interactions, the network complexity and stability beyond just diversity and composition [67]. Some studies have found that molecular ecological networks of microbial communities exhibited clear seasonal patterns, and microbial stability distinctly varied with seasonal variations [42,68]. Water diversion projects have been found to directly impact the stability of the community structure, but a stable state was re-established with the continuation of long-term water diversion [59]. In our study, the robustness results show that the spring network (0.337 $\pm$ 0.023) demonstrated the highest robustness compared to the summer (0.295 $\pm$ 0.030), autumn (0.294 $\pm$ 0.030), and winter (0.287 $\pm$ 0.030), among which there is little difference in robustness, which was in accord with the results of the network complexity. This finding is in agreement with Liu et al. [27], who explored the variations of microbial stability among seasons in lake ecosystems and confirmed that networks with a higher complexity are more likely to be stable. This suggests that stronger interspecific interactions could enhance the stability of microbial communities, leading to better resistance to external environmental disturbances. Moreover, the highest modularity in the summer network, the strongest negative correlation in the autumn network, and the greatest number of key species in the winter network may contribute to the strength of stability [26,69]. Competitive relationships among microorganisms are critical for maintaining a stable ecological system. Although positive correlations within communities can promote mutual assistance among members, cooperation can create dependency and the potential for mutual downfall, leading to oscillations across the network when responding together to environmental changes. Negative connections can help stabilize co-oscillation within communities and contribute to the stability of networks as a high proportion of negative links could better balance the asynchronous dynamics [70,71]. Keystone taxa are important for maintaining microbial community stability, and their removal could lead to the breakdown of networks. Liu et al. [72] found that keystone taxa contributed a lot to maintaining the biological community stability in a seasonal shallow lake, strongly affected by changes in hydrological disturbances and nutrient inputs. The modular organization of species would be beneficial for the local stability of ecological communities by enabling them to recover from small disturbances [73]. Taken together, compared to modularity, negative correlations, and keystone taxa, we speculated that the complexity of the network played a more active and important role in eukaryotic microbial community network stability in the Jihongtan Reservoir in terms of year-round water diversion.

Keystone taxa were identified in the microbial community irrespective of their abundance across space and time, and key taxa were found to be rare species in some studies [18,27]. Recent research on the rare biosphere has been receiving increasing attention. Rare microbes exhibit high genetic diversity and consist of a large number of metabolically active lineages, playing fundamental roles in regulating the ecological function and biogeochemical processes of various aquatic systems [74,75]. In this study, keystone taxa consisted of twelve species with very low abundance rankings (<1%) and three species with high abundance rankings (>1%). In the autumn network, there were two of each rare and abundant species in the keystone taxa. However, in the spring and winter, with the highest water diversions, the key taxa were dominated by rare species. Li et al. posit that within low-stress environmental contexts, the abundant biosphere exerts a substantially more pronounced influence on the stabilization of ecological networks compared to the rare biosphere, but the difference between their relative importance was observed to diminish significantly with the increasing stress [76]. These results suggested that both abundant and rare species exert influence on maintaining eukaryotic microbial community stability in the Jihongtan Reservoir. Additionally, we found no occurrences of crossover species in the four season networks, which is consistent with the findings of previous studies [27,28], suggesting that keystone taxa, serving as the core hubs within a network, may exhibit environment specificity, being confined to particular seasons or time periods.

5. Conclusions

This study investigated the seasonal characteristics of eukaryotic microbial communities in the Jihongtan Reservoir within year-round water diversion. The eukaryotic microbial diversity, community structure, and co-occurrence patterns showed seasonal variations, with significant changes observed in the community structure. The relationship between the microbial diversity and environmental factors was investigated, finding that the water temperature was the primary driver of seasonal succession in eukaryotic microbial communities, both directly and indirectly. The co-occurrence network and robustness results indicated that the complex eukaryotic microbial network showed more stability under the conditions of year-round water diversion. Keystone taxa of the eukaryotic microbes were identified, and it was found that both abundant and rare species play an important role in maintaining eukaryotic microbial community stability in the Jihongtan Reservoir. The impact of the water transfer project on the eukaryotic microbial community in the Jihongtan Reservoir is complex, due to constant variations in the water source and the amount of water diversion and, therefore, the continuous monitoring of aquatic organisms should be implemented. Our study results provide valuable insights for preserving the microbial diversity and aquatic ecosystem of the Jihongtan Reservoir in the future.

Supplementary Materials: The following supporting information can be downloaded at: https://www.mdpi.com/article/10.3390/microorganisms12091873/s1, Table S1: Mantel tests for the correlation between environmental variables and eukaryotic microbes based on Spearman rank correlation; Table S2: Topological parameters of eukaryotic microbial networks in four seasons; Table S3: Taxonomic information of keystone taxa in four season networks.

Author Contributions: Conceptualization, Y.Y., X.Z. and Z.S.; methodology, Y.Y.; formal analysis, Y.Y., F.C. and X.Z.; investigation, F.C., N.D. and N.W.; data curation, Y.Y. and N.D.; writing—original draft preparation, Y.Y. and F.C.; writing—review and editing, A.X. and Z.S.; visualization, Y.Y. and Y.L.; supervision, A.X. and Z.S.; funding acquisition, Z.S. All authors have read and agreed to the published version of the manuscript.

Funding: This research was funded by the National Natural Science Foundation of China, grant number 31570541.

Data Availability Statement: The data presented in this study are openly available in NCBI (https://www.ncbi.nlm.nih.gov/, accessed on 18 August 2024), reference number PRJNA1137302.

Conflicts of Interest: The authors declare no conflicts of interest.

References

1. Yan, H.; Lin, Y.; Chen, Q.; Zhang, J.; He, S.; Feng, T.; Wang, Z.; Chen, C.; Ding, J. A Review of the Eco-Environmental Impacts of the South-to-North Water Diversion: Implications for Interbasin Water Transfers. *Engineering* **2023**, *30*, 161–169. [CrossRef]
2. Chou, Q.; Nielsen, A.; Andersen, T.K.; Hu, F.; Chen, W.; Zhang, X.; Cao, T.; Ni, L.; Jeppesen, E.; Trolle, D. Assessing Impacts of Changes in External Nutrient Loadings on a Temperate Chinese Drinking Water Reservoir. *Front. Environ. Sci.* **2021**, *9*, 108695. [CrossRef]
3. Sun, B.; Wang, G.; Chen, W.; Li, W.; Kong, F.; Li, N.; Liu, Y.; Gao, X. Integrated modeling framework to evaluate the impacts of multi-source water replenishment on lacustrine phytoplankton communities. *J. Hydrol.* **2022**, *612*, 128272. [CrossRef]
4. Zhu, K.; Cheng, Y.; Zhou, Q. China's water diversion carries invasive species. *Science* **2023**, *380*, 1230. [CrossRef] [PubMed]
5. Shen, Z.; Yu, B.; Gong, Y.; Shao, K.; Gao, G.; Tang, X. Unraveling the impact of climatic warming and wetting on eukaryotic microbial diversity and assembly mechanisms: A 10-year case study in Lake Bosten, NW China. *Water Res.* **2024**, *256*, 121559. [CrossRef]
6. Logares, R.; Tesson, S.V.M.; Canbäck, B.; Pontarp, M.; Hedlund, K.; Rengefors, K. Contrasting prevalence of selection and drift in the community structuring of bacteria and microbial eukaryotes. *Environ. Microbiol.* **2018**, *20*, 2231–2240. [CrossRef]
7. Siriarchawatana, P.; Harnpicharnchai, P.; Phithakrotchanakoon, C.; Kitikhun, S.; Mayteeworakoon, S.; Chunhametha, S.; Eurwilaichitr, L.; Ingsriswang, S. Elucidating potential bioindicators from insights in the diversity and assembly processes of prokaryotic and eukaryotic communities in the Mekong River. *Environ. Res.* **2024**, *243*, 117800. [CrossRef]
8. Li, Y.; Zhang, H.; Huo, S.; Zhang, J.; Ma, C.; Weng, N.; Zhang, P.; Shi, Z. Microbial eukaryote community succession over hundreds of years in Chinese lakes. *Ecol. Indic.* **2024**, *158*, 111512. [CrossRef]
9. Yang, J.R.; Yu, X.; Chen, H.; Kuo, Y.-M.; Yang, J. Structural and functional variations of phytoplankton communities in the face of multiple disturbances. *J. Environ. Sci.* **2021**, *100*, 287–297. [CrossRef]

10. Xue, Y.; Chen, H.; Xiao, P.; Jin, L.; Logares, R.; Yang, J. Core taxa drive microeukaryotic community stability of a deep subtropical reservoir after complete mixing. *Environ. Microbiol. Rep.* **2023**, *15*, 769–782. [CrossRef]
11. Mikhailov, I.S.; Galachyants, Y.P.; Bukin, Y.S.; Petrova, D.P.; Bashenkhaeva, M.V.; Sakirko, M.V.; Blinov, V.V.; Titova, L.A.; Zakharova, Y.R.; Likhoshway, Y.V. Seasonal Succession and Coherence Among Bacteria and Microeukaryotes in Lake Baikal. *Microb. Ecol.* **2022**, *84*, 404–422. [CrossRef] [PubMed]
12. Marquardt, M.; Vader, A.; Stübner, E.I.; Reigstad, M.; Gabrielsen, T.M. Strong Seasonality of Marine Microbial Eukaryotes in a High-Arctic Fjord (Isfjorden, in West Spitsbergen, Norway). *Appl. Environ. Microbiol.* **2016**, *82*, 1868–1880. [CrossRef] [PubMed]
13. Wang, H.; Weil, M.; Dumack, K.; Zak, D.; Münch, D.; Günther, A.; Jurasinski, G.; Blume-Werry, G.; Kreyling, J.; Urich, T. Eukaryotic rather than prokaryotic microbiomes change over seasons in rewetted fen peatlands. *FEMS Microbiol. Ecol.* **2021**, *97*, fiab121. [CrossRef] [PubMed]
14. Lv, J.; Yuan, R.; Wang, S. Water diversion induces more changes in bacterial and archaeal communities of river sediments than seasonality. *J. Environ. Manag.* **2021**, *293*, 112876. [CrossRef] [PubMed]
15. Zhang, C.; Zhu, F.; Wang, Y.; Zhu, Y.; Song, G.; Mi, W.; Bi, Y. Assembly processes of eukaryotic plankton communities in the world's largest drinking water diversion project. *Sci. Total Environ.* **2023**, *884*, 163665. [CrossRef]
16. Ouyang, L.; Chen, H.; Liu, X.; Wong, M.H.; Xu, F.; Yang, X.; Xu, W.; Zeng, Q.; Wang, W.; Li, S. Characteristics of spatial and seasonal bacterial community structures in a river under anthropogenic disturbances. *Environ. Pollut.* **2020**, *264*, 114818. [CrossRef]
17. Zhang, H.; Yang, Y.; Liu, X.; Huang, T.; Ma, B.; Li, N.; Yang, W.; Li, H.; Zhao, K. Novel insights in seasonal dynamics and co-existence patterns of phytoplankton and micro-eukaryotes in drinking water reservoir, Northwest China: DNA data and ecological model. *Sci. Total Environ.* **2023**, *857*, 159160. [CrossRef]
18. Yang, Q.; Zhang, P.; Li, X.; Yang, S.; Chao, X.; Liu, H.; Ba, S. Distribution patterns and community assembly processes of eukaryotic microorganisms along an altitudinal gradient in the middle reaches of the Yarlung Zangbo River. *Water Res.* **2023**, *239*, 120047. [CrossRef]
19. Ma, B.; Wang, Y.; Ye, S.; Liu, S.; Stirling, E.; Gilbert, J.A.; Faust, K.; Knight, R.; Jansson, J.K.; Cardona, C.; et al. Earth microbial co-occurrence network reveals interconnection pattern across microbiomes. *Microbiome* **2020**, *8*, 82. [CrossRef]
20. Yuan, M.M.; Guo, X.; Wu, L.W.; Zhang, Y.; Xiao, N.J.; Ning, D.L.; Shi, Z.; Zhou, X.S.; Wu, L.Y.; Yang, Y.F.; et al. Climate warming enhances microbial network complexity and stability. *Nat. Clim. Chang.* **2021**, *11*, 343–348. [CrossRef]
21. Faust, K.; Raes, J. Microbial interactions: From networks to models. *Nat. Rev. Microbiol.* **2012**, *10*, 538–550. [CrossRef] [PubMed]
22. Liu, J.; Sun, X.; Zuo, Y.; Hu, Q.; He, X. Plant species shape the bacterial communities on the phyllosphere in a hyper-arid desert. *Microbiol. Res.* **2023**, *269*, 127314. [CrossRef] [PubMed]
23. Fuhrman, J.A. Microbial community structure and its functional implications. *Nature* **2009**, *459*, 193–199. [CrossRef] [PubMed]
24. Berry, D.; Widder, S. Deciphering microbial interactions and detecting keystone species with co-occurrence networks. *Front. Microbiol.* **2014**, *5*, 219. [CrossRef]
25. Banerjee, S.; Schlaeppi, K.; van der Heijden, M.G.A. Keystone taxa as drivers of microbiome structure and functioning. *Nat. Rev. Microbiol.* **2018**, *16*, 567–576. [CrossRef]
26. Wang, L.; Wang, Z.; Li, Y.; Cai, W.; Zou, Y.; Hui, C. Deciphering solute transport, microbiota assembly patterns and metabolic functions in the hyporheic zone of an effluent-dominated river. *Water Res.* **2024**, *251*, 121190. [CrossRef] [PubMed]
27. Liu, S.W.; Yu, H.; Yu, Y.H.; Huang, J.; Zhou, Z.Y.; Zeng, J.X.; Chen, P.B.; Xiao, F.S.; He, Z.L.; Yan, Q.Y. Ecological stability of microbial communities in Lake Donghu regulated by keystone taxa. *Ecol. Indic.* **2022**, *139*, 108695. [CrossRef]
28. Chen, C.; Li, P.; Yin, M.; Wang, J.; Sun, Y.; Ju, W.; Liu, L.; Li, Z.-H. Deciphering characterization of seasonal variations in microbial communities of marine ranching: Diversity, co-occurrence network patterns, and assembly processes. *Mar. Pollut. Bull.* **2023**, *197*, 115739. [CrossRef]
29. Li, S.; Luo, N.; Li, C.; Mao, S.; Huang, H. Diversity and distribution analysis of eukaryotic communities in the Xiangshan Bay, East China sea by metabarcoding approach. *Mar. Environ. Res.* **2024**, *197*, 106451. [CrossRef] [PubMed]
30. Stoeck, T.; Bass, D.; Nebel, M.; Christen, R.; Jones, M.D.M.; Breiner, H.-W.; Richards, T.A. Multiple marker parallel tag environmental DNA sequencing reveals a highly complex eukaryotic community in marine anoxic water. *Mol. Ecol.* **2010**, *19*, 21–31. [CrossRef]
31. Chen, S.; Zhou, Y.; Chen, Y.; Gu, J. fastp: An ultra-fast all-in-one FASTQ preprocessor. *Bioinformatics* **2018**, *34*, i884–i890. [CrossRef] [PubMed]
32. Magoč, T.; Salzberg, S.L. FLASH: Fast length adjustment of short reads to improve genome assemblies. *Bioinformatics* **2011**, *27*, 2957–2963. [CrossRef] [PubMed]
33. Edgar, R.C. UPARSE: Highly accurate OTU sequences from microbial amplicon reads. *Nat. Methods* **2013**, *10*, 996–998. [CrossRef] [PubMed]
34. Wang, Q.; Garrity, G.M.; Tiedje, J.M.; Cole, J.R. Naïve Bayesian Classifier for Rapid Assignment of rRNA Sequences into the New Bacterial Taxonomy. *Appl. Environ. Microbiol.* **2007**, *73*, 5261–5267. [CrossRef]
35. Gu, Z.; Hübschmann, D. Make Interactive Complex Heatmaps in R. *Bioinformatics* **2021**, *38*, 1460–1462. [CrossRef]
36. Gloor, G.B.; Macklaim, J.M.; Pawlowsky-Glahn, V.; Egozcue, J.J. Microbiome Datasets Are Compositional: And This Is Not Optional. *Front. Microbiol.* **2017**, *8*, 2224. [CrossRef]

37. Zhang, C.; Li, H.; Zeng, Y.; Ding, H.; Wang, B.; Li, Y.; Ji, Z.; Bi, Y.; Luo, W. Diversity and assembly processes of microbial eukaryotic communities in Fildes Peninsula Lakes (West Antarctica). *Biogeosciences* **2022**, *19*, 4639–4654. [CrossRef]
38. Cao, Y.; Dong, Q.; Wang, D.; Zhang, P.; Liu, Y.; Niu, C. microbiomeMarker: An R/Bioconductor package for microbiome marker identification and visualization. *Bioinformatics* **2022**, *38*, 4027–4029. [CrossRef]
39. Zhou, T.; Wang, Z.; Lv, Q.; Zhang, Y.; Tao, S.; Ren, X.; Gao, H.; Gao, Z.; Hu, S. Sulfur dynamics in saline sodic soils: The role of paddy cultivation and organic amendments. *Ecol. Indic.* **2024**, *162*, 112014. [CrossRef]
40. Wen, T.; Xie, P.; Yang, S.; Niu, G.; Liu, X.; Ding, Z.; Xue, C.; Liu, Y.-X.; Shen, Q.; Yuan, J. ggClusterNet: An R package for microbiome network analysis and modularity-based multiple network layouts. *iMeta* **2022**, *1*, e32. [CrossRef]
41. Guimerà, R.; Nunes Amaral, L.A. Functional cartography of complex metabolic networks. *Nature* **2005**, *433*, 895–900. [CrossRef] [PubMed]
42. Fang, W.; Fan, T.; Wang, S.; Yu, X.; Lu, A.; Wang, X.; Zhou, W.; Yuan, H.; Zhang, L. Seasonal changes driving shifts in microbial community assembly and species coexistence in an urban river. *Sci. Total Environ.* **2023**, *905*, 167027. [CrossRef] [PubMed]
43. Yang, Y.; Stenger-Kovács, C.; Padisák, J.; Pettersson, K. Effects of winter severity on spring phytoplankton development in a temperate lake (Lake Erken, Sweden). *Hydrobiologia* **2016**, *780*, 47–57. [CrossRef]
44. Brown, J.H.; Gillooly, J.F.; Allen, A.P.; Savage, V.M.; West, G.B. Toward a metabolic theory of ecology. *Ecology* **2004**, *85*, 1771–1789. [CrossRef]
45. Guo, J.; Zheng, Y.; Teng, J.; Song, J.; Wang, X.; Zhao, Q. The seasonal variation of microbial communities in drinking water sources in Shanghai. *J. Clean. Prod.* **2020**, *265*, 121604. [CrossRef]
46. Tseng, C.H.; Chiang, P.W.; Shiah, F.K.; Chen, Y.L.; Liou, J.R.; Hsu, T.C.; Maheswararajah, S.; Saeed, I.; Halgamuge, S.; Tang, S.L. Microbial and viral metagenomes of a subtropical freshwater reservoir subject to climatic disturbances. *ISME J.* **2013**, *7*, 2374–2386. [CrossRef]
47. Mo, Y.; Peng, F.; Gao, X.; Xiao, P.; Logares, R.; Jeppesen, E.; Ren, K.; Xue, Y.; Yang, J. Low shifts in salinity determined assembly processes and network stability of microeukaryotic plankton communities in a subtropical urban reservoir. *Microbiome* **2021**, *9*, 128. [CrossRef]
48. Dai, J.; Wu, S.; Wu, X.; Lv, X.; Sivakumar, B.; Wang, F.; Zhang, Y.; Yang, Q.; Gao, A.; Zhao, Y.; et al. Impacts of a large river-to-lake water diversion project on lacustrine phytoplankton communities. *J. Hydrol.* **2020**, *587*, 124938. [CrossRef]
49. Liu, Y.; Pan, B.; Zhu, X.; Zhao, X.; Sun, H.; He, H.; Jiang, W. Patterns of microbial communities and their relationships with water quality in a large-scale water transfer system. *J. Environ. Manag.* **2022**, *319*, 115678. [CrossRef]
50. Yang, Y.; Zhang, J.; Yan, W.; Zhang, Y.; Wang, J.; Wang, G.; Yan, F. Impact assessment of water diversion project on urban aquatic ecological environment. *Ecol. Indic.* **2021**, *125*, 107496. [CrossRef]
51. Ziadi, B.; Dhib, A.; Turki, S.; Aleya, L. Factors driving the seasonal distribution of zooplankton in a eutrophicated Mediterranean Lagoon. *Mar. Pollut. Bull.* **2015**, *97*, 224–233. [CrossRef] [PubMed]
52. Min, C.; Johansson, L.S.; Søndergaard, M.; Lauridsen, T.L.; Chen, F.; Sh, T.; Jeppesen, E. Copepods as environmental indicator in lakes: Special focus on changes in the proportion of calanoids along nutrient and pH gradients. *Aquat. Ecol.* **2021**, *55*, 1241–1252. [CrossRef]
53. Hébert, M.-P.; Soued, C.; Fussmann, G.F.; Beisner, B.E. Dissolved organic matter mediates the effects of warming and inorganic nutrients on a lake planktonic food web. *Limnol. Oceanogr.* **2023**, *68*, S23–S38. [CrossRef]
54. Li, Z.; Lu, X.; Fan, Y. Seasonal shifts in assembly dynamics of phytoplankton communities in a humans-affected river in NE China. *J. Oceanol. Limnol.* **2022**, *40*, 1985–2000. [CrossRef]
55. Li, Q.; Wang, G.; Tan, Z.; Wang, H. Succession of phytoplankton in a shallow lake under the alternating influence of runoff and reverse water transfer. *Hydrol. Res.* **2020**, *51*, 1077–1090. [CrossRef]
56. Zhang, X.; Wang, G.; Tan, Z.; Wang, Y.; Li, Q. Effects of ecological protection and restoration on phytoplankton diversity in impounded lakes along the eastern route of China's South-to-North Water Diversion Project. *Sci. Total Environ.* **2021**, *795*, 148870. [CrossRef]
57. Dai, J.; Sha, H.; Wu, X.; Wu, S.; Zhang, Y.; Wang, F.; Gao, A.; Xu, J.; Tian, F.; Zhu, S.; et al. Pulses outweigh cumulative effects of water diversion from river to lake on lacustrine phytoplankton communities. *Environ. Geochem. Health* **2023**, *45*, 3025–3039. [CrossRef]
58. Istvánovics, V.; Honti, M. Efficiency of nutrient management in controlling eutrophication of running waters in the Middle Danube Basin. *Hydrobiologia* **2012**, *686*, 55–71. [CrossRef]
59. Zhang, S.; Pang, Y.; Xu, H.; Wei, J.; Jiang, S.; Pei, H. Shift of phytoplankton assemblages in a temperate lake located on the eastern route of the South-to-North Water Diversion Project. *Environ. Res.* **2023**, *227*, 115805. [CrossRef]
60. Sun, R.; Wei, J.; Zhang, S.; Pei, H. The dynamic changes in phytoplankton and environmental factors within Dongping Lake (China) before and after the South-to-North Water Diversion Project. *Environ. Res.* **2024**, *246*, 118138. [CrossRef]
61. Huo, S.; Zhang, H.; Wang, J.; Chen, J.; Wu, F. Temperature and precipitation dominates millennium changes of eukaryotic algal communities in Lake Yamzhog Yumco, Southern Tibetan Plateau. *Sci. Total Environ.* **2022**, *829*, 154636. [CrossRef] [PubMed]
62. Margesin, R.; Miteva, V. Diversity and ecology of psychrophilic microorganisms. *Res. Microbiol.* **2011**, *162*, 346–361. [CrossRef] [PubMed]

63. Li, C.; Feng, W.; Chen, H.; Li, X.; Song, F.; Guo, W.; Giesy, J.P.; Sun, F. Temporal variation in zooplankton and phytoplankton community species composition and the affecting factors in Lake Taihu—A large freshwater lake in China. *Environ. Pollut.* **2019**, *245*, 1050–1057. [CrossRef] [PubMed]
64. Buchan, A.; LeCleir, G.R.; Gulvik, C.A.; González, J.M. Master recyclers: Features and functions of bacteria associated with phytoplankton blooms. *Nat. Rev. Microbiol.* **2014**, *12*, 686–698. [CrossRef]
65. Singer, D.; Seppey, C.V.W.; Lentendu, G.; Dunthorn, M.; Bass, D.; Belbahri, L.; Blandenier, Q.; Debroas, D.; de Groot, G.A.; de Vargas, C.; et al. Protist taxonomic and functional diversity in soil, freshwater and marine ecosystems. *Environ. Int.* **2021**, *146*, 106262. [CrossRef]
66. Santini, T.C.; Gramenz, L.; Southam, G.; Zammit, C. Microbial Community Structure Is Most Strongly Associated with Geographical Distance and pH in Salt Lake Sediments. *Front. Microbiol.* **2022**, *13*, 920056. [CrossRef]
67. Barberán, A.; Bates, S.T.; Casamayor, E.O.; Fierer, N. Using network analysis to explore co-occurrence patterns in soil microbial communities. *ISME J.* **2012**, *6*, 343–351. [CrossRef]
68. Wang, M.; Liu, X.; Qu, L.; Wang, T.; Zhu, L.; Feng, J. Untangling microbiota diversity and assembly patterns in the world's longest underground culvert water diversion canal. *Environ. Monit. Assess.* **2023**, *195*, 981. [CrossRef]
69. Hernandez, D.J.; David, A.S.; Menges, E.S.; Searcy, C.A.; Afkhami, M.E. Environmental stress destabilizes microbial networks. *ISME J.* **2021**, *15*, 1722–1734. [CrossRef]
70. Coyte, K.Z.; Schluter, J.; Foster, K.R. The ecology of the microbiome: Networks, competition, and stability. *Science* **2015**, *350*, 663–666. [CrossRef]
71. de Vries, F.T.; Griffiths, R.I.; Bailey, M.; Craig, H.; Girlanda, M.; Gweon, H.S.; Hallin, S.; Kaisermann, A.; Keith, A.M.; Kretzschmar, M.; et al. Soil bacterial networks are less stable under drought than fungal networks. *Nat. Commun.* **2018**, *9*, 3033. [CrossRef] [PubMed]
72. Liu, Y.; Zhang, M.; Peng, W.; Wu, N.; Qu, X.; Yu, Y.; Zhang, Y.; Yang, C. The effects of flood pulse on multiple aquatic organisms in a seasonal shallow lake. *Aquat. Ecol.* **2021**, *55*, 379–399. [CrossRef]
73. Grilli, J.; Rogers, T.; Allesina, S. Modularity and stability in ecological communities. *Nat. Commun.* **2016**, *7*, 12031. [CrossRef] [PubMed]
74. Dang, C.; Wang, J.; He, Y.; Yang, S.; Chen, Y.; Liu, T.; Fu, J.; Chen, Q.; Ni, J. Rare biosphere regulates the planktonic and sedimentary bacteria by disparate ecological processes in a large source water reservoir. *Water Res.* **2022**, *216*, 118296. [CrossRef]
75. Xue, Y.; Chen, H.; Yang, J.R.; Liu, M.; Huang, B.; Yang, J. Distinct patterns and processes of abundant and rare eukaryotic plankton communities following a reservoir cyanobacterial bloom. *ISME J.* **2018**, *12*, 2263–2277. [CrossRef]
76. Li, C.; Jin, L.; Zhang, C.; Li, S.; Zhou, T.; Hua, Z.; Wang, L.; Ji, S.; Wang, Y.; Gan, Y.; et al. Destabilized microbial networks with distinct performances of abundant and rare biospheres in maintaining networks under increasing salinity stress. *iMeta* **2023**, *2*, e79. [CrossRef]

microorganisms

MDPI

Article

Rhizosphere Shifts: Reduced Fungal Diversity and Microbial Community Functionality Enhance Plant Adaptation in Continuous Cropping Systems

Jichao Li, Yingmei Zuo * and Jinyu Zhang *

Medicinal Plants Research Institute, Yunnan Academy of Agricultural Sciences, No. 2238 Beijing Road, Kunming 650221, China; ljc@yaas.org.cn

* Correspondence: fairy1003@163.com (Y.Z.); zym@yaas.org.cn (J.Z.); Tel.: +86-13888872457 (J.Z.)

Abstract: Continuous cropping problems constitute threats to perennial plant health and survival. Soil conditioners have the potential to enhance plant disease resistance in continuous cropping systems. However, how microbes and metabolites of the rhizosphere respond to soil conditioner addition remains largely unknown, but this knowledge is paramount to providing innovative strategies to enhance plant adaptation in continuous cropping systems. Here, we found that a biochar conditioner significantly improved plant survival rates in a continuous cropping system. The biochar-induced rhizosphere significantly alters the fungal community, causing a decline in fungal diversity and the downregulation of soil microbial community functionality. Specifically, the biochar-induced rhizosphere causes a reduction in the relative abundance of pathogenic *Fusarium* sp. and phenolic acid concentration, whose variations are the primary causes of continuous cropping problems. Conversely, we observed an unexpected bacterial diversity increase in rhizospheric and non-rhizospheric soils. Our research further identified key microbial taxa in the biochar-induced rhizosphere, namely, *Monographella*, *Acremonium*, *Geosmithia*, and *Funneliformis*, which enhance soil nutrient availability, suppress *Fusarium* sp., mitigate soil acidification, and reduce phenolic acid concentrations. Collectively, we highlight the critical role of regular microbial communities and metabolites in determining plant health during continuous cropping and propose a synthetic microbial community framework for further optimizing the ecological functions of the rhizosphere.

Keywords: rhizosphere microbiome; biochar; continuous cropping obstacles; microbial community functionality; *Fusarium*

1. Introduction

Continuous cropping obstacles (CCOs) have become a staple in the pursuit of agricultural intensification. However, they lead to imbalances in rhizosphere ecology and heightened plant vulnerability to diseases, especially those caused by soil-borne pathogenic *Fusarium* [1]. Root-harvest-targeted medicinal plants (approximately 70%), e.g., *Panax notoginseng* [2], *Panax ginseng*, *Angelica sinensis*, and *Rehmannia glutinosa* [3], experience significant challenges from continuous cropping obstacles (CCOs) that lead to crop failure. CCOs constitute major threats to the yield and properties of medicinal plants and sustainable agriculture. Soil conditioners have shown the potential to improve plant disease resistance in continuous cropping systems. However, it is still unknown how a soil conditioner affects microbial communities and metabolites, which are crucial for plant health [4].

The rhizosphere, a hub of root activities, microorganisms, and metabolic exchanges, is crucial for driving plant growth, resistance, and evolution. It has been described as the second genome of plants and the first line of defense against pathogens [5]. Plants manipulate the soil microbiota by secreting bioactive molecules into the rhizosphere that attract plant-growth-promoting rhizobacteria (PGPR) and perform essential functions, such as nitrogen fixation, nutrient mobilization, and disease suppression, thereby enhancing soil microbial productivity and nutrient availability [6]. However, this process can also inadvertently promote soil-borne diseases, leading to continuous cropping obstacles (CCOs) that threaten plant health and agricultural economics.

In the rhizosphere, metabolites that favor pathogen preferences can enhance their survival and colonization. For instance, cinnamic acid has been shown to increase fusaric acid secretion by *F. oxysporum*, increasing plant susceptibility to infection [7]. Additionally, syringic acid can induce a shift in the tobacco rhizosphere microbial community from beneficial to detrimental, contributing to CCOs [8]. Organic acids can also enhance toxin production and H_2O_2 secretion by pathogens, weakening the antagonistic capabilities of PGPR and promoting pathogenic fungal growth [9]. Pathogens can manipulate host plants to release specific root exudates that amplify their virulence and produce chemo-sensory autotoxic substances. For example, *F. oxysporum* infection alters the expression of genes related to phenolic acid synthesis in host plants, leading to the increased synthesis and accumulation of phenolics in the rhizosphere [10].

Thus, the presence of pathogens and their "preference-type" rhizosphere metabolites in continuous cropping systems potentially accelerates the threat to rhizosphere microbial ecology. Rhizosphere microecological imbalances can significantly impede plant immunity and survival, presenting a substantial challenge in agricultural practices. In this study, we investigate the effects of soil conditioners on fungal and bacterial dynamics in both rhizosphere and bulk soils and their implications for soil health and plant survival. We further identify biomarkers and analyze their relationships with microbial communities, soil nutrients, and autotoxicity, understanding the critical role of rhizosphere microbes and metabolites in enhancing plant adaptation in continuous cropping systems.

2. Materials and Methods

2.1. Experimental Materials

Test soil: The potted experiment commenced in a greenhouse in Kunming County, Yunnan Province, China. The average humidity in the greenhouse ranged from 68 to 75%, with an average temperature of 17 to 21 °C. The continuous cropping soils for the experiment were collected following the harvest of *Panax notoginseng* (Burk.) F.H. Chen after 3-yr continuous cultivation in Wenshan County, Yunnan Province, China. The continuous cropping soil in this region is classified as red soil. Its initial physical and chemical properties included a pH of 5.7 and total N, total P, total K, and organic matter levels of 0.53, 0.61, 5.80, and 12.436 g/kg, respectively. Source of seedlings: We utilized *P. notoginseng* as a model plant, which causes more CCOs than other crops, thereby having a destructive impact on the survival rates of seedlings and leading to the complete loss of seedlings. The *P. notoginseng* seeds were chosen based on vigor, structural integrity, and the absence of insect infestation. The seeds, along with the seedling substrate, underwent a sterilization process. Subsequently, the seeds were sown and cultivated into seedlings, under uniform management, over a period of one year.

2.2. Strategy Design: Growth, Disease Development, and Soil Collection

We designed three strategies, i.e., biochar, grass ash, and corn stover. The biochar selected was rice husk charcoal with a total nano-pore size of 2.1 cm $\times$ 3 g $^{-1}$. Grass ash was derived from the combustion residues of rice straw and had a total ash content exceeding 80 percent. The chosen corn stover was the leaves of the corn plant, which had undergone natural maturation and desiccation. The experimental substrates, biochar, grass ash, and maize stover, were subjected to a sieving process using a sieve with a mesh size of 20.

We used the root bag method for the experiment. Small and large bags were sewn with 25 μm mesh diameter nylon mesh. The small bags were filled with *P. notoginseng* seedlings planted in the test soil. The three strategic substances were added in amounts of 50 g to a large bag (14 $\times$ 10 cm) and a small bag (12 $\times$ 8 cm), and 200 g of test soil with seedlings was placed into the large bag, forming an interlayer with three adsorption effects.

Then, the large bag was placed in a pot (diameter 28 cm, height 40 cm), and the outside was filled with the same test soil (without the added substances). A control was set up where the interlayer was the same test soil (Figure 1). The potting unit was configured to include 30 pots per treatment group. After a particular growth stage, we investigated the plants' growth and seedling survival rates [11]. Ten months after planting, ten soil cores with a depth of 10 cm were randomly selected from each pot, and these ten soil cores were thoroughly mixed into a single composite soil sample as non-rhizosphere soil. Excess soil on the roots was discarded by gently shaking the plants, and the remaining soil particles attached to the root surface were collected as rhizosphere soil. Samples of rhizosphere soil, non-rhizosphere soil, and interlayer material were collected and stored at −80 °C to facilitate further experimental analyses.

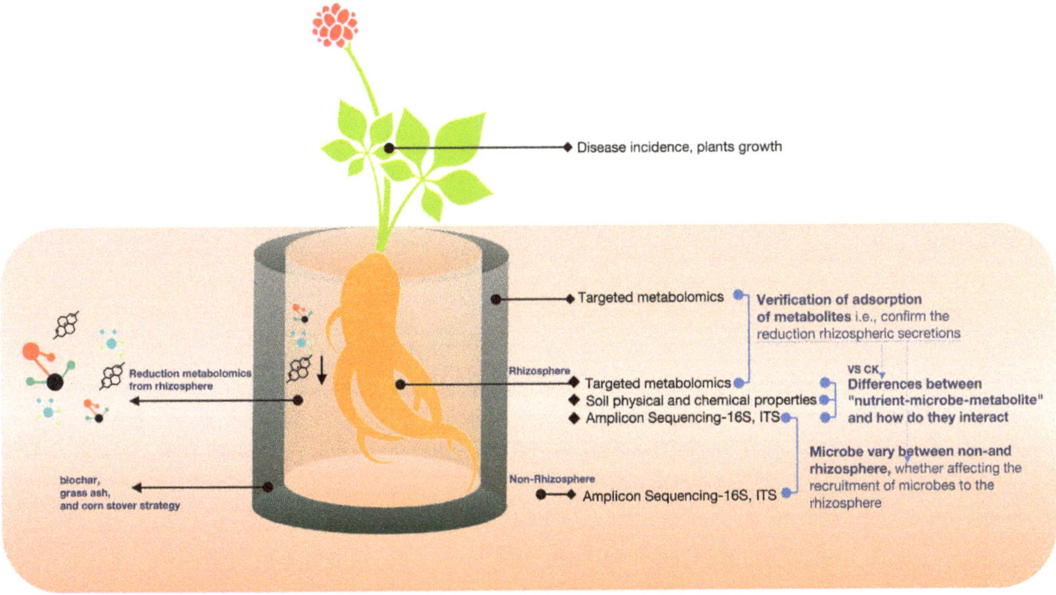

Figure 1. Materials and methods in experimental design.

2.3. Profiling Non-Rhizosphere and Rhizosphere Fungi and Bacteria

Sequencing was performed using Illumina MiSeq, a second-generation high-throughput sequencing platform. The concentration of DNA was verified with a NanoDrop spectrophotometer and agarose gel electrophoresis. The genomic DNA was used as a template for PCR amplification with barcoded primers and Tks Gflex DNA Polymerase (Takara). For bacterial diversity analysis, V3–V4 (or V4–V5) variable regions of 16S rRNA genes were amplified with the universal primers 343 F and 798 R (or 515F and 907R for the V4–V5 region). For fungal diversity analysis, the ITS I variable region was amplified with the universal primers ITS1F and ITS2. The raw sequencing data were in FASTQ format. Paired-end reads were then preprocessed using Trimmomatic software (0.39) to detect and cut off ambiguous bases (N). Low-quality sequences with average quality scores of less than 20 were also removed using the sliding window trimming approach. After trimming, paired-end reads were assembled using FLASH software (2.2.15). The parameters of the assembly were as follows: 10 bp of minimal overlap, 200 bp of maximum overlap, and a maximum mismatch rate of 20 percent. The sequences were further denoised as follows: Reads with ambiguous or homologous sequences or less than 200 bp were discarded. Reads with 75 percent of bases above Q20 were retained. Then, reads with chimeras were detected and removed. These two steps were achieved using QIIME software (version 1.8.0). Clean reads were subjected to primer sequence removal and clustering to generate operational taxonomic units (OTUs) using Vsearch software (2.29.1) with a 97 percent similarity cutoff. The representative read of each OTU was selected using the QIIME package. All representative reads were annotated and blasted against the Silva database Version 138 (16 s/18 s rDNA) using the RDP classifier (the confidence threshold was 70 percent). All representative reads were annotated and blasted against the Unite database (ITS rDNA) using BLAST (2.15.0) [12].

2.4. Quantitative Wide-Target Metabolome Detection of the Rhizosphere Soil and Interlayer Metabolites

Sample preparation and extraction. After thawing the samples from the refrigerator at −80 °C, they were mixed by vortexing for 10 s. After mixing, 2 mL of each sample was added to a centrifuge tube, after which the sample was immersed in liquid nitrogen. The sample was placed into the lyophilizer for freeze-drying after it was completely frozen. After the samples were completely lyophilized, 200 μL of a 70 percent methanol internal standard extract was added. After centrifugation (12,000 r/min, 4 °C) for 3 min, the mixture was swirled for 15 min. The supernatant was filtered through a microporous filter membrane (0.22 μm) and stored in a sample flask for LC–MS/MS analysis.

UPLC conditions. The sample extracts were analyzed using a UPLC-ESI-MS/MS system (UPLC, SHIMADZU Nexera X2 (Shimadzu Corporation, Kyoto, Japan); MS, Applied Biosystems 4500 Q TRAP (AB SCIEX, Framingham, MA, USA)). The analytical conditions were as follows: UPLC: column, Agilent SB-C18 (1.8 μm, 2.1 mm × 100 mm). The mobile phase consisted of solvent A, pure water with 0.1 percent formic acid; solvent B, acetonitrile with 0.1 percent formic acid. Sample measurements were performed with a gradient program that employed starting conditions of 95 percent A, 5 percent B. Within 9 min, a linear gradient to 5 percent A, 95 percent B was programmed, and a composition of 5 percent A, 95 percent B was maintained for 1 min. Subsequently, the composition was adjusted to 95 percent A, 5.0 percent B within 1.1 min, which was maintained for 2.9 min. The flow velocity was set to 0.35 mL/min; the column oven was set to 40 °C; and the injection volume was 4 μL. The effluent was alternatively connected to an ESI-triple quadrupole-linear ion trap (QTRAP)-MS [13].

ESI-Q TRAP-MS/MS. The ESI source operating parameters were as follows: source temperature, 550 °C; ion spray voltage (IS), 5500 V (positive ion mode)/−4500 V (negative ion mode); ion source gas I (GSI), gas II (GSII), and curtain gas (CUR), 50, 60, and 25 psi, respectively; and high collision-activated dissociation (CAD) pressure. Instrument tuning and mass calibration were performed with 10 and 100 μmol/L polypropylene glycol solutions in the QQQ and LIT modes, respectively. QQQ scans were acquired as MRM experiments with the collision gas (nitrogen) set to medium. DP (declustering potential) and CE (collision energy) for individual MRM transitions were determined with further DP and CE optimization. A specific set of MRM transitions was monitored for each period according to the metabolites eluted within this period.

2.5. Effects on Rhizosphere Microbial Community Function

For PICRUSt analysis, we followed the suggested methods for OTU selection with Greengenes 13-5 using Galaxy One UI 6.1.1. PICRUSt (Phylogenetic Investigation of Communities by Reconstruction of Unobserved States) is a bioinformatics tool designed to predict the functional abundances of microbial communities based on marker gene sequences. It infers the metabolic potential of bacteria and archaea. The workflow of PICRUSt2 includes the following steps: The study sequences (OTUs and ASVs) are placed onto a reference tree using tools like HMMER, EPA-ng, and GAPPA. Core hidden state prediction functions are implemented using the R package castor. Metagenome profiles are generated and stratified based on contributing sequences [14]. The predicted gene family abundances were analyzed using the Kyoto Encyclopedia of Genes and Genomes ethology group count level 3, and Story FDR in STAMP software 2.1.3 was used to avoid Type-I errors. KEGG function prediction, enzyme classification number EC, and COG protein prediction were performed; differences were determined according to the Kruskal–Wallis algorithm, and the different results were selected to construct a heatmap map. Linear differential analysis effect size (LEfSe) was used to predict up- and downregulated functions and construct a differential heatmap [15].

2.6. Soil Physicochemical Analysis

The content of organic matter was determined by the potassium dichromate volumetric method. After acid dissolution, total nitrogen was determined by a SEAL-AA3 continuous flow analyzer, total phosphorus was determined by molybdenum–antimony colorimetry, and total potassium and available potassium were determined by a flame photometer. The available nitrogen content was determined via the alkali-hydrolyzed diffusion method, and the available phosphorus content was determined via the sodium bicarbonate method. The cation exchange capacity (CEC) was determined by the neutral ammonium acetate method [16].

2.7. Investigating Potential Interactions Among Biomarkers, Soil Nutrients, Metabolites, and Pathogens

Linear differential analysis effect size (LEfSe) was used to identify the up- and down-regulated microorganisms and to construct a heatmap of the differentially abundant genera. To perform correlation and model prediction analyses, correlations between biomarker microorganisms and other microorganisms, soil nutrients, phenolic acids, and pathogens were evaluated. RDA and mapping were used. Based on two (or three) histological quantitative files, a correlation chord diagram, a cluster heatmap, and a network interaction diagram were drawn [17].

2.8. Statistical Analysis

After data processing, a community composition histogram was drawn to show the community structure distribution. The species richness and distribution uniformity of the samples were evaluated, the alpha diversity was analyzed, and a boxplot plot was drawn. The significance of the differences in the diversity indices among the different groups was calculated through plot analysis (Wilcoxon algorithm). Microbial multivariate statistical analysis: The difference in species (OTU or phylum, genus, and species level) between different groups was calculated through a statistical algorithm (Wilcoxon), the differentially upregulated and downregulated microorganisms were identified through linear differential analysis effect size (LEfSe), and a differential species heatmap was drawn. After the data were processed, a histogram was drawn to show the distribution of the community structure. The species richness and distribution uniformity in the samples were evaluated, and the alpha diversity was analyzed, which was then represented using a boxplot. The significance of the diversity index was calculated for the different groups through plot analysis (using the Wilcoxon algorithm) [18].

3. Results

3.1. Fungal Communities Are More Highly Affected by Rhizosphere Shifts than Bacteria

To explore the impact of soil conditioners on microbial communities and identify strategies for optimizing microbial structure, we utilized three proven approaches: biochar amendment, grass ash application, and corn stover incorporation. Our experimental design contrasts non-rhizosphere and rhizosphere soils.

Plant growth and seedling survival rates differed among the strategies, with the biochar strategy having the most favorable phenotype and seedling survival ($p < 0.01$) (Figure S1). Similarly, we found that the microbial community composition displayed high variability between fungal communities and even between the non-rhizosphere and rhizosphere soils, highlighting the microbiome heterogeneity of soil when the rhizosphere shifts (Figure 2). The bacterial and fungal communities exhibited significant differences; the vast majority of the bacterial communities exhibited structural similarities in response to the different strategies (Figure 2B,D), and the fungal communities were more strongly affected than bacterial communities (Figure 2). In the non-rhizosphere, all the strategies had greater fungal richness than the control group, indicating that rhizosphere shift limits fungal microbial recruitment to the rhizosphere (Figure 2C).

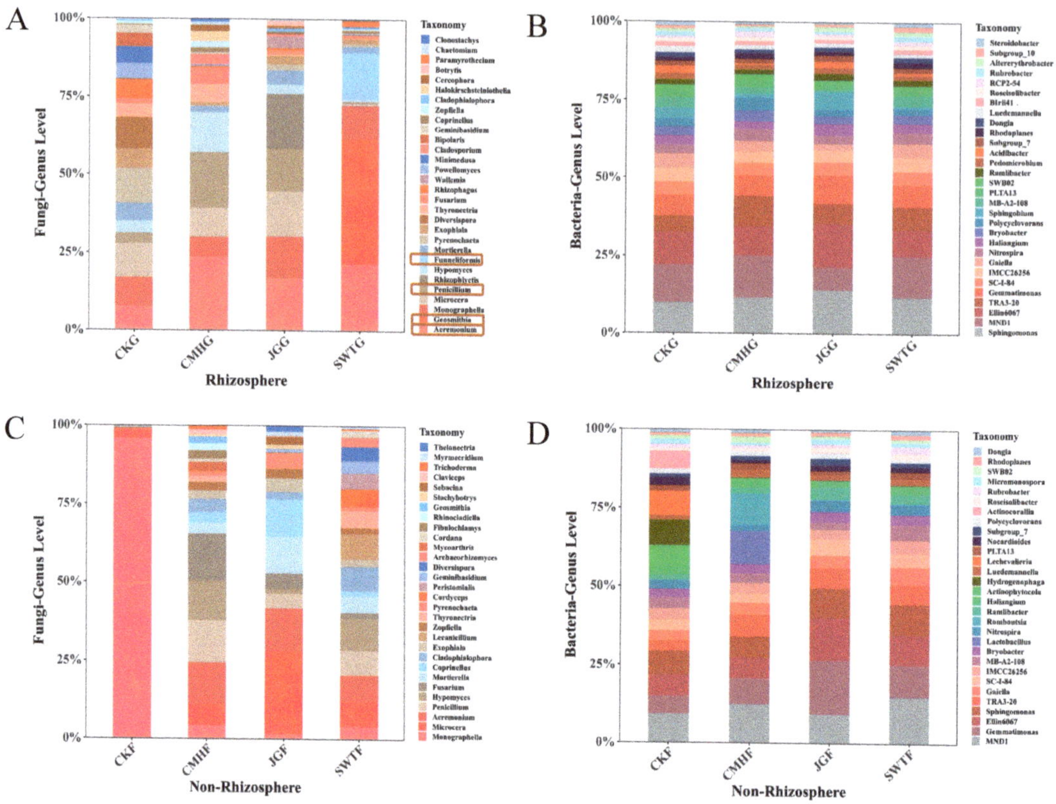

Figure 2. Comparative community structure of fungi and bacteria at the genus level. (**A**,**B**) Fungal and bacterial community structure in the rhizosphere with SWTG (biochar), CMHG (grass ash), JGG (corn stover), and CKG (control). The orange boxes indicate the genera with significant alterations. (**C**,**D**) Fungal and bacterial community structures in the non-rhizosphere. Similarly, SWTF (biochar), CMHF (grass ash), JGF (corn stover), and CKF (control) are shown.

The *Acremonium* genus was highly represented across all three systems, while the richness of the *Penicillium* genus was greatest in the CMHG and corn stover strategies (JGG). Notably, the biochar strategy (SWTG) revealed a distinct trend of specific genera aggregating in the rhizosphere, including *Geosmithia* spp. and *Funneliformis* spp. (Figure 2A).

3.2. Fungal Diversity Reverses Under Soil Conditioner Addition, Decreasing in the Rhizosphere and Increasing in the Non-Rhizosphere

Next, we performed a paired test comparing non-rhizosphere and rhizosphere microbial diversity to examine how microbial communities near or far from the rhizosphere vary. Our findings highlight significant differences in microbial diversity with soil conditioner addition, particularly for fungi, indicating a reversal from the control group. In the control, the rhizosphere exhibited higher fungal diversity than the non-rhizosphere, whereas with soil conditioner addition, the fungal diversity of the rhizosphere was lower than that of the non-rhizosphere, as evidenced by the biochar and grass ash treatments ($p < 0.001$) (Figure 3A,B). Bacterial diversity showed no significant differences between the non-rhizosphere and rhizosphere soils in the control group (Figure 3C,D), but notable variation was observed in the biochar treatment group (Figure 3C).

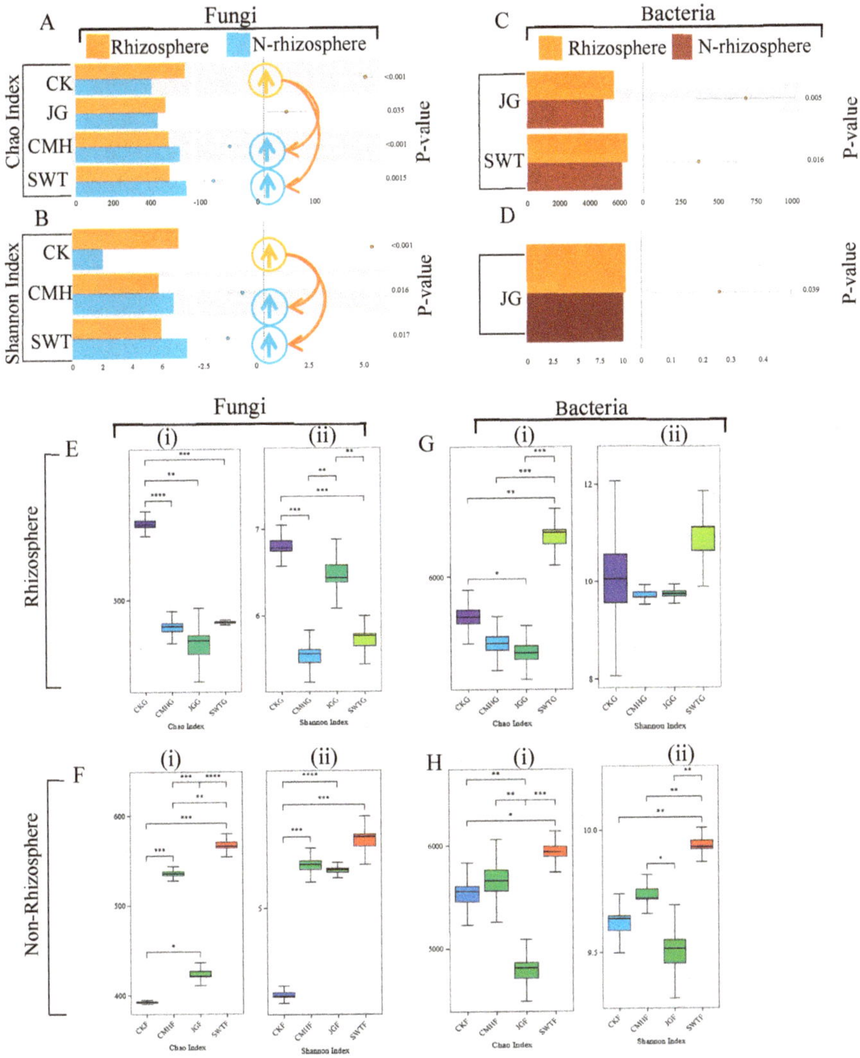

Figure 3. Comparison of rhizosphere and non-rhizosphere fungal and bacterial diversity at the genus level. (**A**,**B**) A fungal Chao and Shannon index paired map between the rhizosphere and non-rhizosphere. It only shows significant differences between strategy pairs. Higher diversity in the rhizosphere than in the non-rhizosphere is indicated by yellow arrows. Higher diversity in the non-rhizosphere than the rhizosphere is indicated by blue arrows. SWT (biochar), CMH (grass ash), JG (corn stover), and CK (control). (**C**,**D**) Bacterial Chao and Shannon index paired map between the rhizosphere and non-rhizosphere. It only shows significant differences between strategy pairs. (**E**,**F**) Fungal Chao (**i**) and Shannon (**ii**) indices with soil conditioners in the rhizosphere and non-rhizosphere. (**G**,**H**) Bacterial Chao (**i**) and Shannon (**ii**) indices with soil conditioners in the rhizosphere and non-rhizosphere. *, $p < 0.05$; **, $p < 0.01$; ***, $p < 0.001$; ****, $p < 0.0001$.

We performed a detailed analysis of the differences in microbial diversity among all the strategies. Fungal diversity followed a similar trend across all strategies for reducing rhizosphere metabolites (Figure 3E,F), while bacterial diversity varied. We observed greater diversity with the biochar strategy (SWTG) in both the rhizosphere and non-rhizosphere and lower diversity with the corn stover strategy (JGG), which was significantly different from the control ($p < 0.05$) and biochar strategy ($p < 0.001$), and almost no change occurred with the grass's ash strategy (CMHG) (Figure 3G(i),H(i)).

Overall, the biochar strategy decreased the fungal diversity and increased the bacterial diversity of the rhizosphere, improving the microbiological environment.

3.3. *Rhizosphere Shifts Downregulated Metabolic Pathways, Enzyme Activity, and Protein Expression*

We further investigated functional variations in the soil microbial community by tagging gene sequences to predict practical abundance using PICRUSt functional prediction, which has a high prediction accuracy that can reach more than 85–90 percent for soil flora. The results reveal that soil microbial community functions tended to decrease with the biochar strategy, and most were downregulated, according to KEGG prediction, protein function, enzyme activity, and metabolic pathway analyses (Figures 4 and S2).

Figure 4. Biochar-induced soil microbial functional abundance variations in the rhizosphere and non-rhizosphere. (**A**) Hierarchical cluster heatmap of KEGG level 2 functional predictions. (**B**) Enzyme activity difference cluster heatmap. (**C**) Protein expression difference cluster heatmap. (**D**) Metabolic pathway difference cluster heatmap. Soil rhizosphere microbial functions, encompassing KEGG predictions, enzyme activities, protein expression, and metabolic pathways, are downregulated with the reduction in rhizosphere metabolites. CKG (rhizosphere of control), SWTG (rhizosphere of biochar), CKF (non-rhizosphere of control), and SWTF (non-rhizosphere of biochar).

Metabolic functions were downregulated under the biochar strategy (SWTG) (Figures 3 and S2). These functions encompassed various processes, such as signaling molecules and interactions, the metabolism of terpenoids and polyketides, transport and catabolism, amino acid metabolism, lipid metabolism, biosynthesis of other secondary metabolites, nucleotide metabolism, and energy metabolism (Figure 4A). Furthermore, cellular processes and genetic information showed signs of downregulation, particularly in processing signaling molecules and interactions, signal transduction, and membrane transport. However, the cellular community of prokaryotes was downregulated, while that of eukaryotes was upregulated (Figure 4A).

However, we detected upregulation during our investigation, specifically in the form of stimulated enzyme expression and activity, e.g., formate dehydrogenase (NADP(+)) and ATP citrate synthase (NADP(+)) (Figure 3B). The expression of proteins, e.g., the hydrogenase-4 membrane subunit HyfE, the SHS2 domain protein implicated in nucleic acid metabolism, formate hydrogenlyase subunit 4, the coenzyme F420-reducing hydrogenase-gamma subunit, glutamate formiminotransferase, and the Perexiredoxin family protein, was upregulated (Figure 4C). Metabolic pathways that were upregulated included sucrose biosynthesis I (from photosynthesis), L-isoleucine brothers IV, the incomplete reductive TCA cycle, and the reductive acetyl coenzyme A pathway (Figure 4D). Overall, reduced rhizosphere metabolites downregulated most microbial community functions in the rhizosphere.

3.4. Phenolic Acids Decrease in the Rhizosphere

We initially speculated that biochar adsorbs root secretions, as previously reported. To validate our hypothesis, we performed targeted metabolomics on the biochar interlayer (biochar-only) strategy and determined its ability to absorb metabolites and which metabolites were most affected.

We found that most metabolites were identified and targeted within the biochar strategy interlayer. In support of our hypothesis, the biochar strategy had fewer metabolites in the rhizosphere but more metabolites in the interlayer of the plants in the biochar strategy (Figure 5). Notably, the metabolites that were adsorbed were phenolic acids with autotoxic effects, e.g., 4-hydroxybenzoic acid, 4-hydroxycinnamic acid, ferulate, stearic acid, and salicylic acid, indicating that biochar can also reduce root autotoxins (Figure 5).

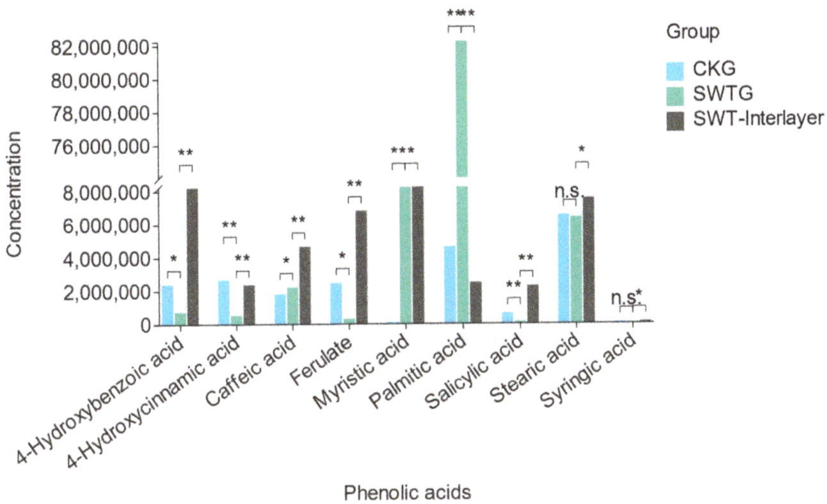

Figure 5. Targeted phenolic acid concentration comparison. *, $p < 0.05$; **, $p < 0.01$; ***, $p < 0.001$.

Consequently, as previously reported, our original prediction was that the experiment would definitely reduce rhizosphere metabolites, accompanied by a decrease in root autotoxins in the rhizosphere soil. As a result, this reduction in autotoxins optimizes the rhizosphere microbial ecology by creating a healthier environment for microorganisms to thrive.

3.5. Fungal Biomarkers Play Competitive Roles in Microbial Interactions

Furthermore, we detected positive correlations between fungi in the biochar-induced rhizosphere, i.e., between *Monographella*, *Acremonium*, *Geosmithia*, and *Funneliformis*. However, they were negatively correlated with 60 percent of the Top 20 differential fungi and were not significantly negatively correlated with fungal biomarkers in the control of rhizosphere.

In addition, the biochar strategy (SWTG) had more bacterial biomarkers than the control (Figure 6A(vii,viii)), which could attract different species of functional bacteria for colonization (Figure 6A(vi,viii)). In addition to the *Proteobacteria* and *Actinobacteria* phyla, the bacterial biomarkers collected in this study (Figure 6A(vii)) included the *Gemmatimonadota* and *Nitrospirota* phyla, which have a wide range of biomarker sources (Figure 6A(vii)).

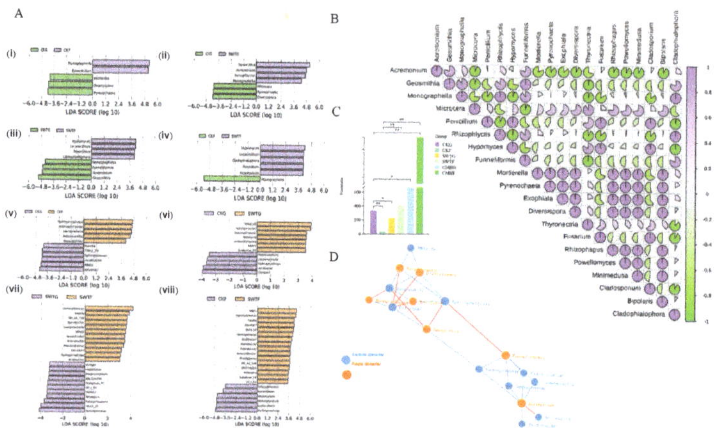

Figure 6. LEfSe analysis and correlation of biomarkers in soil microbiomes. (**A**) (**i–viii**). Discriminating fungal and bacterial microorganisms according to LEfSe analysis. In the biochar strategy, the fungal biomarkers *Monographella*, *Acremonium*, *Geosmithia*, and *Funneliformis* were significantly enriched. (**B**) Correlation analysis of the Top 20 most abundant fungi. *Acremonium*, *Geosmithia*, and *Funneliformis* exhibit negative correlations with more than 60 percent of the Top 20 most abundant fungi. (**C**) Histogram of the pathogen abundance of *Fusarium* sp. A marked decrease in *Fusarium* sp. is observed in the biochar-amended rhizosphere (SWTG). (**D**) The interomics correlation network of fungal and bacterial biomarkers. The fungal biomarkers are positively intercorrelated and negatively correlated with bacterial biomarkers. *, $p < 0.05$; **, $p < 0.01$.

To gain a better understanding of the interactions between biomarkers among community members, an interomics correlation network was constructed between biomarkers found in fungi and bacteria. We found strong negative correlations between fungal biomarkers, i.e., *Acremonium* and *Funneliformis*, and bacterial biomarkers, particularly those found in the control rhizosphere, even stronger than the positive correlations with fungal and bacterial biomarkers within the same biochar-amended rhizosphere.

Taken together, our findings imply that critical fungal biomarkers with reduced rhizosphere metabolites potentially play a competitive role in microbial interactions within the soil rhizosphere.

We conducted a study on the pathogenic fungal *Fusarium* sp. Although not identified as a biomarker, controlling its growth is crucial for restoring the balance of rhizosphere microbial ecology. Our results show that the biochar strategy resulted in a statistically significant decrease in the abundance of *Fusarium* sp. ($p < 0.05$) (Figure 6C); conversely, the grass ash strategy resulted in a significant increase (Figure 6C), which may explain the phenotypic superiority of the biochar strategy to grass ash.

3.6. Ecological Roles of Biomarker Microorganisms in the Rhizosphere

To better understand the role of biomarker microorganisms in optimizing rhizosphere microbial ecology, we also propose a framework for how biomarker microorganisms play roles in three major segments, i.e., microbial networks, metabolites, and soil nutrients.

We found that fungal biomarkers were positively correlated with soil nutrients and elevated pH but negatively correlated with the pathogenic fungus *Fusarium* sp. and phenolic acids, which can be autotoxic to the root system. Therefore, the ecological functions of the fungal biomarkers were more significant in increasing nutrients and decreasing autotoxicity and soil acidification (Figure 7).

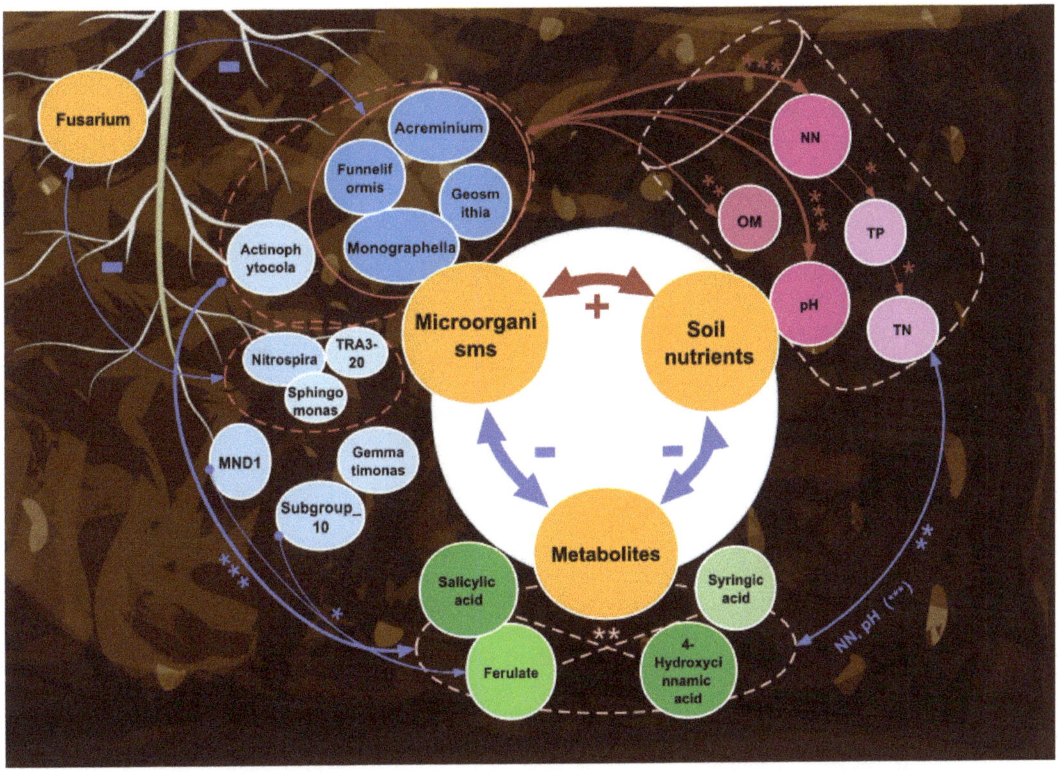

Figure 7. Network interaction maps between microorganisms, metabolites, and soil nutrients. Biomarker microorganisms were positively correlated with soil nutrients and elevated pH but negatively correlated with phenolic acid ($p < 0.001$). Both fungal and bacterial biomarker microorganisms were negatively correlated with the pathogen *Fusarium*. NN, nitrate nitrogen; OM, organic matter; TP, total phosphorus; TN, total nitrogen. *, $p < 0.05$; **, $p < 0.01$; ***, $p < 0.001$; Blue arrows: negative correlation; red arrows: positive correlation; line thickness indicates correlation strength.

Bacteria from the genus *Acreminium* work synergistically with fungi to perform similar functions. Furthermore, nitrate nitrogen and pH were the most critical factors determining the rhizosphere soil environment (Figure 7).

Overall, biomarker microorganisms play a significant ecological role in optimizing the rhizosphere environment by efficiently utilizing soil nutrients, competing with pathogens, and participating in the metabolic function of reducing autotoxic metabolites.

4. Discussion

The biochar-induced rhizosphere shift in fungal diversity from greater to less could redefine plant–microbe symbiosis in the rhizosphere. Typically, the rhizosphere is a hotspot for microbial diversity due to the attractant properties of root exudates [19]. However, our studies show that biochar application triggers a counterintuitive effect: a decrease in fungal diversity in the rhizosphere and a concurrent increase in the surrounding soil. This biochar-induced diversity shift reduces fungal colonization and pathogen invasion, addressing a prevalent challenge in soils prone to continuous cropping obstructions (CCOs), where pathogenic fungi often prevail [20]. Our findings point to a novel strategy for enhancing soil resistance to CCOs, thereby benefiting sustainable agriculture by disrupting pathogenic dominance and rebalancing the soil microbiome to support plant health and disease resistance.

Notably, the biochar-induced rhizosphere would inhibit the synergistic damage of root secretions with pathogens. Our research indicates that biochar not only reduces fungal diversity but also restructures the rhizosphere community, notably by suppressing pathogenic *Fusarium* sp. The secretion of phenolic acids, while a plant defense mechanism, can inadvertently nourish soil-borne fungi, leading to a cycle of acidification and immune suppression [21]. Our findings suggest that reducing phenolic acids can curb pathogen virulence by depriving them of vital nutrients. This targeted approach to soil management disrupts the environment that supports pathogens, offering a new strategy for soil health and agricultural resilience against CCOs. By modulating the composition of root exudates, we can tip the balance in the rhizosphere in favor of beneficial microbes over pathogens, laying the groundwork for innovative soil management practices.

However, the explicit consideration of fundamental ecological processes for developing complex microbial communities is in its infancy. Our work reveals that higher bacterial diversity is indicative of a robust ecosystem that can outcompete pathogens for resources, thus preserving ecosystem health [22]. Soils with greater microbial diversity tend to have more ecological functions and greater resistance to environmental stress. Communities with high diversity have high resource utilization complementarity to use resources in the environment entirely, which leaves fewer resources available for pathogens to invade [23], ultimately helping the ecosystem to remain healthy.

Within the core microbiota, "biomarker microorganisms" can influence the community structure through biotic solid interactions with the host or with other microbial species [24], and these interactions function as the first line of defense against pathogens, as their removal results in the loss of interactions—for example, with Enterobacter cloacae [25]. Identifying "key microorganisms" within this community is essential for understanding their role in enhancing the rhizosphere. We have identified key biomarkers—*Monographella*, *Acremonium*, *Geosmithia*, and *Funneliformis*—that correlate negatively with other microbes, suggesting their role in competitive interactions. These markers optimize microbial ecology by mobilizing soil nutrients, curbing pathogen growth, and reducing phenolic acids. Our results suggest a framework for synthetic microbial communities to fine-tune ecological functions in the rhizosphere.

Our research results show that the rhizosphere shift recruited *Funneliformis* spp., a type of arbuscular mycorrhizal fungi (AMF), enriched the soil environment, and enhanced plant growth by increasing nutrient uptake, particularly phosphorus, and improving soil microbial diversity [26]. This symbiotic relationship boosts plant biomass and health while also mitigating the challenges of continuous cropping, such as soil degradation and disease. By promoting soil enzyme activity and plant resistance to environmental stresses, including heavy metals, *Funneliformis* spp. play a crucial role in fostering soil fertility and supporting sustainable agricultural practices [27].

This research enhances our understanding of leveraging microbial interactions to improve the health and productivity of agro-ecosystems [28]. The development and application of synthetic microbial communities (SynComs) in plants provide a novel avenue for understanding the complex dynamics of plant microbiome interactions, including community assembly and member relationships [29]. Integrating omics-based predictions with SynCom-based microbial interactions illuminates how microbial consortia can be manipulated to combat soil-borne diseases and optimize rhizosphere microbial ecology globally, paving the way for innovative soil management strategies powered by the plant microbiome [30].

In conclusion, a sophisticated understanding of rhizosphere metabolites and their role in shaping microbial communities is pivotal for advancing sustainable agricultural practices and promoting soil and plant health. This knowledge is essential for developing innovative strategies that harness the power of the plant microbiome to enhance soil health and agricultural sustainability.

Supplementary Materials: The following supporting information can be downloaded at https://www.mdpi.com/article/10.3390/microorganisms12122420/s1, Figure S1: Plant growth and seedling survival in different strategies; Figure S2: KEGG prediction difference clustered heatmap; Figure S3: Comparative analysis of physicochemical properties of rhizosphere soil and correlation analysis between differential microbes and soil physicochemical properties.

Author Contributions: Conceptualization, J.Z.; methodology, Y.Z.; software, J.L.; validation, J.L.; formal analysis, J.L.; investigation, J.L.; resources, J.L. and Y.Z.; data curation, J.L.; writing—original draft preparation, J.L.; writing—review and editing, J.L., Y.Z. and J.Z.; visualization, J.L.; supervision, J.Z.; project administration, J.Z.; funding acquisition, J.L. and J.Z. All authors have read and agreed to the published version of the manuscript.

Funding: This work was supported by the National Natural Science Foundation of China (31760360), the Natural Science Foundation of Yunnan Province of China (202301AT070007), and the Natural Science Agricultural Joint Foundation of Yunnan Province of China (202101BD070001-078). The authors declare that no funds, grants, or other support were received during the preparation of this manuscript.

Data Availability Statement: The original contributions presented in the study are included in the article/supplementary material, further inquiries can be directed to the corresponding authors.

Conflicts of Interest: The authors declare no conflicts of interest.

References

1. Yao, J.; Wu, C.; Fan, L.; Kang, M.; Liu, Z.; Huang, Y.; Xu, X.; Yao, Y. Effects of the Long-Term Continuous Cropping of Yongfeng Yam on the Bacterial Community and Function in the Rhizospheric Soil. *Microorganisms* **2023**, *11*, 274. [CrossRef] [PubMed]
2. Tan, Y.; Cui, Y.; Li, H.; Kuang, A.; Li, X.; Wei, Y.; Ji, X. Diversity and composition of rhizospheric soil and root endogenous bacteria in Panax notoginseng during continuous cropping practices. *J. Basic Microbiol.* **2017**, *57*, 337–344. [CrossRef] [PubMed]
3. Yuan, Y.; Zuo, J.; Zhang, H.; Zu, M.; Liu, S. The Chinese medicinal plants rhizosphere: Metabolites, microorganisms, and interaction. *J. Resour. Ecol.* **2022**, *13*, 95–106. [CrossRef]
4. Dong, L.; Xu, J.; Feng, G.; Li, X.; Chen, S. Soil bacterial and fungal community dynamics associated with Panax notoginseng death rate in a continuous cropping system. *Sci. Rep.* **2016**, *6*, 31802. [CrossRef]
5. Vandenkoornhuyse, P.; Quaiser, A.; Duhamel, M.; Le Van, A.; Dufresne, A. The importance of the microbiome of the plant holobiont. *New Phytol.* **2015**, *206*, 1196–1206. [CrossRef]
6. Berihu, M.; Somera, T.S.; Malik, A.; Medina, S.; Piombo, E.; Tal, O.; Cohen, M.; Ginatt, A.; Ofek-Lalzar, M.; Doron-Faigenboim, A.; et al. A framework for the targeted recruitment of crop-beneficial soil taxa based on network analysis of metagenomics data. *Microbiome* **2023**, *11*, 8. [CrossRef]
7. Guo, Y.; Lv, J.; Zhao, Q.; Dong, Y.; Dong, K. Cinnamic Acid Increased the Incidence of Fusarium Wilt by Increasing the Pathogenicity of Fusarium oxysporum and Reducing the Physiological and Biochemical Resistance of Faba Bean, Which Was Alleviated by Intercropping with Wheat. *Front. Plant Sci.* **2020**, *11*, 608389. [CrossRef]
8. Chen, Y.; Du, J.; Li, Y.; Tang, H.; Yin, Z.; Yang, L.; Ding, X. Evolutions and Managements of Soil Microbial Community Structure Drove by Continuous Cropping. *Front. Microbiol.* **2022**, *13*, 839494. [CrossRef]

9. Wu, H.; Wu, L.; Zhu, Q.; Wang, J.; Qin, X.; Xu, J.; Kong, L.; Chen, J.; Lin, S.; Khan, M.U.; et al. The role of organic acids on microbial deterioration in the Radix pseudostellariae rhizosphere under continuous monoculture regimes. *Sci. Rep.* **2017**, *7*, 3497. [CrossRef]
10. Qin, X.; Wu, H.; Chen, J.; Wu, L.; Lin, S.; Khan, M.U.; Boorboori, M.R.; Lin, W. Transcriptome analysis of Pseudostellaria heterophylla in response to the infection of pathogenic Fusarium oxysporum. *BMC Plant Biol.* **2017**, *17*, 155. [CrossRef]
11. Wen, T.; Xie, P.H.; Liu, H.W.; Liu, T.; Zhao, M.L.; Yang, S.D.; Niu, G.Q.; Hale, L.; Singh, B.K.; Kowalchuk, G.A.; et al. Tapping the rhizosphere metabolites for the prebiotic control of soil-borne bacterial wilt disease. *Nat. Commun.* **2023**, *14*, 4497. [CrossRef] [PubMed]
12. Du, S.; Trivedi, P.; Wei, Z.; Feng, J.; Hu, H.W.; Bi, L.; Huang, Q.Y.; Liu, Y.R. The Proportion of Soil-Borne Fungal Pathogens Increases with Elevated Organic Carbon in Agricultural Soils. *Msystems* **2022**, *7*, e01337-21. [CrossRef] [PubMed]
13. Gomes, P.W.P.; Mannochio-Russo, H.; Mao, J.; Zhao, H.N.; Ancira, J.; Tipton, C.D.; Dorrestein, P.C.; Li, M. Co-occurrence network analysis reveals the alterations of the skin microbiome and metabolome in adults with mild to moderate atopic dermatitis. *mSystems* **2024**, *9*, e0111923. [CrossRef] [PubMed]
14. Zhan, Y.; Xu, S.; Hou, Z.; Gao, X.; Su, J.; Peng, B.; Zhao, J.; Wang, Z.; Cheng, M.; Zhang, A.; et al. Co-inoculation of phosphate-solubilizing bacteria and phosphate accumulating bacteria in phosphorus-enriched composting regulates phosphorus transformation by facilitating polyphosphate formation. *Bioresour. Technol.* **2023**, *390*, 129870. [CrossRef] [PubMed]
15. Langille, M.G.I.; Zaneveld, J.; Caporaso, J.G.; McDonald, D.; Knights, D.; Reyes, J.A.; Clemente, J.C.; Burkepile, D.E.; Vega Thurber, R.L.; Knight, R.; et al. Predictive functional profiling of microbial communities using 16S rRNA marker gene sequences. *Nat. Biotechnol.* **2013**, *31*, 814–821. [CrossRef]
16. Guerrieri, E.; Rasmann, S. Exposing belowground plant communication: Root exudation could be harnessed for ecological and applied research. *Science* **2024**, *384*, 272–273. [CrossRef]
17. Lin, H.; Das Peddada, S. Analysis of microbial compositions: A review of normalization and differential abundance analysis. *npj Biofilms Microbiomes* **2020**, *6*, 60. [CrossRef]
18. Jansson, J.; Hofmockel, K.S.; Fuhrman, J.A. Microbiome dynamics on diatoms: Colonization, succession, and regulation of algal blooms. *ISME J.* **2017**, *11*, 637–645. [CrossRef]
19. Xu, J.; Zhang, Y.; Zhang, P.; Trivedi, P.; Riera, N.; Wang, Y.; Liu, X.; Fan, G.; Tang, J.; Coletta-Filho, H.D.; et al. The structure and function of the global citrus rhizosphere microbiome. *Nat. Commun.* **2018**, *9*, 4894. [CrossRef]
20. Chen, D.; Zhou, Y.; Wang, M.; Munir, M.A.M.; Lian, J.; Yu, S.; Dai, K.; Yang, X. Succession Pattern in Soil Micro-Ecology Under Tobacco (*Nicotiana tabacum* L.) Continuous Cropping Circumstances in Yunnan Province of Southwest China. *Front. Microbiol.* **2022**, *12*, 785110. [CrossRef]
21. Debray, R.; Herbert, R.A.; Jaffe, A.L.; Crits-Christoph, A.; Power, M.E.; Koskella, B. Priority effects in microbiome assembly. *Nat. Rev. Microbiol.* **2021**, *20*, 109–121. [CrossRef] [PubMed]
22. Ling, N.; Wang, T.; Kuzyakov, Y. Rhizosphere bacteriome structure and functions. *Nat. Commun.* **2022**, *13*, 836. [CrossRef] [PubMed]
23. Sun, X.; Xie, J.; Zheng, D.; Xia, R.; Wang, W.; Xun, W.; Huang, Q.; Zhang, R.; Kovács, .T.; Xu, Z.; et al. Metabolic interactions affect the biomass of synthetic bacterial biofilm communities. *mSystems* **2023**, *8*, e0104523. [CrossRef] [PubMed]
24. Hamonts, K.; Trivedi, P.; Garg, A.; Janitz, C.; Grinyer, J.; Holford, P.; Botha, F.C.; Anderson, I.C.; Singh, B.K. Field study reveals core plant microbiota and relative importance of their drivers. *Environ. Microbiol.* **2017**, *20*, 124–140. [CrossRef] [PubMed]
25. Niu, B.; Paulson, J.N.; Zheng, X.; Kolter, R. Simplified and representative bacterial community of maize roots. *Proc. Natl. Acad. Sci. USA* **2017**, *114*, E2450–E2459. [CrossRef]
26. Ain, Q.U.; Hussain, H.A.; Zhang, Q.; Maqbool, F.; Ahmad, M.; Mateen, A.; Zheng, L.; Imran, A. Coordinated influence of Funneliformis mosseae and different plant growth-promoting bacteria on growth, root functional traits, and nutrient acquisition by maize. *Mycorrhiza* 2024, *Online ahead of print*. [CrossRef]
27. De Moura, J.B.; Ramos, M.L.G.; Konrad, M.L.d.F.; Júnior, O.J.S.; Lucas, L.d.S.; Junior, W.Q.R. Effects of direct and conventional planting systems on mycorrhizal activity in wheat grown in the Cerrado. *Sci. Rep.* **2024**, *14*, 14322. [CrossRef]
28. Trivedi, P.; Leach, J.E.; Tringe, S.G.; Sa, T.; Singh, B.K. Plant-microbiome interactions: From community assembly to plant health (vol 18, pg 607, 2020). *Nat. Rev. Microbiol.* **2021**, *19*, 72. [CrossRef]
29. Yuan, J.; Zhao, K.; Tan, X.; Xue, R.; Zeng, Y.; Ratti, C.; Trivedi, P. Perspective on the development of synthetic microbial community (SynCom) biosensors. *Trends Biotechnol.* **2023**, *41*, 1227–1236. [CrossRef]
30. Bai, B.; Liu, W.; Qiu, X.; Zhang, J.; Zhang, J.; Bai, Y. The root microbiome: Community assembly and its contributions to plant fitness. *J. Integr. Plant Biol.* **2022**, *64*, 230–243. [CrossRef]

microorganisms

MDPI

Article

Effects of Long-Term Application of Nitrogen Fertilizer on Soil Acidification and Biological Properties in China: A Meta-Analysis

Liqiang Zhang [1], Zehang Zhao [1], Bailing Jiang [1], Bate Baoyin [1], Zhengguo Cui [2], Hongyu Wang [1], Qiuzhu Li [1,*] and Jinhu Cui [1,*]

[1] College of Plant Science, Jilin University, Changchun 130012, China; lqzhang23@mails.jlu.edu.cn (L.Z.); 18846915612@163.com (Z.Z.); jiangbl22@mails.jlu.edu.cn (B.J.); bybt23@mails.jlu.edu.cn (B.B.); hong_yu@jlu.edu.cn (H.W.)

[2] Soybean Research Institute, Jilin Academy of Agricultural Sciences, Changchun 130033, China; 17643346860@163.com

* Correspondence: liqz@jlu.edu.cn or jluzwkx@jlu.edu.cn (Q.L.); cuijinhu@163.com (J.C.)

Abstract: Soil acidification is a global environmental problem with significant impacts on agricultural production, environmental protection, and ecosystem health. Soil acidification is widespread in China, affecting crop yields, agricultural product quality, and biodiversity. Since the 1980s, much work has been done on acidic soils in China, but it is controversial whether excessive nitrogen fertilizer application can lead to soil acidification mechanisms. To address the above issues, we conducted a meta-analysis of 115 published papers to integrate and analyze the effects of N fertilizer application on soil acidification and biological properties from 1980 to 2024. We also quantified the effect of nitrogen fertilization on soil acidification and biological changes under different climatic conditions. The results showed that under long-term application of nitrogen fertilizers in China from 1980 to 2024, soil pH decreased by an average of 15.27%, and the activities of soil urease, nitrate reductase, nitrite reductase, catalase, glutamate dehydrogenase, and glutamate synthetase decreased by an average of 9.82–22.37%. The soil microbial community richness (Chao1 index) increased by 6.53%, but the community diversity (Shannon index) decreased by 15.42%. Among the dominant soil microorganisms, the relative abundance of bacteria decreased by an average of 9.67–29.38% and the abundance of gene expression of *nifH*, *amoA-AOA*, *amoA-AOB*, and *qnorB* decreased by 9.92–19.83%. In addition, we found that the mean annual temperature and rainfall impacted soil acidification via their effect on soil microbial diversity and community composition. This study provides a scientific basis for an in-depth understanding of the spatial and temporal variation of soil acidification and biological properties in China.

Keywords: nitrogen fertilizers; soil acidification; soil enzymes; soil microbial communities; functional genes; climatic conditions

1. Introduction

In the 21st century, China's national food security issues have received increasing attention due to the scarcity of arable land resources and declining soil quality [1]. Acidic soils reduce the bioavailability of nitrogen, phosphorus, and potassium in the soil, thus affecting the uptake and use of nutrients by plants [2], and also reduce soil microbial activity, richness, and diversity, which in turn affects the function and stability of the soil ecosystem [3]. Since the 1980s, with the massive use of nitrogen fertilizers, China's arable land has experienced significant acidification, and by 2023, China's soil acidification area had reached 2.04×10^8 ha, accounting for about 22.7% of the country's total land area [4]. Zhu et al. (2018) found that the pH in the surface layer of 154 arable soils in 35 districts of seven provinces in China decreased by 0.5 units from 1980 to 2010 [5].

Soil acidification can be divided into acidification under natural conditions and acidification due to anthropogenic factors. Natural conditions include natural weathering of the soil, leaching of saline ions, decomposition of plant litter (e.g., leaves and roots) in the soil, plant root secretions, and lightning and rainfall. Anthropogenic factors include soil management practices (e.g., fertilizer application) and anthropogenic discharges (e.g., acid deposition) [6]. Among these, humic acid, produced by the decomposition of soil organic matter, is one of the most important components in acidic soils [7]. Humic acid has strong acidic functional groups, such as carboxylic and aldehyde acids, which can combine with soil particles and cations in the soil to form a stable organic matter–mineral complex. This complex has a good buffering capacity in the soil and prevents the soil pH from changing too quickly [8]. Furthermore, weathering of the parent rock, influenced by environmental factors such as temperature and precipitation, leaches the saline ionic components of the soil, resulting in desaturation of the saline base and a slightly acidic pH response of the soil [9].

Soil temperature has a large impact on soil microbial communities. Soil respiration is positively proportional to soil temperature, with a seasonal dynamic of summer > spring > autumn > winter, peaking in summer. The mechanism of action by which soil temperature affects the rate of soil respiration is through the regulation of microbial and plant respiratory enzyme activities in the soil. An increase in soil air temperature can promote the activity of plant respiratory enzymes and soil microorganisms [7]. A study showed that, when soil temperatures were within 0–35 °C, the number and diversity of microorganisms increased with increasing temperature, but when soil temperature was above 40 °C, microbial activity was significantly suppressed [10]. The number of soil microbial taxa also varies with the seasons, with bacteria being the most abundant, actinomycetes the second most abundant, and fungi the least abundant in different seasons [11]. Soil microbial populations are slightly higher in spring and summer than in autumn and winter, with bacterial populations being highest in summer and fungal populations being highest in spring. Under warmer temperatures, microbial activity is greatly increased, which favors the systematic utilization of nutrients or organic matter that may enter the soil, whereas the ability of surface or subsurface soil microorganisms to take up material is usually reduced under low temperatures [12].

Soil acidification has a direct effect on the growth and metabolic activities of soil microorganisms. Some microorganisms are more tolerant to acidic environments, but most microorganisms (e.g., *Ascomycota*) are sensitive to acidic environments, and microbial abundance and diversity tend to be lower in acidic soils [13]. A decrease in microbial abundance in acidic soils leads to a decrease in soil enzyme activity [14]. Previous studies have shown that soil pH is an important factor influencing soil enzyme activity and nutrient content [15]. Xin et al. (2024) experimentally changed the pH of a soil by increasing litter input and root biomass, which significantly improved soil enzyme activity and nutrient status. As soil microorganisms are very sensitive to changes in soil acidity and alkalinity, moderately lowering soil pH can improve soil enzyme activity, bacterial abundance, and microbial community structure, thereby optimizing the soil ecosystem and increasing soil fertility [16]. Microbial interactions also play an important role in nitrification in acidic soils. The mechanism of action is the oxidation of NO_2^-–N to NO_3^-–N, which is accomplished by nitrifying bacteria [17]. Interactions between different microorganisms can promote or inhibit the activity of nitrifying bacteria, which in turn affects the amount of nitrification. These interactions, which include mechanisms such as symbiosis, competition, and inhibition, play an important role in nitrification in acidic soils [18].

While progress has been made in understanding the effects of soil acidification, it is clear that these effects depend on specific regional environmental and soil factors. Meta-analysis is a statistical method for integrating and analyzing the results of multiple studies of the same type and obtaining consistent conclusions to deal with the heterogeneity of experimental conditions and conflicting results [19]. The aggregated information increases statistical power and provides a more comprehensive analysis of contributing factors than individual studies, ultimately producing more reliable results [20]. In recent years, the use of meta-analysis in agronomy has expanded [21,22]. The effects of time and climatic conditions on the community structure and diversity of soil microorganisms cannot be ignored. Previous studies focused on the effects of anthropogenic factors, such as agricultural practices, on the soil microenvironment, ignoring the importance of their spatial and temporal variations. In this study, we used meta-analysis to (1) clarify the effects of long-term application of nitrogen fertilizers on soil pH and biological properties; (2) determine the relationship between soil acidification, biological properties, and climatic conditions; and (3) analyze the main factors and pathways contributing to soil acidification. This study will provide a scientific basis for mitigating soil acidification and improving soil biological properties, thus providing a reference for sustainable agricultural development in China.

2. Materials and Methods

2.1. Data Collection

A meta-analysis was used to assess the effects of the long-term application of nitrogen fertilizers on soil acidification and biological properties. The Web of Science (WOS, http://apps.webofknowledge.com/, Accessed on 15 June 2024) and China National Knowledge Internet (CNKI, http://www.cnki.net/, Accessed on 24 June 2024) databases were used to find relevant peer-reviewed literature from 1980 to 2024. This was done using Boolean search terms: ("fertilization" OR "soil acidification") AND ("soil pH" OR "soil organic matter" OR "soil enzyme" OR "soil microbial community"). Studies were screened for inclusion in the meta-analysis using the following criteria: 1) all data were obtained from open field cropping; 2) the experimental treatments consisted of both nitrogen fertilizer application (experimental group) and no nitrogen fertilizer application (control group); 3) at least one target variable was included, i.e., data on the effect of long-term nitrogen fertilizer application on soil pH, enzyme activity, or microbial diversity; and 4) the experimental treatments consisted of at least $n = 3$ replicates, for which the mean and standard deviation (SD) could be obtained from a table or from a figure using Engauge Digitizer 12.1 (https://sourceforge.net/projects/digitizer/, Accessed on 30 June 2024). If only the standard error (SE) was reported, the SD was calculated (SD = SE $\times$ $\sqrt{n}$). If neither the SD nor SE was reported, the SD was estimated as 1/10 of the mean [23]. The first screening was done by reading the titles of candidate articles and excluding any reviews and other types of literature (synthesis papers, book chapters, comments/opinions) that did not fit the scope of our meta-analysis. The second screening entailed reading the abstracts and scrutinizing the experimental trials and designs to exclude articles that failed to meet the above criteria [24]. Figure 1 shows a complete flowchart of the screening process.

2.2. Data Categorization

In total, 115 studies fulfilled the criteria (see Supplementary Materials—literature statistics). For each study, the title, author information, literature source, trial location, climate conditions, soil classification, and agricultural management practices were extracted. Climatic conditions included the mean annual temperature (MAT) and mean annual precipitation (MAP). Although it is generally more accurate to use the precipitation and temperature observations during the crop growing season at a given site, these were not available for all studies. Therefore, MAT and MAP were used to facilitate comparison with other reports and ensure data consistency [25]. MAT and MAP were each divided into three groups (MAT: ≤15 °C, 15–20 °C, and ≥20 °C; MAP: ≤1000 mm, 1000–2000 mm, and ≥2000 mm).

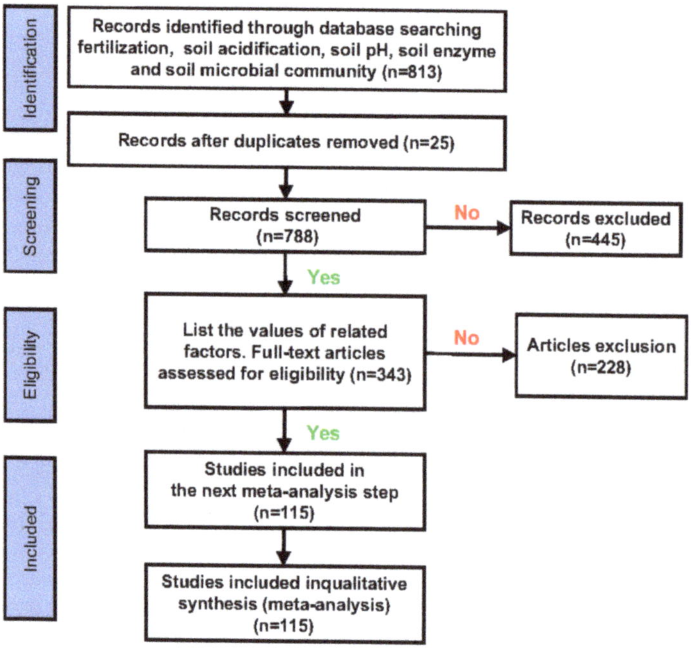

Figure 1. Flowchart showing the screening process for study inclusion in the meta-analysis.

2.3. Meta-Analysis

MetaWin 2.1.5.10 software (https://en.freedownloadmanager.org/, Accessed on 2 July 2024) was used to perform the meta-analysis. The natural logarithm of the response ratio (R) served as the effect size value (lnR), which was calculated as follows:

$$R = X_t / X_c \quad (1)$$

$$lnR = ln(X_t / X_c) = lnX_t - lnX_c \quad (2)$$

where X_t and X_c are respectively the mean values of the data for the test group (N fertilizer input) and the control group (not N fertilizer input) under a saline–alkali land cropping system. lnR is a unitless index with a positive or negative value, respectively indicating an increase or decrease in soil pH, soil enzyme, or soil microbial community.

The variance of the effect size (V_i) was calculated as follows:

$$V_i = SD_t^2 / N_t X_t + SD_c^2 / N_c X_c \quad (3)$$

where SD_t and SD_c are the standard deviations of the target variable in the experimental and control groups, respectively, and N_t and N_c are the sample sizes of the target variable in the experimental and control groups, respectively.

To achieve maximal precision, the weighted mean was used, since statistical precision surely varied considerably among the 115 studies included here. The weighted mean response rate (lnR++) and corresponding weights (values of W_i) were calculated as follows:

$$lnR_{++} = \sum_{i=1}^{k} \frac{lnR_i \times W_i}{\sum_{i=1}^{k} W_i} \quad (4)$$

$$W_i = 1 / V_i \quad (5)$$

where i and k denote the number of comparison and cumulative groups, respectively, and W_i is the weight of a target variable's effect size value.

The 95% CI of the effect size value was calculated via resampling (bootstrapping). If the resulting confidence interval did not overlap with zero, then the effect value was considered significant: when the entire confidence interval was greater or less than zero, this indicated N fertilizer input significantly increased or decreased the target variable, respectively ($p < 0.05$). Conversely, if the confidence interval included zero, then the N fertilizer input had no significant effect on the target variable [26]. The 95% confidence interval for $lnR{+}{+}$ was calculated using the following equation:

$$95\%CI = lnR_{++} \pm 1.96SE_{lnR_{++}} \tag{6}$$

$$SE_{lnR_{++}} = \sqrt{\frac{1}{\sum_{i=1}^{k} W_i}} \tag{7}$$

Based on the comparison between the experimental and control groups, for descriptive purposes, their percentage change (E) in the target variable was also calculated, as follows:

$$E = (exp(lnR_{++}) - 1) \times 100\% \tag{8}$$

2.4. Data Analysis

The response values of the target variables (E) were visualized and plotted using GraphPad Prism 9.1 software (https://www.graphpad.com/, Accessed on 2 July 2024). The relative importance of these variables for soil acidification and biological properties was explored using a random forest model in R v4.3.1 (https://www.r-project.org/, Accessed on 2 July 2024) [27]. Finally, we used structural equation modeling (SEM) to quantify the effects of the N input and environmental variables on the soil acidification and biological properties pathways [28].

3. Results

3.1. Overview of the Dataset

A final total of 115 published papers from China were included in this study (Figure 2). They contained 553 comparative soil pH observations, 2177 comparative soil enzyme (S-UE, urease; S-NR, nitrate reductase; S-NiR, nitrite reductase; S-CAT, catalase; S-SC12, sucrase; S-NPT, glutamate dehydrogenase; S-GS, protease; and S-GDH, glutamate amine synthetase) observations, 579 comparative soil microbial community alpha diversity (Shannon and Chao 1) observations, 1956 comparative soil microbial community composition (phylum level; Actinobacteriota, Proteobacteria, Acidobacteriota, Gemmatimonadetes, Ascomycota, and Basidiomycota) observations, and 765 comparative soil nitrogen cycle functional gene (*nifH*, *amoA-AOA*, *amoA-AOB*, *nirK*, nir*S*, *nosZ*, *qnorB*, *narG*) observations.

Figure 3 shows the effects of long-term nitrogen fertilizer application (1980–2024) in China on soil pH, enzyme activities, abundance of dominant microorganisms, and abundance of functional genes for nitrogen cycling. The application of nitrogen fertilizers decreased the soil pH by 15.27% on average (Figure 3a). Nitrogen fertilizer also increased S-SC12 activity by 9.29%, and decreased S-UE, S-NR, S-NIR, S-CAT, S-GS, and S-GDH activities by an average of 9.82–22.37% (Figure 3a).

Long-term nitrogen fertilizer application increased the soil microbial community richness (Chao1 index) by 6.53% but decreased diversity (Shannon index) by 15.42 (Figure 3b). Furthermore, the relative abundance of Ascomycota and Basidiomycota increased by 12.83% and 20.75%, respectively. However, the application of nitrogen fertilizers decreased the relative abundance of Actinobacteriota, Proteobacteria, Acidobacteriota, and Gemmatimonadetes in the soil, with response percentages of −29.38%, −10.54%, −9.67%, and −25.37%, respectively.

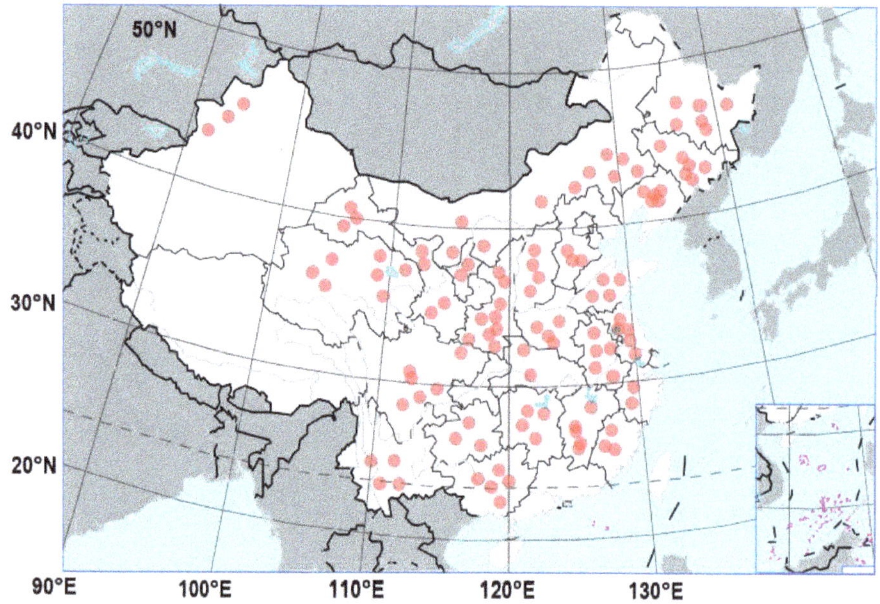

Figure 2. Spatial distribution of the published papers.

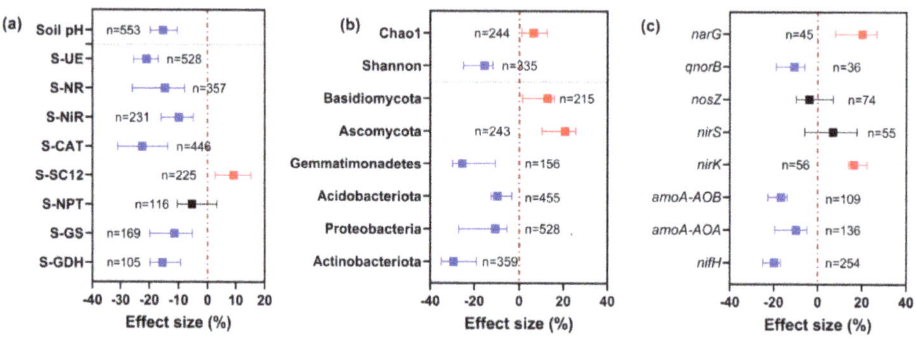

Figure 3. Effects of long-term N application on soil pH and enzyme activity (**a**), microbial community diversity and taxa abundance (**b**), and microbial functional gene abundance (**c**). Symbols and error bars indicate effect sizes and 95% confidence intervals (CIs), respectively. N fertilization treatments were significantly different from no-N-fertilization treatments when the CIs did not overlap with zero (red dashed line) ($p < 0.05$). Red symbols indicate an increasing effect, blue symbols a decreasing effect, and black symbols no significant effect. n is the number of samples for the variable of interest. S-UE, urease; S-NR, nitrate reductase; S-NiR, nitrite reductase; S-CAT, catalase; S-SC12, sucrase; S-NPT, glutamate dehydrogenase; S-GS, protease; S-GDH, glutamate amine synthetase.

As shown in Figure 3c, long-term nitrogen fertilizer application had different effects on different functional genes. The gene expression abundance of *narG* and *nirK* increased by 20.26% and 16.45%, respectively, while the gene expression abundance of *nifH*, *amoA-AOA*, *amoA-AOB*, and *qnorB* decreased by 9.92%–19.83%. The 95% confidence intervals of the response percentages of *nirS* and *nosZ* overlapped with zero, indicating that the long-term application of nitrogen fertilizer had no significant effect on the expression abundance of these two genes.

In summary, these results indicate that long-term N application has a general negative effect on soil pH and soil microbial community diversity and function.

3.2. Effect of Climate Conditions on the Response of Soil PH and Enzyme Activities under Long-Term Nitrogen Fertilizer Application

To further explore the effect of long-term nitrogen fertilizer on soil pH and enzyme activity, we explored the effect under different MAT and MAP categories (Figure 4a,b). The negative effect of nitrogen fertilizer on soil pH weakened with increasing MAT and MAP, from −25.46% at ≤10 °C to −3.54% at ≥20 °C, and −22.56% at ≤1000 mm to −11.43% at ≥2000 mm.

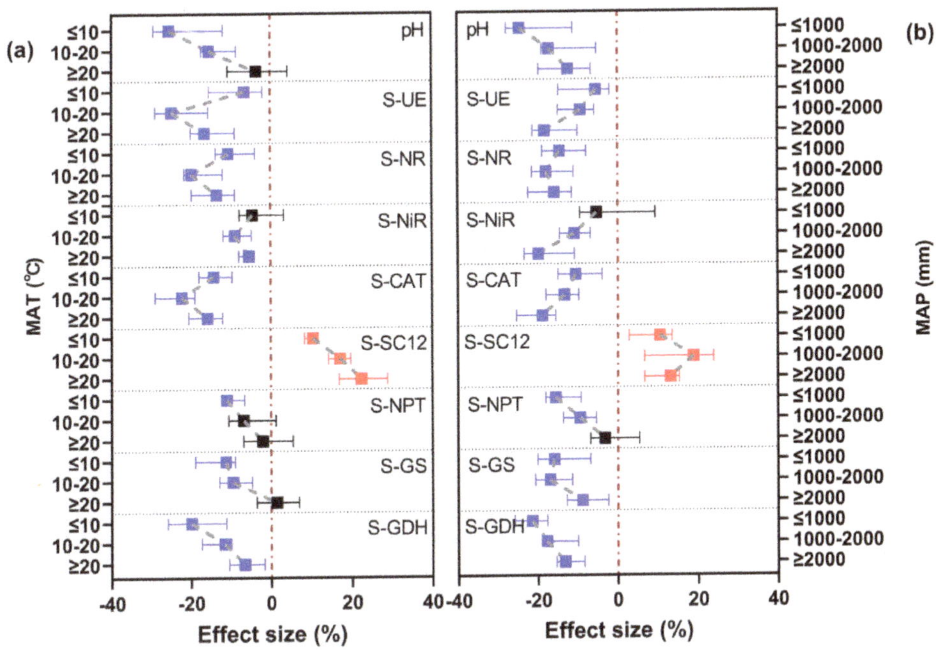

Figure 4. Effects of long-term N fertilization on soil pH and enzyme activities under different (**a**) MAT and (**b**) MAP categories. Symbols and error bars indicate effect sizes and 95% confidence intervals (CIs), respectively. N fertilization treatments were significantly different from no-N-fertilization treatments when the CIs did not overlap with zero (red dashed line) ($p < 0.05$). Red symbols indicate an increasing effect, blue symbols a decreasing effect, and black symbols no significant effect. S-UE, urease; S-NR, nitrate reductase; S-NiR, nitrite reductase; S-CAT, catalase; S-SC12, sucrase; S-NPT, glutamate dehydrogenase; S-GS, protease; S-GDH, glutamate amine synthetase.

Regarding MAT, the negative effects of nitrogen fertilization on S-UE, S-NR, S-NiR, and S-CAT were strongest at 10–20 °C, with reductions ranging from 8.95% to 24.83%. Regarding MAP, the negative effects on S-UE, S-NiR, and S-CAT activities were strongest at ≥2000 mm, with a decrease of 18.29–19.89%, and the negative effect on S-NR activity was strongest at 1000–2000 mm, with a decrease of 17.89%. There was no significant effect of the long-term application of nitrogen fertilizers on S-NiR activity when MAT and MAP were at their lowest levels (≤10 °C and ≤1000 mm).

SC12 activity tended to increase with long-term nitrogen fertilization regardless of temperature and precipitation. The percentage response of S-SC12 activity was highest at ≥20 °C for MAT and at 1000–2000 mm for MAP (22.45% and 18.94%, respectively). The S-NPT, S-GS, and S-GDH activities followed similar trends to soil pH; they all responded negatively to nitrogen fertilizer, with the effects of fertilizer being weakest under MAT ≥ 20 °C and MAP ≥ 2000 mm. Moreover, under MAT ≥ 20 °C and MAP ≥ 2000 mm, there was no effect of nitrogen fertilizer on S-NPT activity.

In summary, MAT and MAP are important factors determining the effect of nitrogen fertilization on soil pH and enzyme activities, with higher MAT and MAP resulting in lower soil acidification, probably due to the increase in S-GS and S-GDH activities.

3.3. *Effect of Climatic Conditions on the Response of Soil Microbial Community Alpha Diversity and Community Composition to Long-Term Nitrogen Fertilizer Application*

The responses of soil microbial alpha diversity and taxa abundance to long-term nitrogen fertilization varied under different MAT and MAP categories (Figure 5). The positive effect of nitrogen fertilization on soil microbial community richness (Chao1 index) gradually decreased with increasing MAT, from 19.35% at ≤10 °C to 5.13% at ≥20 °C, and was highest at a MAP of 1000–2000 mm (19.24%). Microbial community diversity (Shannon's index) responded negatively to nitrogen fertilization regardless of MAT and MAP, with the weakest effect (−14.22% and −10.66%) at the lowest MAT and MAP levels (≤10 °C and ≤1000 mm).

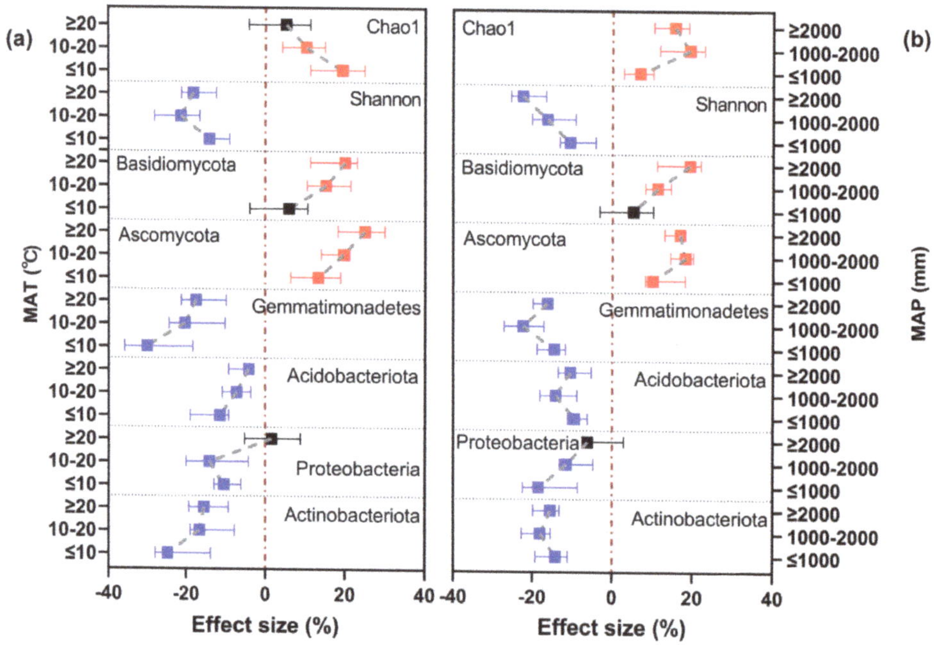

Figure 5. Effects of long-term N fertilization on alpha diversity and community composition of soil microbial communities under different (**a**) MAT and (**b**) MAP categories. Symbols and error bars respectively indicate the values of the effect sizes and their 95% confidence intervals (CIs). N fertilization treatments were significantly different from no-N-fertilization treatments when the CIs did not overlap with zero (red dashed line) ($p < 0.05$), with red symbols indicating an increasing effect, blue symbols a decreasing effect, and black symbols no significant effect. MAT, mean annual temperature; MAP, mean annual precipitation.

In terms of microbial community composition, the positive effect of nitrogen fertilization on the relative abundance of Ascomycota and Basidiomycota was higher under higher MAT and MAP. Fertilizer had the strongest effect on the relative abundance of Basidiomycota under the highest levels of MAT and MAP (effect size: 20.05% and 19.28%, respectively). Conversely, nitrogen fertilizer application had no significant effect on the relative abundance of Basidiomycota at the lowest levels of MAT and MAP (≤10 °C and ≤1000 mm). The relative abundances of Actinobacteriota, Acidobacteriota, and Gemmatimonadetes followed similar trends. The effects of nitrogen fertilizer on these taxa were strongest at MAT ≤ 10 °C and MAP 1000–2000 mm (effect size: 11.47–29.93%). Conversely, at the highest levels of MAT and MAP (≥20 °C and ≥2000 mm), nitrogen fertilizer application had no significant effect on the relative abundance of Proteobacteria. The strongest effect of nitrogen fertilizer on the relative abundance was observed at MAT 10–20 °C and MAP ≤ 1000 mm (effect size: −14.09% and −18.54%, respectively).

In summary, the effects of nitrogen fertilization on the relative abundance of Actinobacteriota, Acidobacteriota, and Gemmatimonadetes and the diversity of soil microbial communities were strongest under higher MAT and MAP.

3.4. Effect of Climatic Conditions on the Response of Functional Genes Related to Nitrogen Cycling under Long-Term Nitrogen Fertilizer Application

As shown in Figure 6a,b, there were differences in the response of functional genes related to nitrogen cycling to nitrogen fertilization between different MAT and MAP categories. Nitrogen fertilizer application increased *narG* and *nirK* regardless of MAT and MAP. However, the effect on *narG* was strongest (effect size: 25.53% and 24.45%) at the lowest MAT and MAP levels (≤10 °C and ≤1000 mm, respectively). The effect on *nirK* was strongest at MAT ≥ 20 °C and MAP 1000–2000 mm (effect size: 19.07% and 17.99%, respectively). In addition, nitrogen application decreased *qnorB*, *amoA-AOA*, *amoA-AOB*, and *nifH* regardless of MAT and MAP, with a similar effect on each among different levels of MAP and MAT. However, the negative effect on *qnorB* and *nifH* was stronger with increasing MAT, and the effect on *amoA-AOA* and *amoA-AOB*, was weaker with increasing MAT. With respect to MAP, the effect of nitrogen on *amoA-AOA* and *amoA-AOB* was weaker with increasing MAP, but the effect of nitrogen on *qnorB* and *nifH* was strongest at a MAP of 1000–2000 mm (effect size: −19.25% and −20.95%, respectively).

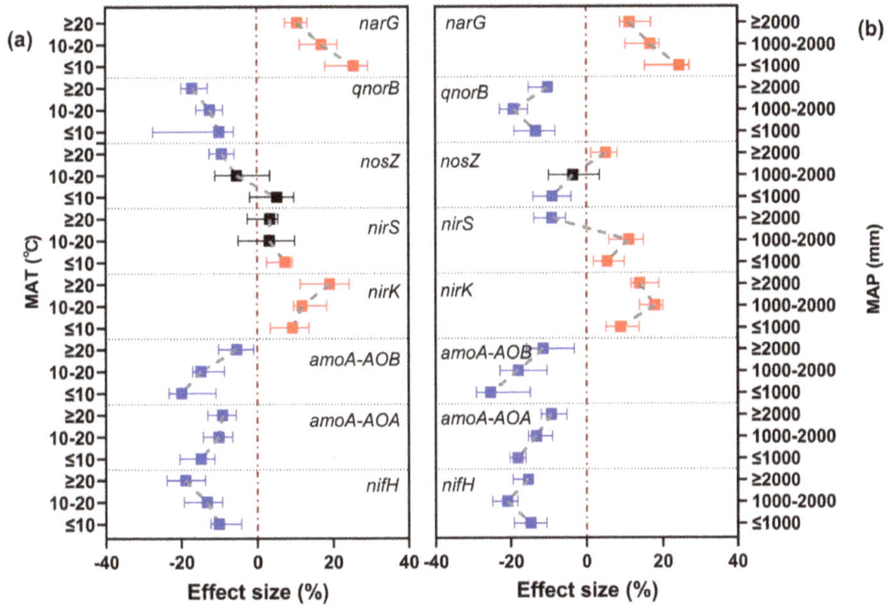

Figure 6. Effects of long-term N fertilization on the abundance of functional genes related to soil nitrogen cycling under different (a) MAT and (b) MAP categories. Symbols and error bars respectively

indicate effect sizes and their 95% confidence intervals (CIs). Nitrogen application treatments were significantly different from no-N-application treatments when the CIs did not overlap with zero (red dashed line) ($p < 0.05$). Red symbols indicate increasing effects, blue symbols indicate decreasing effects, and black symbols indicate no significant effects. MAT, mean annual temperature; MAP, mean annual precipitation.

3.5. Importance of MAT and MAP for Soil PH and Biological Properties

We used the random forest model to determine the relative importance of MAT and MAP for the soil pH and biological properties (Figure 7a–c). MAT had a greater relative effect on soil pH; S-UE, S-NR, S-CAT, and S-SC12 activities; and Shannon diversity. MAP had a greater relative effect on S-NiR, S-NPT, S-GS, and S-GDH activities and Chao 1. Moreover, MAT had a greater relative influence than MAP on the relative abundance of all taxa except Acidobacteriota and Proteobacteria. However, MAP had a greater relative influence on all nitrogen cycling genes except *nirK*. In summary, changes in MAT mainly affected soil pH and enzyme activities, while MAP affected the structure and function of soil microbial communities to a greater extent than MAT.

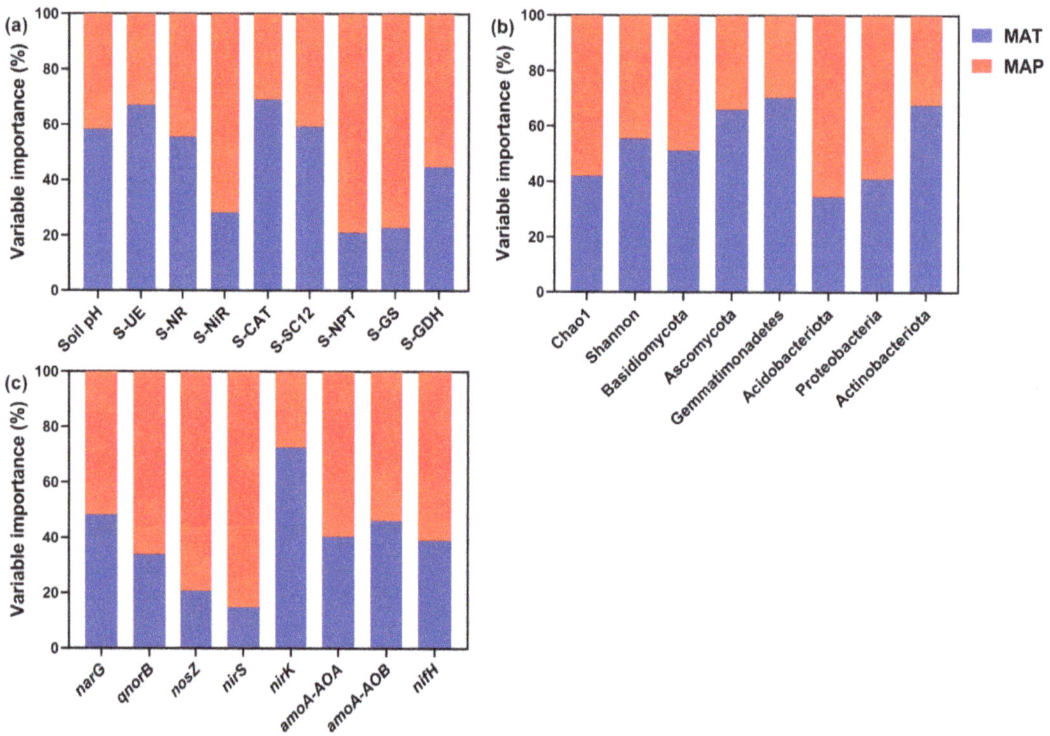

Figure 7. The relative influence of MAT and MAP on soil pH and enzyme activities (**a**), soil microbial community alpha diversity and community composition (**b**), and soil nitrogen cycling functional genes (**c**) based on the random forest model. Red indicates MAP and blue indicates MAT. MAP, mean annual precipitation; MAT, mean annual temperature; S-UE, urease; S-NR, nitrate reductase; S-NiR, nitrite reductase; S-CAT, catalase; S-SC12, sucrase; S-NPT, glutamate dehydrogenase; S-GS, protease; S-GDH, glutamate amine synthetase.

3.6. Structural Equation Modeling Analysis of the Pathways through which Long-Term Nitrogen Fertilizer Application and Climate Conditions Affect Soil Acidification

Structural equation modeling (SEM) was used to determine the relationships of N fertilizer and climate conditions with soil pH, soil enzyme activity, soil microbial community composition, alpha diversity, and N-cycle functional genes. The main pathways and modes by which N fertilizer and climate conditions affect soil acidification were determined (Figure 8). N fertilizer had a direct negative effect on soil pH (loading coefficient: −0.55). N fertilizer also reduced soil pH indirectly by reducing soil enzyme activity, soil microbial community composition, alpha diversity, and N-cycle functional genes, which in turn reduced soil pH. The climatic condition factor had a direct negative effect on soil pH (loading coefficient: −0.36). Climatic conditions also increased the composition and alpha diversity of soil microbial communities (loading coefficient: 0.46), which in turn mitigated the negative effect on soil pH. We also found that soil enzyme activities were positively correlated with soil nitrogen-cycle functional genes, and both were significantly positively correlated with soil pH. Overall, N input reduces soil pH by decreasing soil enzyme activities, microbial community composition, alpha diversity, and nitrogen-cycle functional genes, but appropriate temperature and moisture conditions can mitigate the risk of soil acidification.

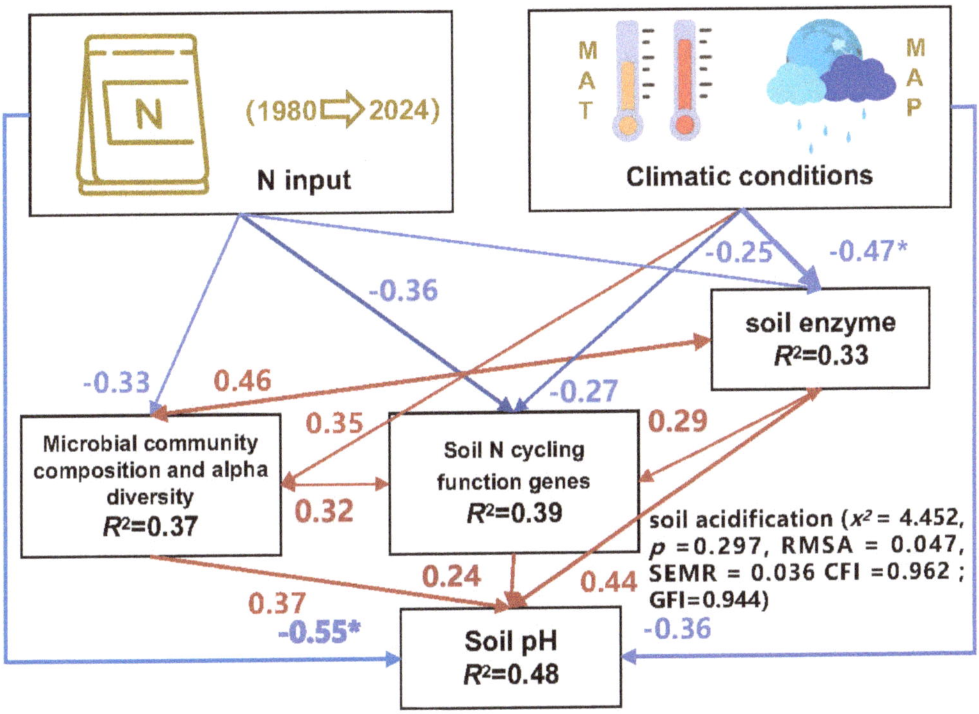

Figure 8. Structural equation modeling (SEM) describing the effects of N input, climatic conditions, and related factors (soil enzyme activity, soil microbial community composition, alpha diversity, and soil N-cycle functional genes) on soil acidification. Blue and red indicate negative and positive correlations, respectively (* $p < 0.05$). MAT, mean annual temperature; MAP, mean annual precipitation.

4. Discussion

4.1. Effects of Long-Term Nitrogen Application on Soil PH in China

Nitrogen fertilizer has been identified as an important trigger for farmland soil acidification [29]. From the 1980s to the beginning of the 21st century, the pH of farmland soils in China declined. The contribution of nitrogen application to this decline was 70% for major crops such as wheat, maize, and rice, and 90% for fruit and vegetable crops [30]. In this study, we found that with the long-term application of nitrogen fertilizer, soil pH in China decreased by an average of 15.27% from 1980 to 2024. If soil was a closed system, the processes could be described as follows: the mineralization of molecular organic nitrogen in the soil consumes 1 H^+ to produce 1 NH_4^+, 1 NH_4^+ nitrifies to 1 NO_3^- to produce 2 H^+, the plant absorbs 1 NO_3^- to consume 1 H^+, and the whole process of H^+ production and excretion is balanced and does not lead to soil acidification [31–33]. However, soil is not a closed system, and the soil frequently exchanges substances with the atmosphere, biosphere, and hydrosphere. If the amount of nitrogen applied exceeds the amount of nitrogen taken up by plants, nitrate nitrogen will remain in the soil or be leached out, resulting in a relative decrease in nitrate nitrogen taken up by plants, a decrease in H^+ consumed, and an increase in the amount of net H^+, which can lead to soil acidification [34].

4.2. Effects of Long-Term Nitrogen Application on Soil Enzyme Activities in China

In this study, we found that long-term nitrogen application in China increased S-SC12 activity by 9.29%, whereas it decreased activities of S-UE, S-NR, S-NIR, S-CAT, S-GS, and S-GDH, with an average decrease of 9.82–22.37%. The nitrogen inputs on which plants depend are mainly nitrate and ammonium, and the nitrate absorbed by plants must be reduced to ammonium before it can be further assimilated into amino acids. This process is catalyzed by S-NR and S-NiR, while S-GS and S-GDH are mainly involved in the process of ammonia assimilation in soil [35]. S-UE, S-SC12, S-GS, and S-CAT are of vital importance for soil nitrogen accumulation and plant growth [36]. In addition, S-CAT mitigates hydrogen peroxide toxicity in crops and can promote nutrient uptake and utilization [37]. However, S-CAT activity decreased under long-term N fertilizer application, indicating that continuous N fertilizer application not only led to soil acidification but also weakened the resistance of the soil to toxicity.

4.3. Effects of Long-Term Nitrogen Application on the Structure and Function of Soil Microbial Communities in China

Soil microorganisms are the engine of terrestrial ecosystem functioning. Microorganisms are involved in almost all known soil matter transformation processes and are important drivers of aboveground plant productivity and nutrient use, which determine the sustainability of terrestrial ecosystems. In addition, soil microbial communities are highly sensitive to soil acidification and are important indicators of the degree of soil acidification [38]. In this study, soil microbial diversity was found to decrease by 15.42% under long-term nitrogen fertilizer application. Among the dominant microbial phyla included in this study, the relative abundance of fungi increased by an average of 12.83–20.75%, while the relative abundance of bacteria decreased by 9.67–29.38%. In addition, the expression abundance of functional genes related to nitrogen cycling such as *nifH*, *amoA-AOA*, *amoA-AOB*, and *qnorB* decreased by 9.92–19.83%. This is likely because numerous functions of soil microorganisms, such as the activities of enzymes involved in nitrogen fixation, nitrification, denitrification, and phosphorus solubilization, as well as their functional genes, are closely related to soil pH [39]. Soil acidification and its altered nutrient efficiency are key drivers of changes in the composition, structure, and diversity of soil microbial communities, leading to a decline in ecological functions in acidic soils [40]. Long-term high inputs of chemical nitrogen fertilizers are not only the main cause of acidification of agricultural soils, but also significantly alter the composition and structure of soil microbial communities, which in turn negatively affects soil microbial functions and soil health [41]. Regulations that focus on enhancing soil microbial function, such as those that improve nitrogen fixation, can reduce the need for nitrogen fertilizer application and thus mitigate soil acidification. [42]. Rare and dominant bacteria in soil microbial communities contribute differently to soil functions, with rare bacteria being more important in maintaining soil ecological functions [43]. Soil microbial keystone species play an important role in improving acid soil fertility, plant nutrient uptake, and productivity [44]. As the so-called second genome of plants, the rhizosphere microbiome has become a hot research topic in agricultural science, with studies promoting its effect on plant growth, plant stress resistance, and plant health [45]. The rhizosphere is important for the structure and function of microbial communities in acid soils under different fertilizer schemes and crops [46]. Therefore, microbial function in acid soils plays an important role in improving plant nutrition, increasing nutrient use efficiency, mitigating soil acidification, and reducing fertilizer application.

4.4. Climatic Conditions as a Major Influence on Soil Acidification and Changes in Biological Properties in China under Long-Term N Application

Temperature is the main factor affecting soil respiration and microbial growth, and soil temperature is the main driver of seasonal changes in microbial communities [47]. In this study, the effect of MAT on the soil microbial community was more dominant than that of MAP. This is likely due to the significant positive correlation of bacteria, fungi, and actinomycetes with soil temperature. Soil microbial abundance increased gradually from spring to summer, peaked in summer, and then declined; bacterial abundance peaked in July and fungal diversity in September [48]. Seasonal freezing and thawing affect 70% of the global land area. Soil microbial activity and community composition are strongly responsive to seasonal freezing and thawing; a sharp decrease in temperature leads to structural changes in the soil microbial community, resulting in a decrease in the number, abundance, and diversity of bacterial taxa but a significant increase in the number of fungal species and fungal/bacterial ratios. Moreover, the freezing and thawing cycle has a strong effect on the microbial metabolic community, which is predominantly bacterial in summer and fungal-dominated in winter [49]. Dominant populations at higher temperatures can metabolize substrates not used by members of the microbial community at lower temperatures. Moreover, the freezing process can directly or indirectly kill microorganisms by limiting the availability of soil water and nutrients, thereby reducing microbial abundance, with fungi being more resistant to low temperatures. Surviving species have greater tolerances and adaptations to low temperatures [50]. The main factors influencing the seasonal variation of microorganisms vary due to differences in dominant vegetation types and climatic hydrothermal conditions in different regions, with precipitation having the greatest influence on microbial abundance in the tropics, whereas air temperature plays a dominant role in the subtropics [51].

Soil water content is one of the dominant factors affecting soil microbial activities, and soil microbial basal respiration is closely related to precipitation [52]. In this study, the positive effect of nitrogen fertilizer application on soil microbial community richness was strongest at a MAP of 1000–2000 mm. However, excessive moisture can reduce the permeability of wetland soils, impede the diffusion and circulation of gases in the soil, and lead to a decrease in the abundance of soil microorganisms [53]. Changes in MAP can reduce the diversity of soil microbial communities and shift them towards more fungal-dominated systems [54]. Seasonal and temporal variations in rainfall, particularly in ecosystems where organisms may be at or near the limits of their physiological tolerance, can have a major impact on the diversity, abundance, and responsiveness of soil microbial communities. In drylands, fungal abundance is particularly low during the pre-dry season and increases significantly during the rainy season, when soil moisture is 4.7 times higher than in drylands, and bacteria are more affected by short-term changes in rainfall [55].

5. Conclusions

Soil acidification has become one of the most serious land degradation problems in global agricultural systems. By analyzing 4864 sets of data comparisons from 115 studies, this meta-analysis revealed that under long-term N fertilizer application from 1980 to 2024, soil pH decreased by 15.27%, soil enzyme activities decreased by 9.82–22.37%, the relative abundance of bacteria decreased by an average of 9.67–29.38%, and the gene expression abundance of *nifH*, *amoA-AOA*, *amoA-AOB*, and *qnorB* decreased by 9.92–19.83%. We also found that the mean annual temperature and rainfall could affect the rate of soil acidification via changes to the diversity and composition of soil microbial communities. To reduce the effect of nitrogen fertilizer on soil acidification, we recommend the development of efficient, environmentally friendly, and low-cost slow/controlled-release fertilizers. These findings improve our understanding of soil pH–microbe interactions. This study provides insights for the sustainable development of Chinese agriculture.

Supplementary Materials: The following supporting information can be downloaded at: https://www.mdpi.com/article/10.3390/microorganisms12081683/s1, Supplementary Materials—literature statistics.

Author Contributions: Investigation, B.J., B.B. and Z.C.; writing—original draft preparation, L.Z. and Z.Z.; writing—review and editing, L.Z. and Q.L.; visualization—L.Z. and H.W.; supervision, Q.L. and J.C. All authors have read and agreed to the published version of the manuscript.

Funding: The work was supported by the science and technology development plan of Jilin Province, China (project nos. 20220508095RC and 20230302003NC) and the Key Technologies Research and Development Program (project no. 2023YFD1501104).

Data Availability Statement: The datasets generated and analyzed during the current study are available from the corresponding author upon reasonable request.

Acknowledgments: We wish to thank Wuliang Shi, Bin Li, and Yubin Zhang for valuable and encouraging discussions. We appreciate the efforts of the anonymous reviewers and the editors' valuable suggestions in earlier versions of this manuscript.

Conflicts of Interest: The authors declare no conflicts of interest. The funders had no role in the design of the study; in the collection, analyses, or interpretation of data; in the writing of the manuscript; or in the decision to publish the results.

References

1. Wang, W.; Chen, Y.L.; Liu, K.; Dang, Y.C.; Li, G.L.; Wen, L.Y.; Cao, Y. Optimization and classification control of permanent basic farmland based on quality classification. *Front. Environ. Sci.* **2024**, *12*, 14. [CrossRef]
2. Wang, F.W.; Gao, Y.; Li, X.; Luan, M.D.; Wang, X.Y.; Zhao, Y.W.; Zhou, X.H.; Du, G.Z.; Wang, P.; Ye, C.L.; et al. Changes in microbial composition explain the contrasting responses of glucose and lignin decomposition to soil acidification in an alpine grassland. *Sci. Total Environ.* **2024**, *930*, 8. [CrossRef]
3. Lin, S.X.; Liu, Z.J.; Wang, Y.C.; Li, J.Y.; Wang, G.G.; Ye, J.H.; Wang, H.B.; He, H.B. Soil metagenomic analysis on changes of functional genes and microorganisms involved in nitrogen-cycle processes of acidified tea soils. *Front. Plant Sci.* **2022**, *13*, 13. [CrossRef]
4. Liang, F.; Li, B.Z.; Vogt, R.D.; Mulder, J.; Song, H.; Chen, J.S.; Guo, J.H. Straw return exacerbates soil acidification in major Chinese croplands. *Resour. Conserv. Recycl.* **2023**, *198*, 8. [CrossRef]
5. Zhu, Q.C.; de Vries, W.; Liu, X.J.; Hao, T.X.; Zeng, M.F.; Shen, J.B.; Zhang, F.S. Enhanced acidification in Chinese croplands as derived from element budgets in the period 1980–2010. *Sci. Total Environ.* **2018**, *618*, 1497–1505. [CrossRef]
6. Jin, L.; Hua, K.K.; Zhan, L.C.; He, C.L.; Wang, D.Z.; Nagano, H.; Cheng, W.G.; Inubushi, K.; Guo, Z.B. Effect of Soil Acidification on Temperature Sensitivity of Soil Respiration. *Agronomy* **2024**, *14*, 16. [CrossRef]
7. Fujii, K.; Hayakawa, C. Fluxes of dissolved organic matter and nitrate and their contribution to soil acidification across changing permafrost landscapes in northwestern Canada. *Geoderma* **2023**, *430*, 13. [CrossRef]
8. Cui, T.T.; Zhang, J.B.; Luo, W.Q. The Quantity and Quality of Humic Substances following Different Land Uses in Karst Peak-Cluster Depression in Guangxi, China. *Agriculture* **2023**, *13*, 14. [CrossRef]
9. Ma, C.; Tu, Q.; Zheng, S.M.; Deng, S.H.; Xia, Y.H.; Mao, W.Q.; Gao, W.; Hu, L.N.; Kuzyakov, Y.; Hu, Y.J.; et al. Soil acidification induced by intensive agricultural use depending on climate. *J. Soils Sediments* **2022**, *22*, 2604–2607. [CrossRef]
10. Jimma, T.B.; Chemura, A.; Spillane, C.; Demissie, T.; Abera, W.; Ture, K.; Terefe, T.; Solomon, D.; Gleixner, S. Coupled Impacts of Soil Acidification and Climate Change on Future Crop Suitability in Ethiopia. *Sustainability* **2024**, *16*, 17. [CrossRef]
11. Jansson, J.K.; Hofmockel, K.S. Soil microbiomes and climate change. *Nat. Rev. Microbiol.* **2020**, *18*, 35–46. [CrossRef]

12. Alkorta, I.; Epelde, L.; Garbisu, C. Environmental parameters altered by climate change affect the activity of soil microorganisms involved in bioremediation. *FEMS Microbiol. Lett.* **2017**, *364*, 9. [CrossRef]
13. Shi, R.Y.; Ni, N.; Wang, R.H.; Nkoh, J.N.; Pan, X.Y.; Dong, G.; Xu, R.K.; Cui, X.M.; Li, J.Y. Dissolved biochar fractions and solid biochar particles inhibit soil acidification induced by nitrification through different mechanisms. *Sci. Total Environ.* **2023**, *874*, 8. [CrossRef] [PubMed]
14. Filimon, M.N.; Roman, D.L.; Bordean, D.M.; Isvoran, A. Impact of the Herbicide Oxyfluorfen on the Activities of Some Enzymes Found in Soil and on the Populations of Soil Microorganisms. *Agronomy* **2021**, *11*, 19. [CrossRef]
15. Wade, J.; Li, C.Y.; Vollbracht, K.; Hooper, D.G.; Wills, S.A.; Margenot, A.J. Prescribed pH for soil β-glucosidase and phosphomonoesterase do not reflect pH optima. *Geoderma* **2021**, *401*, 11. [CrossRef]
16. Gu, S.S.; Wu, S.L.; Zeng, W.A.I.; Deng, Y.; Luo, G.W.; Li, P.F.; Yang, Y.S.; Wang, Z.Q.; Hu, Q.L.; Tan, L. High-elevation-induced decrease in soil pH weakens ecosystem multifunctionality by influencing soil microbiomes. *Environ. Res.* **2024**, *257*, 14. [CrossRef] [PubMed]
17. Xin, P.Q.; Zhang, Y.L.; Jiang, N.; Chen, Z.H.; Chen, L.J. Neutral soil pH conditions favor the inhibition of phenol on hydrolase activities and soil organic carbon mineralization. *Eur. J. Soil Biol.* **2024**, *121*, 10. [CrossRef]
18. Waheed, A.; Li, C.; Muhammad, M.; Ahmad, M.; Khan, K.A.; Ghramh, H.A.; Wang, Z.W.; Zhang, D.Y. Sustainable Potato Growth under Straw Mulching Practices. *Sustainability* **2023**, *15*, 16. [CrossRef]
19. He, Z.J.; Cao, H.X.; Qi, C.; Hu, Q.Y.; Liang, J.P.; Li, Z.J. Straw management in paddy fields can reduce greenhouse gas emissions: A global meta-analysis. *Field Crop. Res.* **2024**, *306*, 15. [CrossRef]
20. Zhang, L.Q.; Wang, Y.L.; Lou, Z.X.; Hsu, L.F.; Chen, D.; Piao, R.Z.; Zhao, H.Y.; Cui, Z.J. Meta-Analysis of Factors Affecting C-N Fractions and Yield of Paddy Soils by Total Straw Return and N Fertilizer Application. *Agronomy* **2022**, *12*, 21. [CrossRef]
21. Dang, C.R.; Kong, F.L.; Li, Y.; Jiang, Z.X.; Xi, M. Soil inorganic carbon dynamic change mediated by anthropogenic activities: An integrated study using meta-analysis and random forest model. *Sci. Total Environ.* **2022**, *835*, 11. [CrossRef] [PubMed]
22. Franco, H.H.S.; Guimaraes, R.M.L.; Tormena, C.A.; Cherubin, M.R.; Favilla, H.S. Global applications of the Visual Evaluation of Soil Structure method: A systematic review and meta-analysis. *Soil Tillage Res.* **2019**, *190*, 61–69. [CrossRef]
23. Fan, X.P.; Yin, C.; Yan, G.C.; Cui, P.Y.; Shen, Q.; Wang, Q.; Chen, H.; Zhang, N.; Ye, M.J.; Zhao, Y.H.; et al. The contrasting effects of *N*-(*n*-butyl) thiophosphoric triamide (NBPT) on N_2O emissions in arable soils differing in pH are underlain by complex microbial mechanisms. *Sci. Total Environ.* **2018**, *642*, 155–167. [CrossRef] [PubMed]
24. Mei, K.; Wang, Z.F.; Huang, H.; Zhang, C.; Shang, X.; Dahlgren, R.A.; Zhang, M.H.; Xia, F. Stimulation of N_2O emission by conservation tillage management in agricultural lands: A meta-analysis. *Soil Tillage Res.* **2018**, *182*, 86–93. [CrossRef]
25. Philibert, A.; Loyce, C.; Makowski, D. Assessment of the quality of meta-analysis in agronomy. *Agric. Ecosyst. Environ.* **2012**, *148*, 72–82. [CrossRef]
26. Hedges, L.V.; Gurevitch, J.; Curtis, P.S. The meta-analysis of response ratios in experimental ecology. *Ecology* **1999**, *80*, 1150–1156. [CrossRef]
27. Yan, X.J.; Chen, X.H.; Ma, C.C.; Cai, Y.Y.; Cui, Z.L.; Chen, X.P.; Wu, L.Q.; Zhang, F.S. What are the key factors affecting maize yield response to and agronomic efficiency of phosphorus fertilizer in China? *Field Crop. Res.* **2021**, *270*, 12. [CrossRef]
28. Zhang, W.Q.; Dong, A.H.; Liu, F.L.; Niu, W.Q.; Siddique, K.H.M. Effect of film mulching on crop yield and water use efficiency in drip irrigation systems: A meta-analysis. *Soil Tillage Res.* **2022**, *221*, 11. [CrossRef]
29. Feijoó, C.; Hegoburu, C.; Messetta, M.L.; Guerra-López, J.; Rigacci, L.; Anselmo, J.; Di Franco, L.; Marcé, R. Acidification and increase of phosphorus levels in Pampean streams after 12 years of agricultural intensification. *Aquat. Sci.* **2023**, *85*, 14. [CrossRef]
30. Lu, X.Q.; Zhang, X.Y.; Zhan, N.; Wang, Z.; Li, S.F. Factors contributing to soil acidification in the past two decades in China. *Environ. Earth Sci.* **2023**, *82*, 12. [CrossRef]
31. Zhang, H.; Wang, L.; Fu, W.G.; Xu, C.; Zhang, H.; Xu, X.J.; Ma, H.B.; Wang, J.D.; Zhang, Y.C. Soil Acidification Can Be Improved under Different Long-Term Fertilization Regimes in a Sweetpotato-Wheat Rotation System. *Plants* **2024**, *13*, 13. [CrossRef]
32. Zhang, Y.J.; Ye, C.; Su, Y.W.; Peng, W.C.; Lu, R.; Liu, Y.X.; Huang, H.C.; He, X.H.; Yang, M.; Zhu, S.S. Soil Acidification caused by excessive application of nitrogen fertilizer aggravates soil-borne diseases: Evidence from literature review and field trials. *Agric. Ecosyst. Environ.* **2022**, *340*, 10. [CrossRef]
33. Zhu, X.; Ros, G.H.; Xu, M.; Xu, D.; Cai, Z.; Sun, N.; Duan, Y.; de Vries, W. The contribution of natural and anthropogenic causes to soil acidification rates under different fertilization practices and site conditions in southern China. *Sci. Total Environ.* **2024**, *934*, 172986. [CrossRef] [PubMed]
34. Shi, R.Y.; Li, J.Y.; Ni, N.; Xu, R.K. Understanding the biochar's role in ameliorating soil acidity. *J. Integr. Agric.* **2019**, *18*, 1508–1517. [CrossRef]
35. Gálvez, S.; Lancien, M.; Hodges, M. Are isocitrate dehydrogenases and 2-oxoglutarate involved in the regulation of glutamate synthesis? *Trends Plant Sci.* **1999**, *4*, 484–490. [CrossRef] [PubMed]
36. Zheng, B.C.; Zhou, Y.; Chen, P.; Zhang, X.N.; Du, Q.; Yang, H.; Wang, X.C.; Yang, F.; Xiao, T.; Li, L.; et al. Maize-legume intercropping promote N uptake through changing the root spatial distribution, legume nodulation capacity, and soil N availability. *J. Integr. Agric.* **2022**, *21*, 1755–1771.
37. Lu, M.; Zhao, J.X.; Lu, Z.R.; Li, M.J.; Yang, J.F.; Fullen, M.; Li, Y.M.; Fan, M.P. Maize-soybean intercropping increases soil nutrient availability and aggregate stability. *Plant Soil* **2023**. [CrossRef]

38. Zhou, Z.H.; Wang, C.K.; Zheng, M.H.; Jiang, L.F.; Luo, Y.Q. Patterns and mechanisms of responses by soil microbial communities to nitrogen addition. *Soil Biol. Biochem.* **2017**, *115*, 433–441. [CrossRef]
39. Zhang, M.Y.; Xu, Z.H.; Teng, Y.; Christie, P.; Wang, J.; Ren, W.J.; Luo, Y.M.; Li, Z.G. Non-target effects of repeated chlorothalonil application on soil nitrogen cycling: The key functional gene study. *Sci. Total Environ.* **2016**, *543*, 636–643. [CrossRef]
40. Wang, W.B.; Chen, D.S.; Sun, X.M.; Zhang, Q.; Koide, R.T.; Insam, H.; Zhang, S.G. Impacts of mixed litter on the structure and functional pathway of microbial community in litter decomposition. *Appl. Soil Ecol.* **2019**, *144*, 72–82. [CrossRef]
41. Cui, H.Y.; Vitousek, P.M.; Reed, S.C.; Sokoya, B.; Bamigboye, A.R.; Mukherjee, A.; Peñaloza-Bojacá, G.F.P.; Teixido, A.L.; Trivedi, P.; He, J.Z.; et al. Environmental filtering controls soil biodiversity in wet tropical ecosystems. *Soil Biol. Biochem.* **2022**, *166*, 9. [CrossRef]
42. Gandois, L.; Perrin, A.S.; Probst, A. Impact of nitrogenous fertiliser-induced proton release on cultivated soils with contrasting carbonate contents: A column experiment. *Geochim. Cosmochim. Acta* **2011**, *75*, 1185–1198. [CrossRef]
43. Morrow, J.L.; Sa, P.T.; Beattie, G.A.C.; Milham, P.J.; Riegler, M.; Spooner-Hart, R.N.; Holford, P. Additions of sugar and nitrogenous fertiliser affect plant nitrogen status and soil microbial communities. *Appl. Soil Ecol.* **2019**, *139*, 47–55. [CrossRef]
44. Hu, X.J.; Gu, H.D.; Liu, J.J.; Wei, D.; Zhu, P.; Cui, X.A.; Zhou, B.K.; Chen, X.L.; Jin, J.; Liu, X.B.; et al. Metagenomics reveals divergent functional profiles of soil carbon and nitrogen cycling under long-term addition of chemical and organic fertilizers in the black soil region. *Geoderma* **2022**, *418*, 10. [CrossRef]
45. Hu, Q.Y.; Liu, T.Q.; Ding, H.N.; Li, C.F.; Tan, W.F.; Yu, M.; Liu, J.; Cao, C.G. Effects of nitrogen fertilizer on soil microbial residues and their contribution to soil organic carbon and total nitrogen in a rice-wheat system. *Appl. Soil Ecol.* **2023**, *181*, 9. [CrossRef]
46. Wang, T.T.; Cao, X.X.; Chen, M.M.; Lou, Y.H.; Wang, H.; Yang, Q.G.; Pan, H.; Zhuge, Y.P. Effects of Soil Acidification on Bacterial and Fungal Communities in the Jiaodong Peninsula, Northern China. *Agronomy* **2022**, *12*, 11. [CrossRef]
47. Tang, Z.X.; Sun, X.L.; Luo, Z.K.; He, N.P.; Sun, O.J. Effects of temperature, soil substrate, and microbial community on carbon mineralization across three climatically contrasting forest sites. *Ecol. Evol.* **2018**, *8*, 879–891. [CrossRef] [PubMed]
48. Guo, Z.B.; Liu, C.A.; Hua, K.K.; Wang, D.Z.; Wu, P.P.; Wan, S.X.; He, C.L.; Zhan, L.C.; Wu, J. Changing soil available substrate primarily caused by fertilization management contributed more to soil respiration temperature sensitivity than microbial community thermal adaptation. *Sci. Total Environ.* **2024**, *912*, 13. [CrossRef]
49. Ballhausen, M.B.; Hewitt, R.; Rillig, M.C. Mimicking climate warming effects on Alaskan soil microbial communities via gradual temperature increase. *Sci. Rep.* **2020**, *10*, 10. [CrossRef]
50. Barreiro, A.; Lombao, A.; Martín, A.; Cancelo-Gonzalez, J.; Carballas, T.; Díaz-Raviña, M. Soil Heating at High Temperatures and Different Water Content: Effects on the Soil Microorganisms. *Geosciences* **2020**, *10*, 17. [CrossRef]
51. Huang, G.; Li, Y.; Su, Y.G. Effects of increasing precipitation on soil microbial community composition and soil respiration in a temperate desert, Northwestern China. *Soil Biol. Biochem.* **2015**, *83*, 52–56. [CrossRef]
52. Yu, H.Y.; Li, L.; Ma, Q.H.; Liu, X.D.; Li, Y.B.; Wang, Y.H.; Zhou, G.S.; Xu, Z.Z. Soil microbial responses to large changes in precipitation with nitrogen deposition in an arid ecosystem. *Ecology* **2023**, *104*, 14. [CrossRef] [PubMed]
53. Li, G.; Kim, S.; Han, S.H.; Chang, H.; Du, D.L.; Son, Y. Precipitation affects soil microbial and extracellular enzymatic responses to warming. *Soil Biol. Biochem.* **2018**, *120*, 212–221. [CrossRef]
54. Liu, Y.C.; Tian, H.M.; Li, J.R.; Wang, H.; Liu, S.R.; Liu, X.J. Reduced precipitation neutralizes the positive impact of soil warming on soil microbial community in a temperate oak forest. *Sci. Total Environ.* **2022**, *806*, 8. [CrossRef] [PubMed]
55. Zhou, Z.H.; Wang, C.K.; Luo, Y.Q. Response of soil microbial communities to altered precipitation: A global synthesis. *Glob. Ecol. Biogeogr.* **2018**, *27*, 1121–1136. [CrossRef]

microorganisms

Article

Maize–Soybean Rotation and Intercropping Increase Maize Yield by Influencing the Structure and Function of Rhizosphere Soil Fungal Communities

Liqiang Zhang †, Yuhan Yang †, Zehang Zhao, Yudi Feng, Baoyin Bate, Hongyu Wang, Qiuzhu Li * and Jinhu Cui *

College of Plant Science, Jilin University, Changchun 130012, China; lqzhang23@mails.jlu.edu.cn (L.Z.); wakk47640887@163.com (Y.Y.); 18846915612@163.com (Z.Z.); fengyudi1@163.com (Y.F.); bybt23@mails.jlu.edu.cn (B.B.); hong_yu@jlu.edu.cn (H.W.)

* Correspondence: liqz@jlu.edu.cn (Q.L.); cuijinhu@163.com (J.C.)

† These authors contributed equally to this work.

Abstract: Soil-borne diseases are exacerbated by continuous cropping and negatively impact maize health and yields. We conducted a long-term (11-year) field experiment in the black soil region of Northeast China to analyze the effects of different cropping systems on maize yield and rhizosphere soil fungal community structure and function. The experiment included three cropping systems: continuous maize cropping (CMC), maize–soybean rotation (MSR), and maize–soybean intercropping (MSI). MSI and MSR resulted in a 3.30–16.26% lower ear height coefficient and a 7.43–12.37% higher maize yield compared to CMC. The richness and diversity of rhizosphere soil fungi were 7.75–20.26% lower in MSI and MSR than in CMC. The relative abundances of *Tausonia* and *Mortierella* were associated with increased maize yield, whereas the relative abundance of *Solicoccozyma* was associated with decreased maize yield. MSI and MSR had higher proportions of wood saprotrophs and lower proportions of plant pathogens than CMC. Furthermore, our findings indicate that crop rotation is more effective than intercropping for enhancing maize yield and mitigating soil-borne diseases in the black soil zone of Northeast China. This study offers valuable insights for the development of sustainable agroecosystems.

Keywords: continuous maize cropping; maize–soybean rotation; maize–soybean intercropping; fungal community diversity; fungal community function; soil-borne diseases

1. Introduction

Northeast China, as one of the four major black soil areas worldwide, has become an important grain production area [1]. The grain production capacity of the northeastern region has increased in recent years; however, continuous cropping still dominates [2]. Continuous cropping is unsustainable and has led to soil degradation in the region [3], limiting crop yields. It is vital to find sustainable intensification methods that will ensure food security at a lower cost to resources and the environment [4]. Systems that incorporate crop diversity could solve the above problems.

In recent years, high-throughput sequencing technology has opened up a new era of environmental microbial research. This technology allows us to obtain a large amount of genetic information without the need for culturing samples [5]. Microbial diversity and function are essential for maintaining soil quality and ecosystems [6]. As soil microbial systems are large and complex, they play a key role in the decomposition of soil organic matter, the cycling of important elements such as C and N, the degradation of soil pollutants, and the formation and stabilization of soil structure [7]. In a study comparing maize–soybean rotation with continuous maize cropping, it was found that the soil under maize–soybean rotation had a higher relative abundance of the *nifB* gene, which encodes a nitrogen-fixing enzyme, higher bacterial diversity, and altered bacterial community structure [8]. Specifically, there was a higher relative abundance of Acidobacteria at the phylum level, and a higher and lower relative abundance of *Rhizobiaceae* and *Sphingomonadaceae*, respectively, at the family level [9]. A study comparing maize–soybean intercropping with maize monocropping found that intercropping resulted in a higher number of soil bacteria and actinomycetes, and a higher relative abundance of beneficial bacteria such as Proteobacteria and Acidobacteria, which are dominant soil bacteria [10].

A meta-analysis on the effect of crop rotation on soil biological properties at the global scale found that crop rotation significantly increased soil microbial biomass carbon and nitrogen in areas with a mean annual temperature (MAT) $\geq$ 8 °C and increased the Shannon diversity of soil bacterial communities in colder areas (MAT < 8 °C) [11]. For different soil types, rotation performs better in coarse or medium textured soil, with medium levels of initial SOC (7–10 g ka^{-1}) and low TN (<1.2 g kg^{-1}) [12]. Furthermore, a crop rotation increased the Shannon diversity by 6.55% in a region with a MAT of 15 °C [13]. In terms of pests and diseases, one study found that intercropping reduced nematode damage to the target crop by 40% and reduced disease incidence by 55% [14].

Based on the above literature, both intercropping and crop rotation with maize and soybean can increase yields by improving biological characteristics compared to continuous maize cropping. However, these studies focused on microbiological communities. For example, Li et al. (2023) found that crop rotation and intercropping plantings altered bacterial and fungal community composition and improved the abundance of potentially plant-beneficial fungi, such as *Mortierella*, while reducing the abundance of potentially plant-pathogenic fungi, such as *Fusarium* [15]. There are few reports on the role of fungal community structure and function in crop growth and development. We hypothesized that cropping patterns may lead to changes in soil fungal communities and that such changes would inevitably affect crop agronomic traits and yields. To test this hypothesis, we sampled maize rhizosphere soils from an 11-year field experiment that included continuous maize cropping, maize–soybean rotation, and maize–soybean intercropping. We characterized fungal community structure and function using high-throughput sequencing. The objectives of this study were (1) to compare fungal community structure and function in maize rhizosphere soil under the three cropping systems and (2) clarify the relationships between the rhizosphere soil fungal community and maize agronomic traits and yield. This study provides a theoretical basis for promoting the sustainable development of maize cultivation.

2. Materials and Methods

2.1. Study Area

A long-term field experiment was established in May 2012 in Sijian Village, Chengxi Town, Changchun City, central Jilin Province. The study site was located at the Agricultural Experimental Base of the College of Plant Science, Jilin University (125°26.83′ E, 43°90.77′ N). The mean monthly rainfall and mean air temperature during the crop-growing season are shown in Figure 1. There is a frost-free period of 140 days and a mean annual effective cumulative temperature of 3442.3 °C. The soil is black calcium soil. Before the experiment (25 April 2012), the soil (0–20-cm layer) had the following characteristics: pH, 5.64; organic matter content, 1.89%; total nitrogen content, 1.05 g kg^{-1}; total phosphorus content, 0.35 g kg^{-1}; total potassium content, 9.75 g kg^{-1}; available nitrogen content, 145.82 mg kg^{-1}; available phosphorus content, 23.65 mg kg^{-1}; available potassium content, 160.03 mg kg^{-1}; soil capacity, 1.13 g cm^3; soil temperature, 14.8 °C; and relative soil water content, 75.83%.

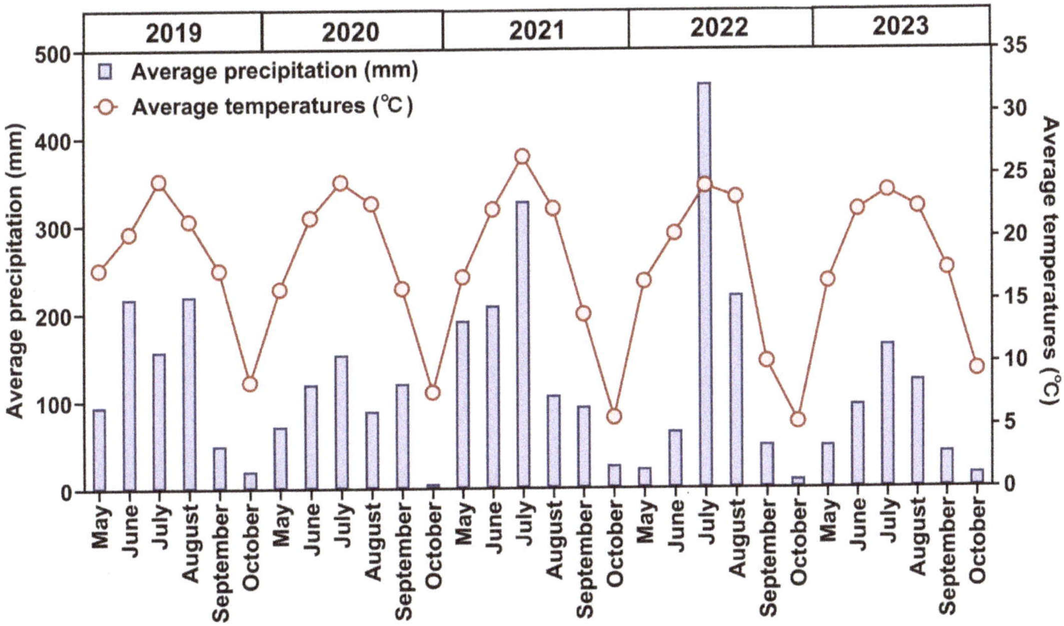

Figure 1. Monthly average rainfall and average temperature during the sampling period (2019–2023).

2.2. Experimental Design

The maize variety Fumin 985 (128 days growth period, late-maturing, compact variety, 2012–2023) and the soybean variety Changnong 39 (129 days growth period, medium maturity branched variety with an unlimited podding habit, 2012–2023) were used. The experiment was designed as a large-area production trial with three cropping systems, namely, continuous maize cropping (CMC), maize–soybean rotation (maize: 20 rows, soybean: 20 rows; MSR), and maize–soybean intercropping (maize: 20 rows, soybean: 20 rows; MSI). Before the experiment, the site was planted with maize. Four replicate plots were established for MSI and MSR, and one plot for CMC. Soybean was planted at a density of 200,000 plants per hectare and maize at 75,000 plants per hectare (Figure 2), with each row being 50 m in length and spaced 65 cm apart. Fertilizer was applied once on May 12. The nitrogen application rate was 200 kg N ha^{-1} for maize and 80 kg N ha^{-1} for soybean. Phosphate was applied as heavy superphosphate (P_2O_5, 46%) for maize and soybean, at a rate of 120 kg P ha^{-1}. Potassium was applied as potassium sulfate (K_2O, 50%) at a rate of 90 kg K ha^{-1} for maize and soybean. Before seedlings emerged, Acetochlor ($C_{14}H_{20}ClNO_2$, 50%, JINQIU, Changchun, China) was manually sprayed at 1200 mL ha^{-1} during sunny weather from 3 to 5 May each year. In addition, maize was manually weeded twice during the 6- to 12-leaf stage, and maize was manually sprayed with Chlorantraniliprole ($C_{18}H_{14}BrCl_2N_5O_2$, 4.5%, JINQIU, Changchun, China) at the jointing stage (during sunny weather from 10 to 15 June each year) at 350 mL ha^{-1}. Soybean herbicide and insecticide rates and timing were consistent with maize. At the end of each growing season, maize straw was crushed, plowed, and returned to the field. All other field management methods were consistent across each treatment during the growth period.

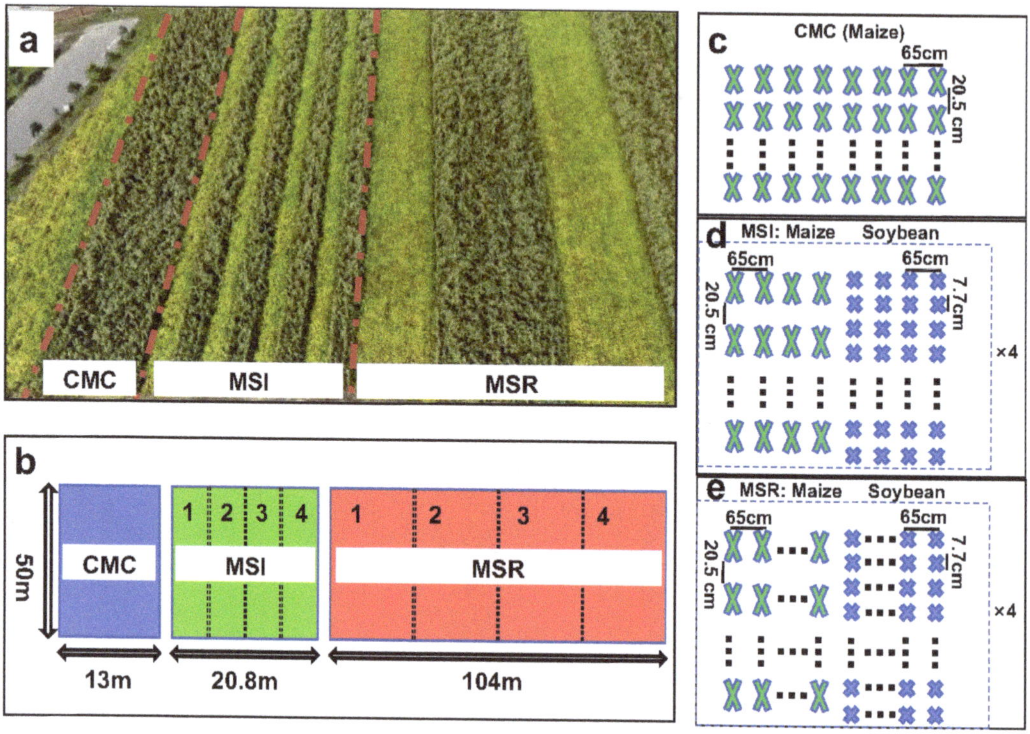

Figure 2. Experimental design. (**a**) Aerial photograph of the long-term field experiment; (**b**) diagram of the field distribution of the three cropping patterns; 1–4 indicate replicate plots; (**c–e**) diagram of the distribution of crops planted in each treatment; 65 cm refers to the space between rows, while 20.5 cm and 7.7 cm refer to the space between plants within a given row. Continuous maize cropping (CMC), maize–soybean rotation (MSR), maize–soybean intercropping (MSI).

2.3. Sample Collection

Maize plants were sampled at maturity each year from 2019 to 2023. First, the plant height, ear height, and stem diameter of 10 maize plants in each treatment were measured. The ear position coefficient (%) was calculated as: ear height (cm)/plant height (cm) × 100%. Then, the maize in a 20 m^2 area in each treatment was harvested. All the grain was threshed, and the kernel weight was measured and converted to yield per hectare based on the area harvested. In addition, we measured ear length, ear diameter, grain number per ear, bald tip length and 100-grain weight. The seed moisture content was determined using a PM8188 (PM-8188-A, KETT, Tokyo, Japan) moisture meter with three replications, and the final yield data were determined using a seed moisture content of 14%.

Maize rhizosphere soil was sampled in 2023. Five soil samples were collected from each plot. Five plants were removed from the ground, and large pieces of soil were removed with a sterile brush so that only about 1 mm of soil was still attached to the roots. The roots were placed in a sterile tube, 50 mL of 1× PBS buffer was added, and the roots were stirred vigorously with sterile forceps for 5 min. Roots were removed and centrifuged at 13,000× *g* for 30 s (TGL-16 s, SHUKE, Chengdu, China). The centrifugal precipitate was considered to be rhizosphere soil and was stored at −80 °C until DNA extraction.

2.4. DNA Extraction and Sequencing

Genomic DNA was extracted from rhizosphere soil samples using the E.Z.N.A. Soil DNA Kit (Omega Bio-tek, Inc., Norcross, GA, USA) and stored at −20 °C for subsequent experiments. A 30 ng DNA sample was taken for PCR amplification. Specific primers with barcodes were used to amplify DNA with an ABI 9700 PCR instrument (Applied Biosystems, Inc., Foster City, CA, USA). The quality of DNA extraction was assessed by 1% agarose gel electrophoresis. The ITS1F (5′-CTTGGTCATTTAGAGGAAGTAA-3′) and ITS2 (5′-GCTGCGTTTCTTTCATCGATGC-3′) primers were used to amplify the ITS1 region. The amplification system and procedures are detailed in Table 1.

Table 1. PCR amplification system and procedure for fungi in soil.

PCR Amplification System		PCR Amplification Procedure	
Reagent Composition	**Volume**	**Reaction**	
DNA template	30 ng	① Pre-denaturation	94 °C 2 min
Forward Primer (5 uM)	1 μL	② Denaturation	94 °C 15 s
Reverse Primer (5 uM)	1 μL	③ Anneal	55 °C 30 s
BSA (2 ng μL^{-1})	3 μL	④ Extension	72 °C 60 s
2x Taq Plus Master Mix	12.5 μL	⑤ Final extension	72 °C 7 min
ddH_2O	7.5 μL	③–⑤ Number of cycles	30
Total	25 μL	⑥ Reaction termination	10 °C ∞

The library construction comprised the following steps: (1) ligation of the "Y" junction; (2) removal of junction self-associated fragments by magnetic bead screening; (3) enrichment of the library template by PCR amplification; (4) denaturation by sodium hydroxide to produce single-stranded DNA fragments [16]. Sequencing methods and hardware and software information are shown in Table 2. All sequence data were uploaded to the National Center for Biotechnology Information (NCBI; https://www.ncbi.nlm.nih.gov/) (URL, accessed on 1 August 2024) under accession number: PRJNA1123432.

Table 2. Statistics of sequencing instruments and reagents.

Step	Instrument/Reagents	Manufacturer	Specification/Model/Lot Number
Amplicon extraction	MoBio PowerSoil DNA Isolation Kit (100)	QIAGEN (Frankfurt, Germany)	100 times
Amplifier amplification	KAPA 2G Robust Hot Start Ready Mix	KAPA (Boston, MA, USA)	
	ABI 9700 PCR	ABI (Guangzhou, China)	
Amplicon purification	Agencourt® AMPure® XP	Beckman Coulter (Shanghai, China)	Dispense 45 mL/bottle, total 450 mL/bottle
Amplicon building	NEBNext Ultra II DNA Library Prep Kit	NEB (Beijing, China)	96 reactions
	Agencourt® AMPure® XP	Beckman Coulter	Dispense 45 mL/bottle, total 450 mL/bottle
	ABI 9700 PCR	ABI	
Library quality control (instruments)	Bioanalyzer (Agilent 2100)	Agilent (Palo Alto, CA, USA)	DE13806339
	Biomolecule Analyzer (Labchip GX)	PerkinElmer (Shanghai, China)	
	ABI Qpcr	ABI	Step One Plus (Shanghai, China) (www.PuDi.cn) (accessed on 1 August 2024)
Library quality control (reagents)	Agilent DNA 1000 Kit	Agilent (Palo Alto, CA, USA)	300 samples
	HT DNA-Extended Range LabChip	PerkinElmer (Shanghai, China)	
	KAPA Library Quantification Kit	KAPA	500 times

Table 2. *Cont.*

Step	Instrument/Reagents	Manufacturer	Specification/Model/Lot Number
Sequencing (equipment)	High-throughput second-generation sequencer	Illumina (Beijing, China)	MiSeq (Shanghai, China) (www.PuDi.cn) (accessed on 1 August 2024)
Sequencing (reagents)	MiSeq® Reagent Kit v3 (600 cycle) (PE300)	Illumina	
	MiSeq Reagent Kit v2 (500 cycle)	Illumina	

2.5. Statistical Analysis

The following statistical analyses were performed using SPSS 22.0 (https://www.ibm.com/spss) (accessed on 6 August 2024) (Norman H. Nie, C. Hadlai Hull and Dale H. Ben, Stanford University, Stanford, CA, USA) software: one-way ANOVA (https://www.ibm.com/spss) (accessed on 6 August 2024) was used to compare the effects of cropping system on maize agronomic traits and yield component factors. After evaluating the significant differences between the sample means via one-way ANOVA, Duncan's test, which is a post hoc test, was used to determine which specific group means were significantly different from each other. Pearson's correlation was used to evaluate the relationships between the relative abundance of dominant genera, soil chemical properties, and yield. Beta diversity was determined using the coefficient of variation of the Aitchison distance and was plotted using non-metric multidimensional scaling (NMDS) to compare the degree of similarity between samples.

A soil fungal genera-level network was constructed using the Hmisc package (5.1-1, Frank Harrell, University of Alabama at Birmingham: Birmingham, AL, USA) and the igraph package (2.0.3, Kirill Müller, Karlsruhe Institute of Technology: Karlsruhe, DE, USA) in R version 4.3.1 (MathSoft, 2023; https://www.r-project.org/, accessed on 14 July 2024). The network was analyzed using Gephi 0.9.2 (Netbeans, 2022) [17] and plotted using Java (SE 8u 171, Netbeans, 2022). The model was constructed using R based on the following equations:

$$Cb(n) = \sum_{s \neq n \neq t} \left(\frac{\sigma_{st}(n)}{\sigma_{st}} \right) \tag{1}$$

$$T_n = \frac{avg\left(J_{(n,m)}\right)}{k_n} \tag{2}$$

where n is the destination node, s and t are nodes in the network other than n, σ_{st} represents the number of shortest paths from node s to node t, and $\sigma_{st}(n)$ denotes the number of shortest paths from node s to node t that must pass through node n. $J_{(n,m)}$ is the number of all nodes adjacent to both nodes n and m, and the value of $J_{(n,m)}$ is increased by 1 if n is directly adjacent to m. k_n is the number of all neighbours of the node.

Structural equation modeling (SEM) of the effects of different cropping modes on maize yield pathways was implemented using the 'Lavaan' package (0.6–18, Yves Rosseel, Ghent University, BEL) in R v4.3.1 (https://lavaan.ugent.be, accessed on 1 July 2024).

3. Results

3.1. Maize Plant Growth

MSR and MSI had lower maize ear position coefficients than CMC in 2019–2023 (Figure 3), which indicated a better growth structure and higher lodging resistance. In 2019–2020, the plant height and ear height under CMC were lower than those under MSR and MSI, by 4.21–13.97% and 6.75–8.98%, respectively. The ear position coefficients under CMC were significantly lower than that under MSR, by 4.94–7.6%. In 2021, the plant height under MSR was 4.88–8.52% higher than that under MSI and CMC, and the ear height under MSR was 10.85% higher than that under CMC. However, in 2022–2023, plant height followed the order CMC > MSI > MSR, with the plant height under CMC being 3.31–13.01%

higher than that under MSI and MSR, and the plant height under MSI being 9.39% higher than that under MSR. In 2022–2023, the ear position coefficient under MSR was significantly lower than that under MSI and CMC, by 12.96–16.26%.

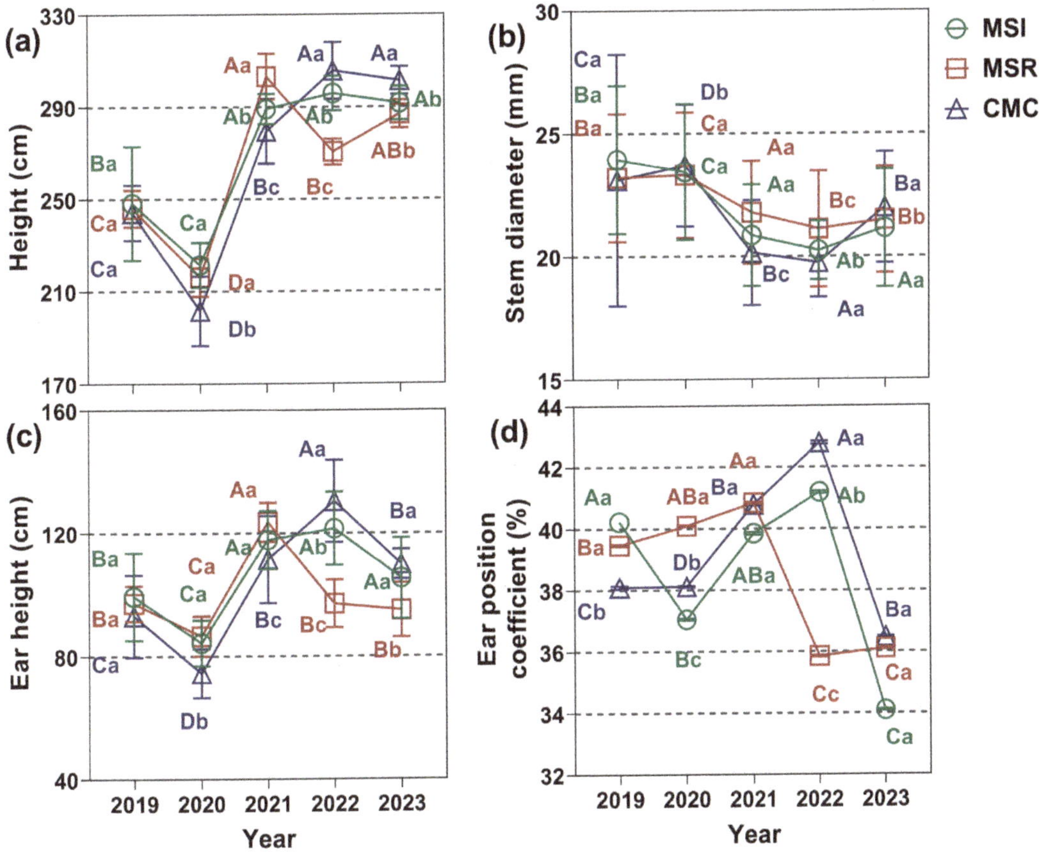

Figure 3. Effects of cropping system on maize growth in 2019–2023. (**a**) Plant height, (**b**) stem diameter, (**c**) ear height, and (**d**) ear position coefficient. Values are means ± SE, n = 10 replicates. The lowercase letters (a–c) indicate differences among cropping systems within the same year, while the uppercase letters (A–D) indicate differences among years under the same cropping system ($p < 0.05$). MSI, maize–soybean intercropping; MSR, maize–soybean rotation; CMC, continuous maize cropping.

3.2. Maize Yield and Yield Components

The effects of cropping system on maize yield components are shown in Figure 4. Maize ear length and ear diameter varied considerably under CMC from 2019 to 2023 and were lowest in 2022, when they were 9.58% and 11.33% lower than those under MSI and MSR, respectively. From 2019 to 2023, the ear diameter, ear length, and number of grains per ear were similar between MSI and MSR (Figure 4a–c), but in general, they were higher under MSR than under MSI. From 2019 to 2023, the maize bald tip length under each cropping system showed a downward trend (Figure 4d), with the average maize bald tip length decreasing from 0.60 cm in 2019 to 0.06 cm in 2023. However, the maize bald tip length was always higher under CMC than under MSI and MSR, and the 100-grain weight was higher under MSR than under CMC and MSI, by an average of 3.07% (Figure 4e). Overall, the maize yield in each year followed the order MSR > MSI > CMC (Figure 4f). Except for in 2020, the maize yield under MSR generally increased over time. The maize yield under MSI stayed similar over time, with the difference between the highest and lowest yield year being only 1259.7 kg ha^{-1}. The maize yield under CMC decreased with time, and the maize yield in 2020–2023 was 5.03–10.99% lower than that in 2019. In summary, compared with CMC, MSR and MSI generally increased maize ear length, ear diameter, and number of grains per ear, reduced maize bald tip length, and thus increased maize yield. Moreover, MSR resulted in a higher maize yield than MSI.

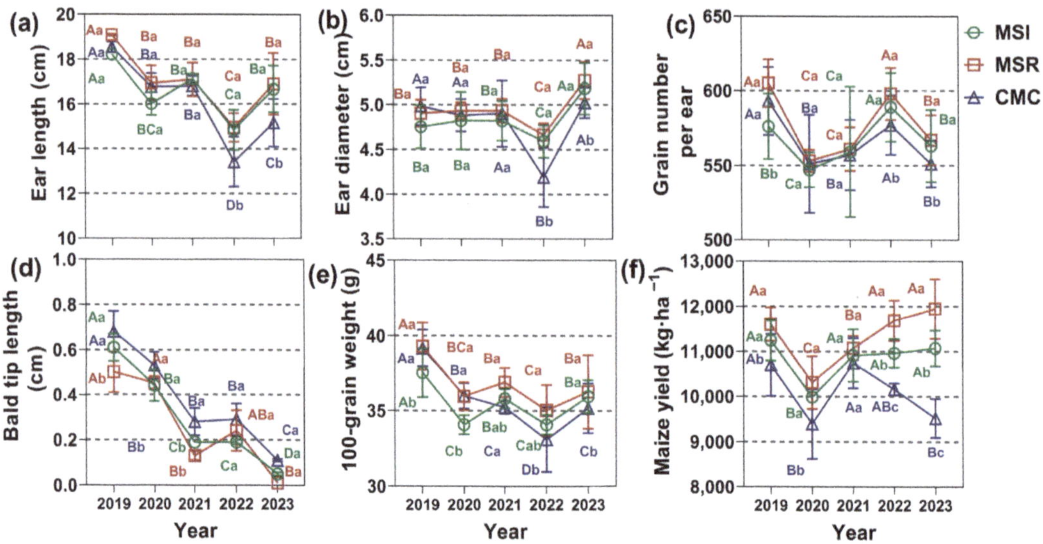

Figure 4. Effects of cropping system on maize yield and yield components in 2019–2023. (**a**) Ear length, (**b**) ear diameter, (**c**) grain number per ear, (**d**) bald tip length, (**e**) 100-grain weight, and (**f**) yield. Values are mean ± SE, n = 10 replicates. The lowercase letters (a–c) indicate differences among cropping systems within the same year, while the uppercase letters (A–D) indicate differences among years under the same cropping pattern ($p < 0.05$). MSI, maize–soybean intercropping; MSR, maize–soybean rotation; CMC, continuous maize cropping.

3.3. *Rhizosphere Soil Fungal Community*

3.3.1. Alpha Diversity of Maize Rhizosphere Soil Fungal Community

To investigate soil fungal alpha diversity under different cropping systems, two soil fungal community richness indices (Chao1 and observed number of species; Figure 5a,b) and two community diversity indices (Faith's phylogenetic diversity (PD) and Shannon diversity; Figure 5c,d) were calculated. The Chao1 and number of observed species under MSR and CMC were higher than those under MSI. The Chao1 and number of observed species under MSR were 48.30% and 29.48% higher, respectively, compared with those under MSI, and the Chao1 and number of observed species under CMC were 62.58% and 40.07% higher, respectively, compared with those under MSI. Faith's PD under CMC was 14.79% and 10.43% higher compared to that under MSI and MSR, respectively. Shannon diversity under CMC was significantly higher than that under MSI, by 10.07%. In summary, compared with under CMC, the richness and diversity of soil fungal communities under MSI and MSR were lower, with MSI having lower richness and diversity than MSR.

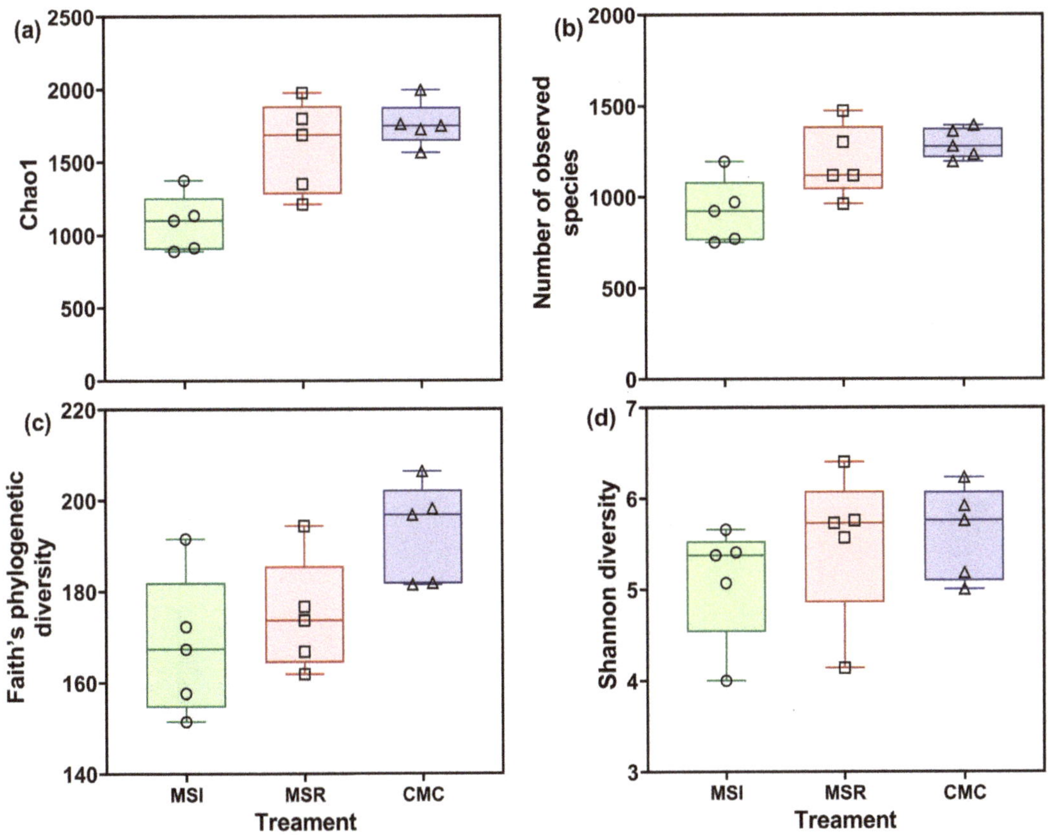

Figure 5. Effects of cropping system on maize rhizosphere soil fungal richness and diversity. (**a**) Chao1, (**b**) number of observed species, (**c**) Faith's phylogenetic diversity (PD), and (**d**) Shannon diversity. The two ends of the box plot are the upper and lower quartiles, the horizontal line in the middle indicates the median, and the lines connecting the two ends are the minimum and maximum values (n = 5). MSI, maize–soybean intercropping; MSR, maize–soybean rotation; CMC, continuous maize cropping.

3.3.2. NMDS of Maize Rhizosphere Soil Fungal Community

NMDS was used to explore differences in soil fungal communities among cropping systems (Figure 6). Each cropping system was separated along NMDS2, indicating different fungal community compositions among the cropping systems. MSI and MSR were close to each other in the NMDS plot and far from CMC, indicating that intercropping and rotation systems affected the fungal community structure.

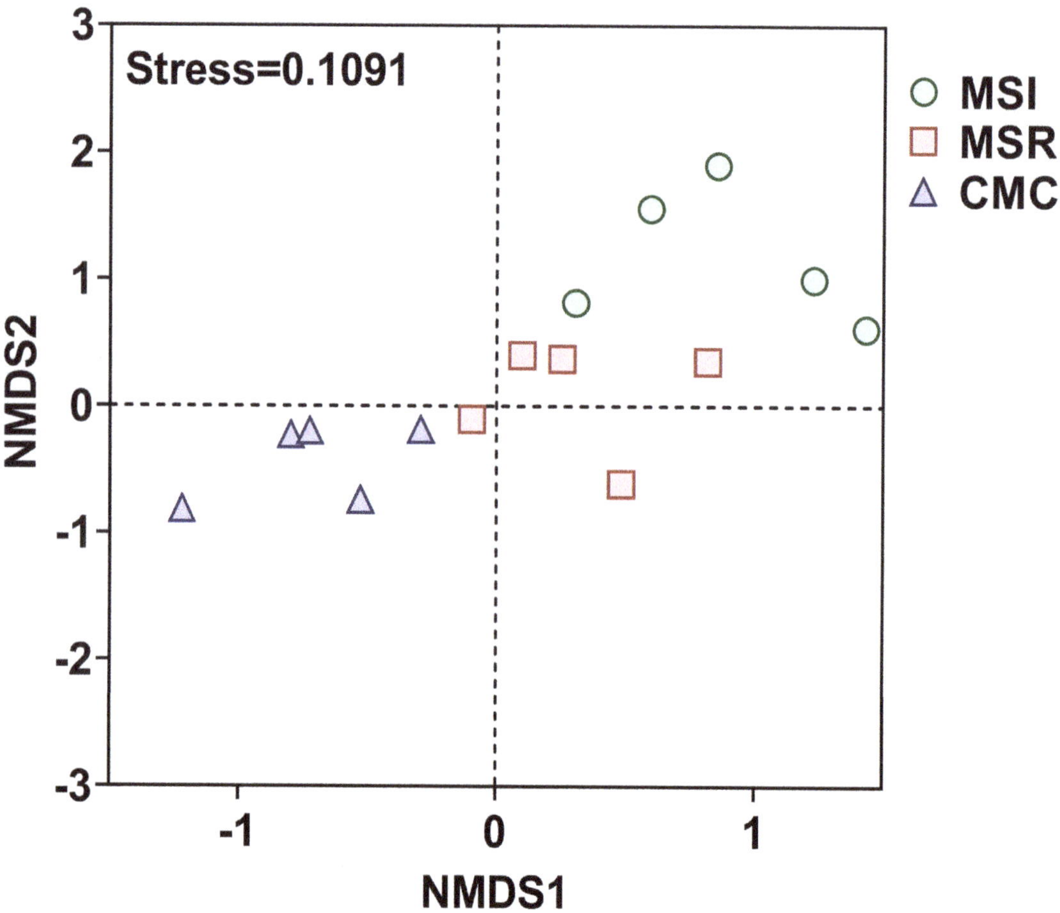

Figure 6. NMDS plot of maize rhizosphere soil fungal communities under different cropping systems. Stress = 0.1091. MSI, maize–soybean intercropping; MSR, maize–soybean rotation; CMC, continuous maize cropping.

3.3.3. Fungal Community Composition in Maize Rhizosphere Soil under Different CropPing Systems

In total, 707 genera of fungi were annotated. The unnamed genera were removed and the genera with a relative abundance greater than 0.1% were retained, resulting in 18 genera (Figure 7). For MSI and CMC, *Tausonia*, *Ceratobasidium*, *Chaetomium*, and *Mycena* were the dominant genera, accounting for 37.21% and 27.47% of the total relative abundance, respectively. For MSR, the dominant genera were *Tausonia*, *Chaetomium*, and *Solicoccozyma*, accounting for 48.12% of the total relative abundance. Other genera included *Mortierella*, *Crocicreas*, and *Exophiala*. Further analysis showed that the relative abundance of *Tausonia* under MSR was 147.78% and 89.89% higher than that under MSI and CMC, respectively, and the differences were significant. The relative abundance of *Ceratobasidium* under MSI was higher than that under MSR and CMC, by 378.41% compared with CMC. The relative abundance of *Chaetomium* under MSI and MSR was 37.76% and 75.20% higher than that under CMC, respectively. Similarly, the relative abundance of *Mortierella* under MSI and MSR was 40.54% and 47.11% higher compared with that under CMC, respectively. The relative abundance of *Solicoccozyma* was 3.46% under CMC, 1.63% under MSI, and 0.33% under MSR.

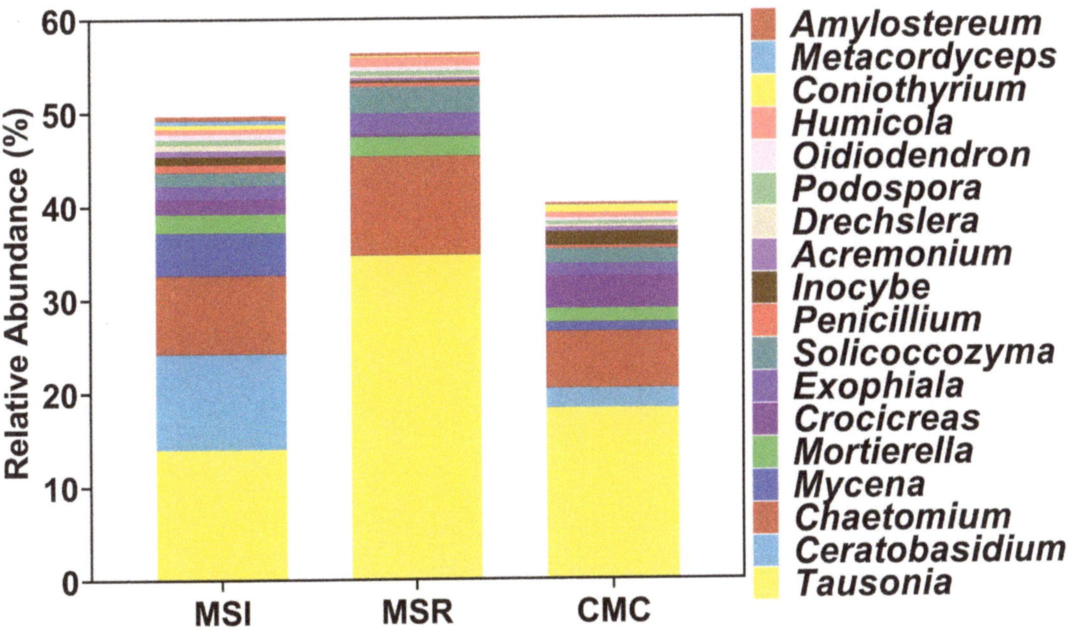

Figure 7. Composition of fungal genera in maize rhizosphere soil under different cropping systems. MSI, maize–soybean intercropping; MSR, maize–soybean rotation; CMC, continuous maize cropping.

3.3.4. Co-Occurrence Network of the Maize Rhizosphere Soil Fungal Community

To determine relationships among genera, we constructed a co-occurrence network of the top 200 fungal genera in maize rhizosphere soil under different cropping systems (Figure 8). Topological parameters of the network are shown in Table 3. All genera in the three cropping systems were from Ascomycota, Basidiomycota, and Mortierellomycota. Under the same network node, there were 17.50% and 3.40% fewer edges under MSI and MSR, respectively, compared with under CMC. In addition, the proportion of positive correlations was 2.01% and 24.92% lower, respectively. The average degree, average weighted degree, average clustering coefficient, and modularity under CMC were higher than those under MSI and MSR. The CMC co-occurrence network had more connections between network nodes, and the connections were stronger and more complex. These results indicate that the rotation and intercropping systems can reduce the complexity of the fungal community network, weakening the positive relationships among genera and increasing competition. This could in turn inhibit pathogens.

Table 3. Topological characteristics of maize rhizosphere soil fungal co-occurrence networks.

Treatment	Node	Edge	Positive (%)	Negative (%)	Average Degree	Average Weighting	Modularity	Cluster Coefficient
MSI	201	1114	57.95	42.05	11.14	10.36	0.68	0.65
MSR	199	1266	47.32	52.68	12.74	11.77	0.65	0.62
CMC	200	1309	59.11	40.89	13.03	12.04	0.63	0.61

Note: MSI, maize–soybean intercropping; MSR, maize–soybean rotation; and CMC, continuous maize cropping.

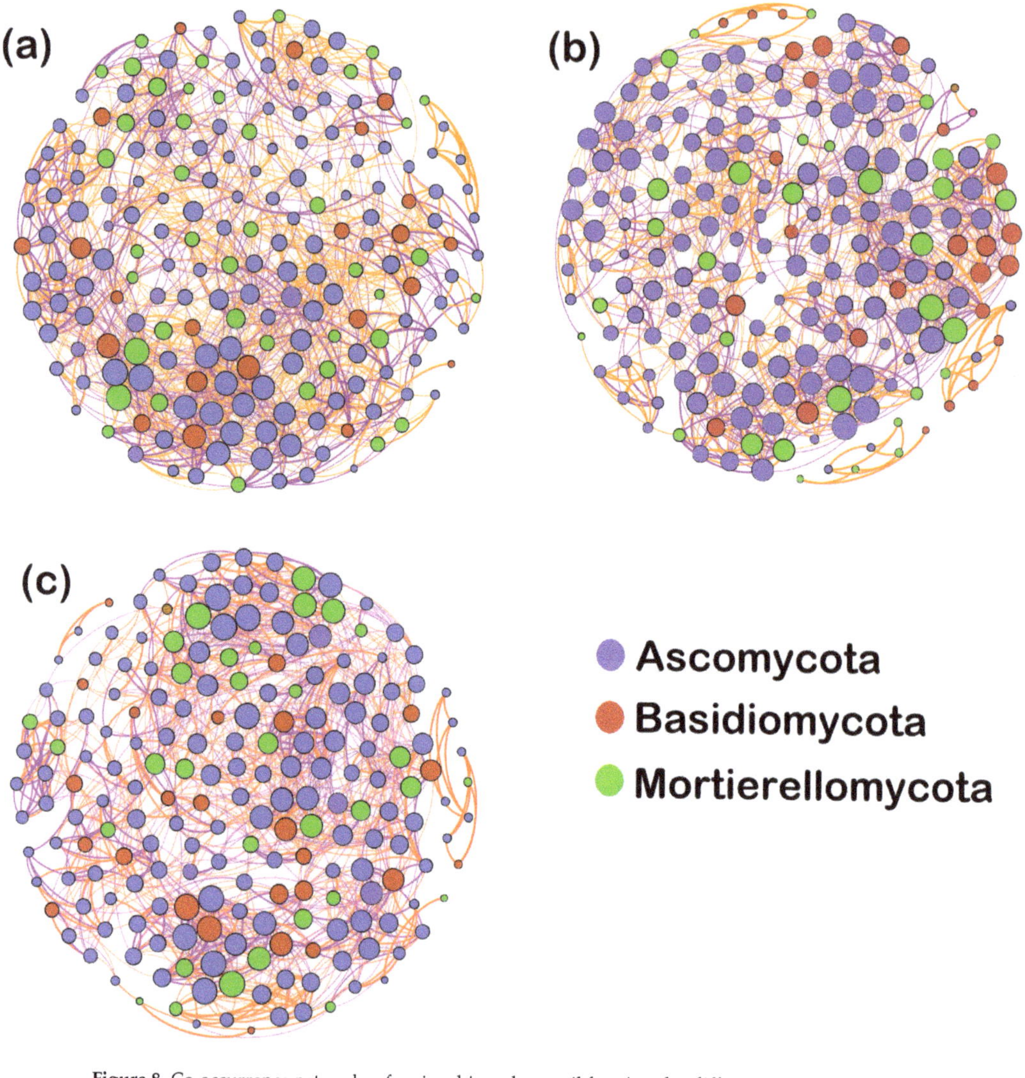

Figure 8. Co-occurrence networks of maize rhizosphere soil fungi under different cropping systems. (**a**) Maize–soybean intercropping, (**b**) maize–soybean rotation, and (**c**) continuous maize cropping. Circles indicate different genera. Genera from the same phylum are labeled with the same color. The size of the circles indicates the average abundance of the genera, lines indicate correlations between two genera, and the line thickness indicates the strength of the correlation. Orange lines indicate positive correlations and purple lines indicate negative correlations.

3.3.5. Prediction of Fungal Function in Rhizosphere Soil

A total of seven metabolic pathways were annotated across the treatments (Figure 9a–c). The saprotroph pathway was dominant across treatments, ranging from 57.15% to 79.78%. The saprotroph pathway was 30.59% and 37.01% more abundant under MSR than under MSI and CMC, respectively. Other fungal pathways showed the opposite trend (CMC > MSI > MSR). To explore the fungal function further, different functional guilds were annotated (Figure 9d). The average proportion of endophytes under MSR was 22.32–54.13% lower than that under MSI and CMC, while the average proportion of wood

10. Zheng, B.C.; Chen, P.; Du, Q.; Yang, H.; Luo, K.; Wang, X.C.; Yang, F.; Yong, T.W.; Yang, W.Y. Soil Organic Matter, Aggregates, and Microbial Characteristics of Intercropping Soybean under Straw Incorporation and N Input. *Agriculture* **2022**, *12*, 1362. [CrossRef]
11. Liu, Q.; Zhao, Y.X.; Li, T.; Chen, L.; Chen, Y.Q.; Sui, P. Changes in soil microbial biomass, diversity, and activity with crop rotation in cropping systems: A global synthesis. *Appl. Soil Ecol.* **2023**, *186*, 9. [CrossRef]
12. Zhao, J.; Yang, Y.D.; Zhang, K.; Jeong, J.; Zeng, Z.H.; Zang, H.D. Does crop rotation yield more in China? A meta-analysis. *Field Crop. Res.* **2020**, *245*, 107659. [CrossRef]
13. Li, S.J.; Ye, S.; Liu, Z.Q.; Hassan, M.U.; Huang, G.Q.; Zhou, Q. How does intercropping contribute to soil organic carbon accumulation? A global synthesis. *Agric. Ecosyst. Environ.* **2024**, *374*, 10. [CrossRef]
14. Xu, Z.; Li, C.J.; Zhang, C.C.; Yu, Y.; van der Werf, W.; Zhang, F.S. Intercropping maize and soybean increases efficiency of land and fertilizer nitrogen use; A meta-analysis. *Field Crop. Res.* **2020**, *246*, 107661. [CrossRef]
15. Li, Q.M.; Zhang, D.; Zhang, J.Z.; Zhou, Z.J.; Pan, Y.; Yang, Z.H.; Zhu, J.H.; Liu, Y.H.; Zhang, L.F. Crop rotations increased soil ecosystem multifunctionality by improving keystone taxa and soil properties in potatoes. *Front. Microbiol.* **2023**, *14*, 14. [CrossRef] [PubMed]
16. Cui, J.J.; Li, S.; Baoyin, B.; Feng, Y.D.; Guo, D.Y.; Zhang, L.Q.; Gu, Y. Maize/Soybean Intercropping with Straw Return Increases Crop Yield by Influencing the Biological Characteristics of Soil. *Microorganisms* **2024**, *12*, 1108. [CrossRef] [PubMed]
17. Faust, K. Open challenges for microbial network construction and analysis. *Isme J.* **2021**, *15*, 3111–3118. [CrossRef] [PubMed]
18. Jin, W.; Liu, Z.T.; Cheng, Z.R.; Wang, Q.; Hu, W.; Chen, B.L.; Meng, Y.L.; Zhou, Z.G. The trade-off between root growth redundancy and premature senescence under different straw returning modes affects boll formation and seedcotton yield. *Eur. J. Agron.* **2024**, *156*, 12. [CrossRef]
19. Mwila, M.; Silva, J.V.; Kalala, K.; Simutowe, E.; Ngoma, H.; Nyagumbo, I.; Mataa, M.; Thierfelder, C. Do rotations and intercrops matter? Opportunities for intensification and diversification of maize-based cropping systems in Zambia. *Field Crop. Res.* **2024**, *314*, 13. [CrossRef]
20. Tarekegne, A.; Wegary, D.; Cairns, J.E.; Zaman-Allah, M.; Beyene, Y.; Negera, D.; Teklewold, A.; Tesfaye, K.; Jumbo, M.B.; Das, B.; et al. Genetic gains in early maturing maize hybrids developed by the International Maize and Wheat Improvement Center in Southern Africa during 2000-2018. *Front. Plant Sci.* **2024**, *14*, 17. [CrossRef]
21. Li, B.B.; Chen, X.M.; Deng, T.; Zhao, X.; Li, F.; Zhang, B.C.; Wang, X.; Shen, S.; Zhou, S.L. Timing effect of high temperature exposure on the plasticity of internode and plant architecture in maize. *J. Integr. Agric.* **2024**, *23*, 551–565. [CrossRef]
22. Bian, D.H.; Jia, G.P.; Cai, L.J.; Ma, Z.Y.; Eneji, A.E.; Cui, Y.H. Effects of tillage practices on root characteristics and root lodging resistance of maize. *Field Crop. Res.* **2016**, *185*, 89–96. [CrossRef]
23. Benitez, M.S.; Ewing, P.M.; Osborne, S.L.; Lehman, R.M. Rhizosphere microbial communities explain positive effects of diverse crop rotations on maize and soybean performance. *Soil Biol. Biochem.* **2021**, *159*, 13. [CrossRef]
24. Liu, L.; Han, Q.L.; Liu, C.M.; Yang, J.; Liu, L.; Wu, J.R.; Li, C.Y. The Dynamics of Soil Microbe Metabolic Function Diversity in the Root-Zone of Maize-Soybean Intercropping. *Int. J. Agric. Biol.* **2019**, *21*, 639–647. [CrossRef]
25. Nurgi, N.; Tana, T.; Dechassa, N.; Alemayehu, Y.; Tesso, B. Effects of planting density and variety on productivity of maize-faba bean intercropping system. *Heliyon* **2023**, *9*, 12. [CrossRef] [PubMed]
26. Faghihinia, M.; Jansa, J.; Halverson, L.J.; Staddon, P.L. Hyphosphere microbiome of arbuscular mycorrhizal fungi: A realm of unknowns. *Biol. Fertil. Soils* **2023**, *59*, 17–34. [CrossRef]
27. He, D.X.; Yao, X.D.; Zhang, P.Y.; Liu, W.B.; Huang, J.X.; Sun, H.X.; Wang, N.; Zhang, X.J.; Wang, H.Y.; Zhang, H.J.; et al. Effects of continuous cropping on fungal community diversity and soil metabolites in soybean roots. *Microbiol. Spectr.* **2023**, *11*, e01786-23. [CrossRef]
28. Gahagan, A.C.; Shi, Y.C.; Radford, D.; Morrison, M.J.; Gregorich, E.; Aris-Brosou, S.; Chen, W. Long-Term Tillage and Crop Rotation Regimes Reshape Soil-Borne Oomycete Communities in Soybean, Corn, and Wheat Production Systems. *Plants* **2023**, *12*, 2338. [CrossRef]
29. Wang, Q.F.; Zhou, D.P.; Chu, C.B.; Zhao, Z.; Ma, M.C.; Wu, S.H. The choice of rice rotation system affects the composition of the soil fungal community and functional traits. *Heliyon* **2024**, *10*, 12. [CrossRef]
30. Niewiadomska, A.; Majchrzak, L.; Borowiak, K.; Wolna-Maruwka, A.; Waraczewska, Z.; Budka, A.; Gaj, R. The Influence of Tillage and Cover Cropping on Soil Microbial Parameters and Spring Wheat Physiology. *Agronomy* **2020**, *10*, 200. [CrossRef]
31. Shi, Y.X.; He, Y.S.; Zheng, Y.X.; Liu, X.X.; Wang, S.Z.; Xiong, T.E.; Wen, T.; Duan, H.; Liao, X.L.; Cui, Q.R.; et al. Characteristics of the phyllosphere microbial community and its relationship with major aroma precursors during the tobacco maturation process. *Front. Plant Sci.* **2024**, *15*, 14. [CrossRef]
32. Wang, Q.; Liang, A.Z.; Chen, X.W.; Zhang, S.X.; Zhang, Y.; McLaughlin, N.B.; Gao, Y.; Jia, S.X. The impact of cropping system, tillage and season on shaping soil fungal community in a long-term field trial. *Eur. J. Soil Biol.* **2021**, *102*, 10. [CrossRef]
33. Nieves-Campos, E.I.; Méndez-Bravo, A.; Pérez-Bautista, Y.; Llanderal-Mendoza, J.; Guevara-Avendano, E.; Solís-García, I.A.; Diyarza-Sandoval, N.A.; Contreras-Ramos, S.M.; Rodriguez-Campos, J.; Méndez-Bravo, A.; et al. Anti-oomycete activity and plant growth promoting properties of avocado fungal endophytes. *Rhizosphere* **2024**, *31*, 10. [CrossRef]
34. Peltoniemi, K.; Velmala, S.; Lloret, E.; Ollio, I.; Hyvönen, J.; Liski, E.; Brandt, K.K.; Campillo-Cora, C.; Fritze, H.; Iivonen, S.; et al. Soil and climatic characteristics and farming system shape fungal communities in European wheat fields. *Agric. Ecosyst. Environ.* **2024**, *370*, 12. [CrossRef]

35. Nicola, L.; Landínez-Torres, A.Y.; Zambuto, F.; Capelli, E.; Tosi, S. The Mycobiota of High Altitude Pear Orchards Soil in Colombia. *Biology-Basel* **2021**, *10*, 1002. [CrossRef]
36. Lu, Q.S.; Hu, C.L.; Cai, L.N.; Wu, C.F.; Zhang, H.Q.; Wei, L.; Zhang, T.Y.; Hu, H.C.; Liu, S.; Lei, J.J.; et al. Changes in soil fungal communities after onset of wheat yellow mosaic virus disease. *Front. Bioeng. Biotechnol.* **2022**, *10*, 13. [CrossRef]

microorganisms

MDPI

Article

Driving Factors Influencing Soil Microbial Community Succession of Coal Mining Subsidence Areas during Natural Recovery in Inner Mongolia Grasslands

Dongqiang Lu, Zhen Mao *, Yan Tang, Bo Feng and Liang Xu

School of Environment Science and Spatial Informatics, China University of Mining and Technology, Xuzhou 221116, China; ludongqiang@cumt.edu.cn (D.L.); tangyan@cumt.edu.cn (Y.T.); fengbo@cumt.edu.cn (B.F.); xuliang0911@cumt.edu.cn (L.X.)

* Correspondence: maozhen@cumt.edu.cn

Abstract: Soil microorganisms significantly influence the energy flow and material cycle of soil ecosystems, making them highly susceptible to environmental changes, such as those induced by mining activities. Studying the succession of soil microbial communities after mining subsidence is crucial for comprehending the significance of soil microbes in the natural recovery process following subsidence. Therefore, the soil properties, vegetation communities, and soil microbial communities of the subsidence area, as well as unexploited areas, were analyzed during the natural restoration process (1, 2, 5, 10, and 15 years). The results demonstrate that mining subsidence has a significant impact on the aboveground vegetation community, soil properties, and microbiological community. Following an extended period of natural recovery, a new stable state has emerged, which differs from that observed in non-subsidence areas. The total nitrogen, nitrate nitrogen, and ammonium nitrogen amounts may be key factors driving the natural recovery of bacterial communities, and total potassium and available potassium may be key factors driving the natural recovery of fungal communities. The natural recovery mechanism of soil microorganisms was analyzed along with the changes related to vegetation and soil physicochemical properties. The mechanism was explained from three perspectives, namely, plant-led, soil-led, and soil-microbial-led, which could provide a theoretical basis for the natural restoration of grassland ecosystems and provide guidance for the treatment of coal mining subsidence areas.

Keywords: coal mining subsidence areas; driving factors; grassland; microbial community diversity; natural recovery

1. Introduction

As one of the most important material energy sources, coal accounts for about 74% of China's primary energy consumption. The rapid development of the coal mining industry plays a vital role in promoting the development of human society and civilization [1,2]. As the world's largest coal producer, statistically, China mines about 4.56 billion tons of coal annually, accounting for more than half of the world's total output [3]. A total of 96% of China's coal mining is underground mining, and 4% is open-pit mining. However, coal hollowing causes serious disturbances to the ecosystem surrounding the mining area, and environmental factors change accordingly; these disturbances introduce a series of geological disasters and environmental problems, such as surface subsidence, ground fissures, ecological landscape degradation, and groundwater pollution [4–6]. As the demand for minerals and energy continues and as mining activities continue to evolve, coal mining subsidence areas continue to increase at a rapid rate [7–9].

With the emergence and development of high-volume sequencing technology, the role of the soil microbial community in mine remediation has gradually attracted attention [10].

By comparing the vegetation characteristics, microbial diversity, and microbial community structure in the coal mining subsidence area with the unexploited area, the recovery status of the coal mining subsidence area in the mining area can be evaluated [11,12]. At present, there have been many studies on the ecological recovery of coal mining subsidence sites around the world. Some studies have shown that with the restoration of vegetation in the collapse area, the community structure and function of soil bacteria will change with changes in vegetation; the physical and chemical properties, as well as the microbial community diversity of soil, are positively correlated with soil nutrients but negatively correlated with soil physical properties and soil pH values at various developmental stages [13,14]. Relevant studies have shown that the organic matter and nutrients in soil from disturbed areas have been greatly reduced [15]. The assessment of the microbial community structure in coal mining subsidence areas shows that when the microbial community structure is damaged, the relevant material cycle of the soil and the sustainability of the ecosystem will be affected to varying degrees [16]. Disturbances cause a decline in soil microbial diversity and affect the activity of related microorganisms, thereby affecting energy flow and material cycle and ultimately reducing soil nutrients [17].

Studies have shown that after 5 years of vegetation reconstruction, the diversity of bacterial communities in artificial and natural restoration areas does not reach pre-subsidence index values. However, after 15 years of vegetation reconstruction, the diversity of bacterial communities can reach pre-subsidence levels, and compared with natural restoration, artificial remediation can accelerate the restoration of soil variables [18]. Different research has demonstrated variations in microbial diversity, abundance, and functions between soils that have been reclaimed and those that have undergone natural restoration in subsidence areas [19–21]. However, over time, microbial diversity and ecosystem sustainability can be restored to be as close to the pre-disturbance state as possible through artificial restoration [22]. However, most of these studies focus on artificial restoration in collapse areas and less on the differences in soil microbial diversity in coal mining collapse areas under natural recovery conditions across different recovery years. Grassland ecosystems have remarkable resilience; studying the natural restoration of grassland vegetation and soil microorganisms after subsidence is significant in guiding artificial restoration.

Therefore, this project focuses on a grassland mining area. Soil samples with different subsidence years (1, 2, 5, 10, and 15 years) that have not been artificially restored and unexploited areas were selected as the research objects. This research mainly focuses on two aspects: (1) investigating the mechanism of the vegetation community, soil properties, and the soil microbial community during natural recovery after subsidence and (2) examining the main driving factors of the natural recovery mechanism in the soil microbial community.

2. Materials and Methods

2.1. Study Area

The grassland mining area selected by this institute is the Zhalainuoer mining area, located in the west of Hulunbuir City, Inner Mongolia Autonomous Region, administratively under the jurisdiction of Hulunbuir City and Manzhouli City. The mining area is about 23.8 km long from north to south, 22.83 km wide from east to west, and covers an area of 543.43 km^2. The geographical coordinates are 49°19′~49°46′ north latitude and 117°12′~117°53′ east longitude (Figure 1).

This area belongs to the moderate temperate semi-arid continental monsoon climate, with cold winters and hot summers, temperature changes, a freezing period of up to 7 months, and a summer of only about 5 months. According to Manzhouli City and Zhaqu meteorological station data, the lowest temperature is −42.7 °C, the highest temperature is +37.8 °C, and the annual average temperature is generally 0.2 to −2 °C; the evaporation is the largest from April to September, and the annual evaporation is generally 1200~1500 mm. The southwest wind is the most common throughout the whole year, followed by the northwest wind in June, July, and August. The wind speed is generally 2–5 m/s, and the maximum wind speed is 20 m/s. The wind is the lowest in January, February, and December. The wind is the strongest in March, April, May, and November [23].

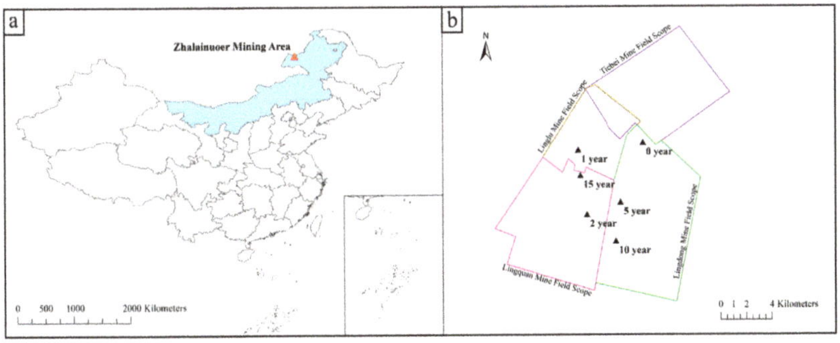

Figure 1. (**a**,**b**) Schematic map of the study area.

2.2. Research Methods

In July 2022, under the leadership and guidance of staff familiar with the situation of the mining area, areas with five subsidence years of 1, 2, 5, 10, and 15 years without artificial restoration and one unexploited area were selected as the sampling group.

Before sampling, the site was investigated, and related data were collected. The sampling range was divided into sampling units according to soil type, topography, and other factors, and the soil of each sampling unit was kept as consistent as possible. According to the size of the subsidence plot on site, we set the number of sampling points to 6 to ensure the greatest possible representativeness of the sampling. At the same time, sampling was carried out according to the principle of "multi-point mixing". All samples were taken in duplicate for further analysis. One fresh sample was stored cryogenically in sterile centrifuge tubes to determine soil microorganisms, and the other was used to determine soil physicochemical properties. The removed surface plants were weighed fresh in sterile bags, then taken back to the laboratory to dry to a constant weight, and their dry weight was measured.

2.3. Sample Collection and Processing Methods

2.3.1. Plant Samples

To clarify the effect of plant community development on soil properties and bacterial community changes, two duplicate quadrats (1×1 m grassland communities) were randomly selected in each plot to investigate the component and diversity of the plant community. We recorded all herb species, names, quantity, coverage, and maximum/average height in each quadrat (Table S1). The plant coverage was estimated visually by observers. A single plant was used to calculate the number of individuals of each species in each plot, and the plant density of each plot was calculated. Additionally, aboveground portions of all vegetation were cut and dried at 70 °C for 48 h to obtain the aboveground biomass. The species Pielou uniformity (E) and the Shannon–Wiener index (H) of the plant communities were calculated as a characterization of plant community diversity.

2.3.2. Soil Samples

Soil moisture (SW) and pH of samples were measured simultaneously on-site using the soil temperature, moisture, salt, and pH velocity tester (model: HM-WSYP). After removing the topsoil, the soil sample was collected with a depth of 0~10 cm, and the mixed soil sample was made for subsequent analysis. The soil sample was sealed with a ziplock bag and returned to the laboratory for pretreatment to determine physical and chemical properties. Soil organic matter (SOM) was detected by potassium dichromate oxidation–external heating. The semi-trace Kjeldahl method was used to detect total nitrogen (TN) in soil. Total phosphorus (TP) and total potassium (TK) were tested separately using the sodium hydroxide melt–molybdenum antimony anti-colorimetric method and sodium hydroxide molten flame photometry [24]. Alkali-hydrolyzable nitrogen (AN), available phosphorus (AP), and available potassium (AK) were determined according to the methods of Olsen [25] and Colwell [26], respectively.

2.3.3. Soil Microbial Samples

(1) DNA extraction and PCR products' amplification

Total microbial genomic DNA was extracted from soil samples using the E.Z.N.A.® soil DNA Kit (Omega Bio-tek, Norcross, GA, USA) following the product manuals. Before being used further, the DNA was stored at −80 °C. Its quality and concentration were assessed using a 1.0% agarose gel electrophoresis and a NanoDrop2000 spectrophotometer (Thermo Scientific, Waltham, MA, USA). The hypervariable region V3-V4 of the bacterial *16S rRNA* gene was amplified with primer pairs 338F (5′-ACTCCTACGGGAGGCAGCAG-3′) and 806R (5′-GGACTACHVGGGTWTCTAAT-3′) [27] using a T100 Thermal Cycler PCR thermocycler (BIO-RAD, Hercules, CA, USA). The ITS1-1F-F (CTTGGTCATTTAGAGGAAGTAA) and ITS1-1F-R (GCTGCGTTCTTCATCGATG C) were used for a PCR of the fungi ITS. The PCR reaction mixture included 0.8 μL of each primer (5 μM), 4 μL 5 × Fast Pfu buffer, 2 μL 2.5 mM dNTPs, 10 ng of template DNA, 0.4 μL Fast Pfu polymerase, and ddH_2O to a final volume of 20 μL. The following conditions were used for the PCR amplification cycle: a 3-min initial denaturation at 95 °C, 27 cycles of denaturing at 95 °C for 30 s, annealing at 55 °C for 30 s, extending at 72 °C for 45 s, a single extension at 72 °C for 10 min, and concluding at 4 °C. Following the manufacturer's instructions, the PCR product was extracted from a 2% agarose gel and purified using the PCR Clean-Up Kit (YuHua, Shanghai, China) and quantified using Qubit 4.0 (Thermo Fisher Scientific, Waltham, MA, USA).

(2) Illumina PE300/PE250 sequencing

Using an Illumina PE300/PE250 platform (Illumina, San Diego, CA, USA) and standard techniques from Majorbio Bio-Pharm Technology Co. Ltd. (Shanghai, China), purified amplicons were pooled in equimolar quantities and paired-end sequenced.

(3) Amplicon sequence processing and analysis

Fastp (0.19.6) was utilized for quality filtering the resultant sequences, and FLASH (v1.2.11) was employed for their combination [28]. Subsequently, the DADA2 plugin in the Qiime2 pipeline [29] (version 2020.2) was utilized for the de-noising of the high-quality sequences, resulting in a single-nucleotide resolution based on error patterns within samples. These variants of amplicon sequences are commonly referred to as DADA2-denoised sequences (ASVs). To minimize the impact of sequencing depth on alpha and beta diversity measurements, the number of sequences from each sample was filtered to 20,000. This process yielded an average Good's coverage of 97.90%. Taxonomic assignment of ASVs was carried out using the SILVA 16S rRNA database (v138), and a consensus taxonomy classifier built in Qiime2.

2.4. Data Processing and Analysis

SPSS and Origin were used to process and draw the plant-related and soil physical and chemical property data.

The Majorbio Cloud platform (https://cloud.majorbio.com, accessed on 1 March 2023) was used for bioinformatic analysis. To calculate rarefaction curves and alpha diversity indices based on the data from the ASVs, including observed ASVs, Chao1 richness, the Shannon index, and Good's coverage, Mothur v1.30.1 [30] was utilized. Principal coordinate analysis (PCoA) based on the Bray–Curtis dissimilarity was conducted using the Vegan v2.5-3 software to assess the similarities between the microbial communities in various samples. Subsequently, the statistical significance of the variation explained by the treatment and its percentage was determined by conducting the PERMANOVA test using the Vegan v2.5-3 package. Additionally, LefSe [31] (http://huttenhower.sph.harvard.edu/

LEfSe, accessed on 1 March 2023) was utilized to perform a linear discriminant analysis effect size (LDA > 2, $p < 0.05$) in order to identify bacterial taxa with substantial differences in phylum to genus abundance amongst various groups.

3. Results

3.1. Natural Recovery of Vegetation after Subsidence in Grassland Mining Areas

As shown in Figure 2, the species composition of the study area was mainly Asteraceae, legumes, and amaranth plants. Compared to one or two annual herbaceous species, there were substantially more perennial herbaceous species. The number of sample species after subsidence was greater than that of non-subsidence sample species; it first increased and then decreased with the increase in subsidence time. As the subsidence time increases, the vegetation cover and biomass recover year by year. There were significant differences in coverage and biomass between the non-subsidence and the subsidence plot after 1, 2, and 5 years, and there was no significant difference between the non-subsidence plot and the subsidence plot after 10 and 15 years. Compared with the non-subsidence group, the vegetation cover after 10 years of subsidence exceeded that of the non-subsidence group, but its biomass did not show the same regularity. Subsidence significantly reduced vegetation species diversity; both the H and E indexes increased first and then slowly decreased with the increase in subsidence time, indicating that the community gradually reached a relatively stable stage. The difference was that the H index had exceeded the non-subsidence group after 2 years of subsidence, but the E index in the subsidence group was lower than that of the non-subsidence group. This finding shows that, during the natural recovery process, it is difficult for the plant community to recover after subsidence to the same state it was in when it was not submerged.

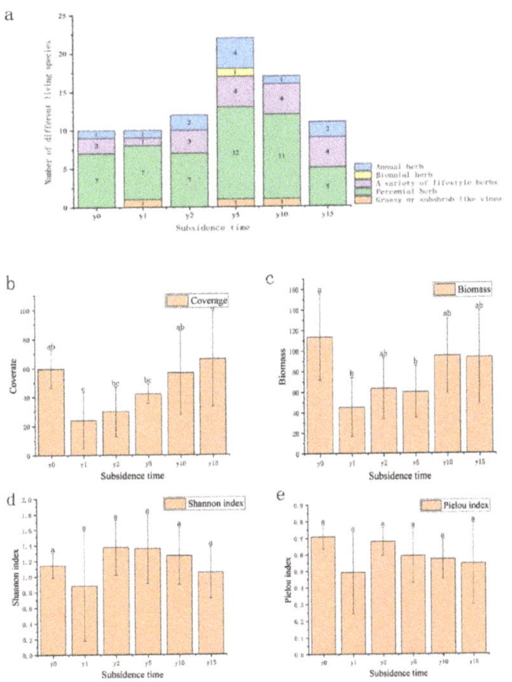

Figure 2. (**a**) Variation diagram of species quantity and subsidence time of different lifeforms. (**b**) Variation diagram of plant coverage and subsidence time. (**c**) Variation diagram of biomass and subsidence time. (**d**) Variation diagram of Shannon index under different subsidence times. (**e**) Variation diagram of Pielou index under different subsidence times. Different letters indicate significant differences.

3.2. Changes in Soil Physical–Chemical Properties after Subsidence in Grassland Mining Areas

As shown in Figure 3, the area was mainly composed of sand grains (62.90%) before being disturbed, with a fine sand content of 27.58%, a powder content of 4.15%, and a clay content of 5.37%. The coarse sand grain content decreased significantly due to subsidence; with the increase in subsidence time, the soil texture gradually changed from clay to sand, but it was sandy loam after 15 years of subsidence and did not recover to sand. The soil water content of most samples was between 15% and 30%; it exhibits a pattern of initially increasing before subsequently decreasing, reaching a maximum of 27.33% at 2 years of subsidence and then decreasing year by year, except for 10 years of subsidence with a soil water content of 60%. Soil water content after 10 years of subsidence was significantly different from other subsidence years and the non-subsidence soil. The soil pH in this area is mainly between 6 and 8 and is classified as neutral soil. Compared with non-subsidence soil (7.43), the soil pH was relatively stable after subsidence.

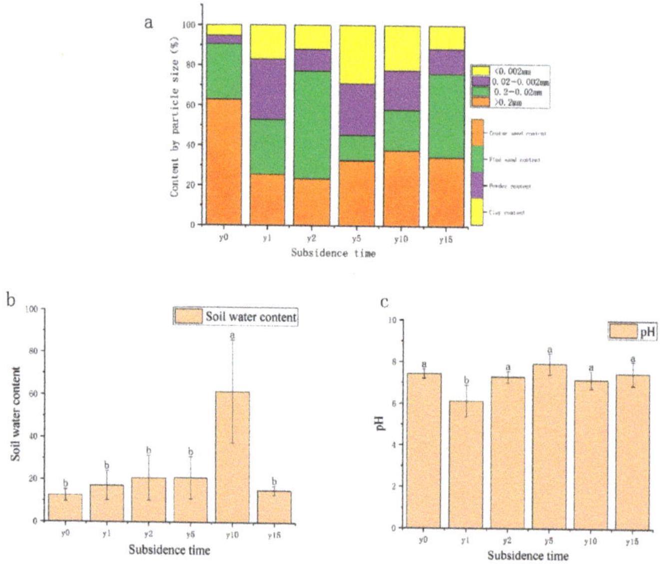

Figure 3. (**a**) Variation diagram of mechanical composition and subsidence time. (**b**) Variation diagram of water content and subsidence time. (**c**) Variation diagram of soil pH and settlement time. Different letters indicate significant differences.

As shown in Figure 4, compared with the non-subsidence soil, the soil organic matter (SOM) content after settlement was significantly higher than that of the non-subsidence soil and then decreased roughly year by year, except for subsidence 2 years. There was little overall change in total soil total nitrogen (TN) and total phosphorus (TP) content, except for a significant decrease in total nitrogen and a significant increase in total phosphorus in the second year of subsidence. For total potassium, there was a significant increase in total potassium content in the soil after subsidence.

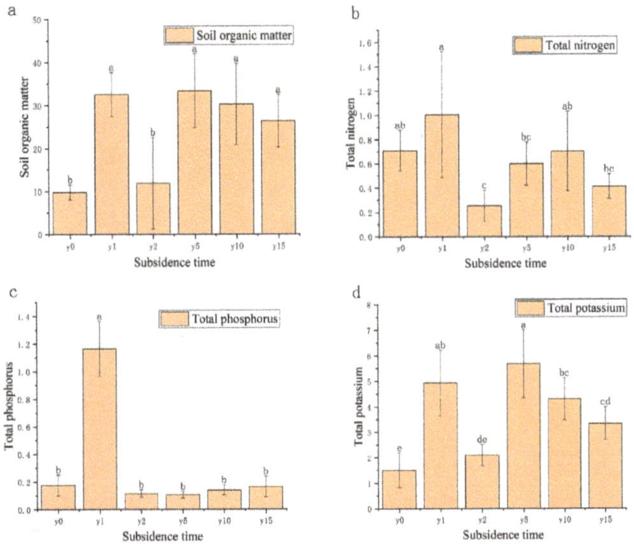

Figure 4. (**a**) Variation diagram of soil organic matter and subsidence time. (**b**) Relationship between soil total nitrogen, nitrate nitrogen, and ammonium nitrogen and subsidence time. (**c**) Relationship between soil total phosphorus, available phosphorus, and subsidence time. (**d**) Relationship between soil total potassium, available potassium, and subsidence time. Different letters indicate significant differences.

3.3. Changes in the Soil Microbial Community after Subsidence in Grassland Mining Areas

As presented in Table 1, subsidence will lead to an increase in the number of bacteria and fungi. However, the number of bacteria is lower than that in the non-subsidence group after 5 years of subsidence, and the number of fungi is greater than that in the non-subsidence group. The number of bacterial sequences in the early stage of subsidence was the largest, and then it showed a downward trend year by year. However, while the number of fungal sequences increased after subsidence, there was no obvious trend with the subsidence time.

Table 1. ASV quantity in high-throughput sequencing.

Constituencies	y0	y1	y2	y5	y10	y15
Bacteria	25567 ± 1677^{ab}	27869 ± 5611^{a}	25807 ± 3405^{ab}	22675 ± 2630^{ab}	24452 ± 1858^{ab}	22298 ± 3293^{b}
Fungi	38866 ± 1682^{b}	53242 ± 4685^{a}	40946 ± 5386^{b}	45588 ± 7322^{ab}	43380 ± 6523^{ab}	49403 ± 4410^{a}

Note: a,b: in the same column, values with different superscript letters differed significantly ($p < 0.05$).

As shown in Figure 5, subsidence is a form of negative feedback for bacteria; that is, subsidence weakens the activity intensity of bacterial communities and reduces their diversity as a whole. In contrast, subsidence is a form of positive feedback for fungi; that is, subsidence enhances the activity intensity of fungi and increases the diversity of fungal communities. The different subsidence times affected the community structure of soil bacteria and fungi, but the difference was not obvious. For bacterial communities, the similarity between bacterial communities at 15 years of subsidence and non-subsidence soil microbial communities was higher. Bacterial microbial communities with 1 year of subsidence were unstable, and their sample sites were relatively scattered; meanwhile, the soil bacterial communities of 2 and 5 years of subsidence were more similar, but the bacterial communities at 10 years of subsidence were significantly different and differentiated from those of other years to a certain extent, indicating that the bacterial community had a large change after 10 years of subsidence. However, bacterial communities showed no obvious trend with the increase in subsidence time. For fungal communities, the soil fungal community after subsidence was very different from the non-subsidence soil fungal community, and the fungal community after 15 years of subsidence was closer to the microbial communities of 1, 2, and 5 years of subsidence, indicating that with the advancement of natural recovery time, the fungal community did not change greatly.

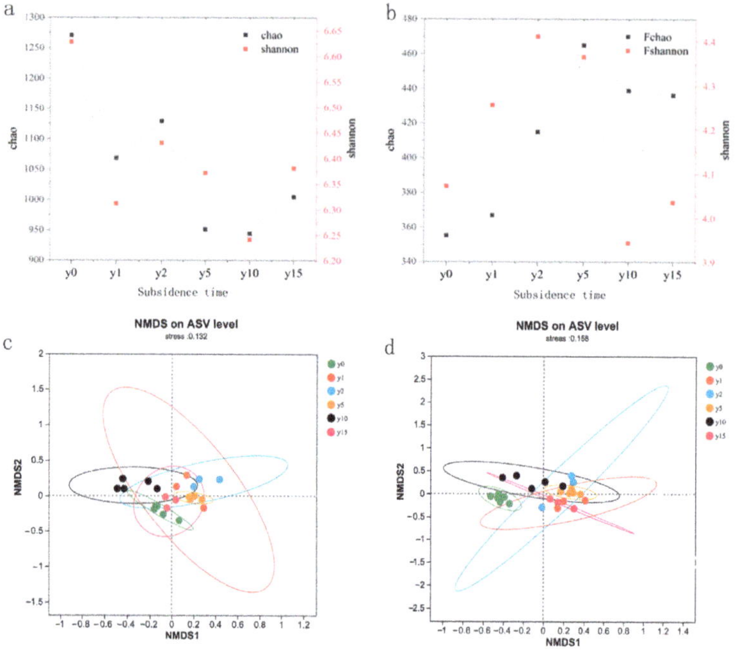

Figure 5. (**a**) Bacterial alpha diversity index of different subsidence times. (**b**) Fungal alpha diversity index of different subsidence times. (**c**) NMDS analysis of soil bacteria at different subsidence times. (**d**) NMDS analysis of soil fungi at different subsidence times.

As shown in Figure 6, for bacteria, at the phylum level, the dominant bacterial phylum are *Actinobacteria*, *Proteobacteria*, *Acidobacteria*, *Chloroflexi*, and *Firmicutes*, which together account for more than 90% of the bacterial population. Among them, *Actinobacteria* is the most dominant phylum, accounting for about 35%; *Proteobacteria* is the subdominant phylum, accounting for about 20%; *Acidobacteria* is slightly lower than *Proteobacteria* overall, accounting for about 15%; *Acidobacteria* is equal to or slightly higher than Proteobacteria in some subsidence year groups; the proportion of *Chloroflexi* fluctuates around 10%; and *Firmicutes* differed significantly in different groups, ranging from 2% to 20%. The *Actinobacteria* phylum showed the greatest changes during natural recovery after subsidence. In the first year just after subsidence, the *Actinobacteria* phylum increased dramatically with its proportion accounting for almost half of all bacteria. Subsequently, it recovered to almost the same level as it had before subsidence. In contrast, the *Firmicutes* phylum gradually decreases during the 1–2 years of subsidence, after which it gradually increases with natural recovery, ultimately returning to a level almost identical to that before the subsidence.

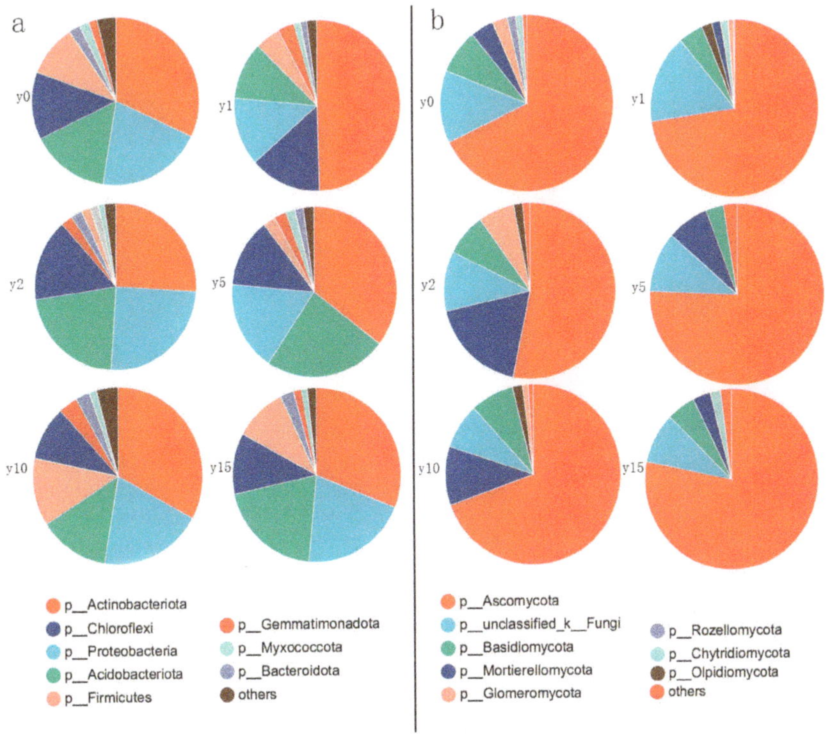

Figure 6. Relative abundance of bacterial phyla (**a**) and fungal phyla (**b**) at different subsidence times.

For fungi, at the phylum level, the dominant phyla are *Ascomycota*, *Mortierellomycota*, *Basidiomycota*, and *Glomeromycota*. *Ascomycetes* in all groups were above 67%, except for the soil 2 years after subsidence, where the percentage of *Ascomycetes* was only 53%. The percentage of *Ascomycota* after 15 years of natural restoration reached 78%, which is even higher than the level before subsidence. During the period of subsidence, the percentage of *Mortierellomycota* initially experienced a decrease in levels, followed by an increase, and ultimately returned to levels similar to those before the subsidence occurred. *Basidiomycota* accounts for an average of about 8% and did not change much during the subsidence and natural restoration process. *Glomeromycota* can account for up to 7% at 2 years of subsidence, but it is almost impossible to find after 15 years of natural recovery.

3.4. Key Factors in the Natural Recovery of the Soil Microbial Community

The results of the RDA analysis of 14 environmental factors, namely, biomass, species number (SN), coverage, E, SW, pH, mechanical composition, coarse sand content, and soil chemical properties (i.e., organic matter, total nitrogen, nitrate nitrogen, ammonium nitrogen, available phosphorus, total potassium, and available potassium), are shown in Figure 7. For bacteria, the explanatory degrees of the first and second axes of RDA were 27.54% and 22.08%, respectively. The first axis was mainly composed of SW, pH, SOM, NH_4-N, TK, AK, E, coverage, and biomass, and the second axis mainly comprised fine sand content and AP. For fungi, the explanatory degrees of the first and second axes of the RDA were 47.75% and 7.73%, respectively. The first axis was mainly composed of SOM, TK, and AK, and the second axis was mainly composed of SW, E, SN, and biomass.

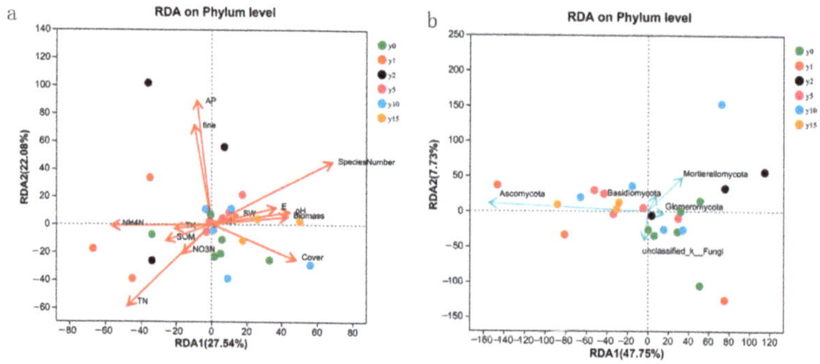

Figure 7. RDA analysis of bacteria (**a**) and fungi (**b**) samples and environmental factors at different subsidence times.

The SN, coverage, TN, NH_4-N, fine sand content (fine), and AP had a large impact on bacterial samples, among which AP was positively correlated with fine sand content, NH_4N was positively correlated with TN, SN was positively correlated with coverage, and AP/fine sand content had no correlation with SN/NH_4N and was negatively correlated with coverage/TN. The influence of all environmental factors on fungi was more significant. The sample points with different subsidence times were scattered and did not have obvious characteristics with environmental factors.

Soil microorganisms were analyzed for the Spearman correlation between species and environmental factors at the phylum classification level, and correlation heat maps were obtained. As shown in Figure 8, the dominant bacterial phylum, *Actinobacteria*, was significantly positively correlated with TN, very positively correlated with NO_3-N and NH_4-N, and negatively correlated with the number of species. There was no obvious correlation between *Proteobacteria* and environmental factors; *Acidobacteria* was significantly positively correlated with biomass; and *Chloroflexi* was significantly positively correlated with NO_3-N and negatively correlated with AK and species number. There was no obvious correlation between *Firmicutes* and environmental factors. There was no significant correlation between *Firmicutes* and environmental factors in the phylum, with significantly different phylum abundances observed between different groups. *Gemmatimonadetes* were significantly correlated with several environmental factors, such as being highly positively correlated with SW, SOM, TK, and AP and positively correlated with AK and SN. They were negatively correlated with fine mechanical composition, and E. *Desulfobacterota* was significantly negatively correlated with SOM and AK, very negatively correlated with TK, significantly positively correlated with NH_4-N, and highly positively correlated with biomass. Strong positive correlations were found between *Bdellovibrionota* and NH_4-N and strong negative correlations were found between RCP2-54 and TK and AK. There was a substantial negative correlation found with coverage and biomass and a significant positive correlation with SOM, TK, and AK for *Deinococcota*. Significantly adversely connected with SOM, TK, AK, and AP, extremely negatively correlated with NH_4-N and NO_3-N, and significantly positively linked with E and biomass were the relationships observed between *Spirochaetota* and these variables. *Margulisbacteria* showed a strong positive correlation with NO_3-N and NH_4-N and a substantial negative correlation with SW, TK, and AP.

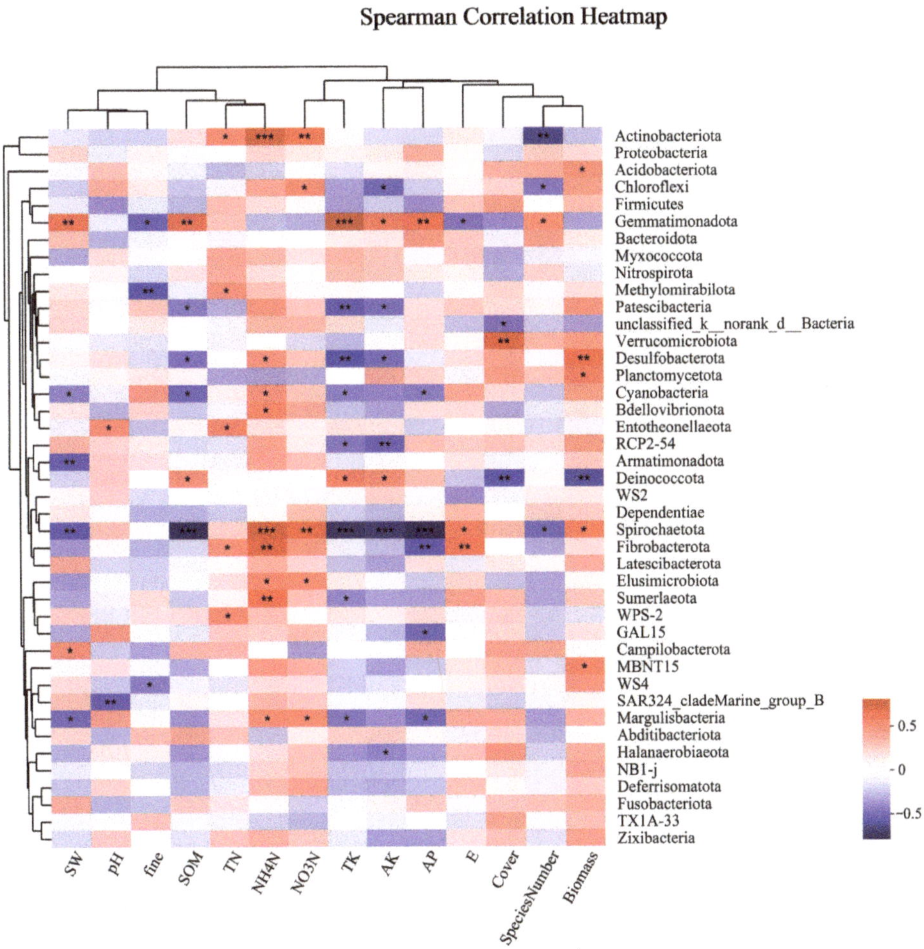

Figure 8. Cluster thermogram of horizontal abundance of soil bacteria microbial community and environmental factors. *: The correlation is significant at the level of 0.05; **: The correlation is significant at the level of 0.01; ***: The correlation is significant at the level of 0.001.

As shown in Figure 9, the dominant phylum of fungi, *Ascomycota*, was significantly positively correlated with TK and AK. *Unclassified_k_Fungi* had no obvious correlation with environmental factors. *Mortierellomycota* was positively correlated with SW, AP, and SN. *Basidiomycota* and *Glomeromycota* were not significantly correlated with environmental factors. *Mortierellomycota* was significantly positively correlated with SW, AP, and SN among the phyla with significantly different phylum abundance. There was a substantial negative correlation between *Rozellomycota* and SOM, while there was a positive correlation between AK and TK and plant biomass and coverage. *Aphelidiomycota* exhibited a substantial negative correlation with NH_4-N, while demonstrating positive correlations with TK, AK, and SN, with AK showing the most positive correlation.

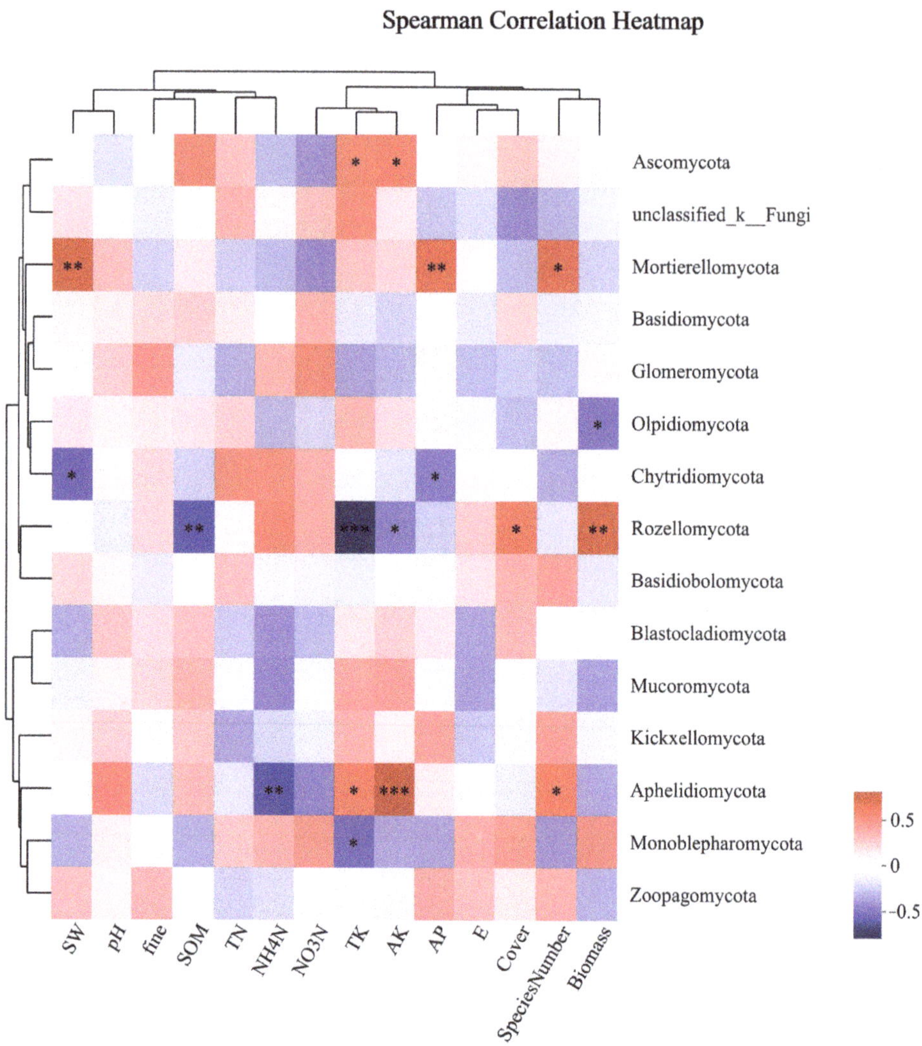

Figure 9. Cluster thermogram of horizontal abundance of soil fungal microbial community and environmental factors. *: The correlation is significant at the level of 0.05; **: The correlation is significant at the level of 0.01; ***: The correlation is significant at the level of 0.001.

4. Discussion

4.1. The Regulation of the Natural Restoration of Vegetation

Vegetation coverage, biomass, and diversity decreased significantly in the early stage of subsidence. These factors gradually increased with the increase in subsidence years, with coverage and diversity basically returning to the pre-subsidence level. However, there was still a certain gap between biomass before and after subsidence, which was consistent with the results of Zhang et al. [32] on vegetation restoration in different subsidence areas. From the perspective of species population, the number of species gradually increased after subsidence, reached a maximum value at 5 years of subsidence, and then gradually decreased. This may be related to the degradation of grasslands caused by subsidence, resulting in the spread of a large number of weeds. The number of species in the subsidence period of 1 year was the lowest, indicating that the grassland was seriously degraded and biodiversity was seriously lost [33]. With the increase in sedimentation time, the dominant species changed from one or two annual herbs to perennial herbs and from single life forms to multiple life forms. Perennial plants have a stronger ability to resist environmental disturbances and maintain population stability than annual plants, so this change in the composition of species also reflects the changes in ecosystem structure and functioning during the process of vegetation restoration; the community structure tends to be stable, and the ecological function is enhanced [34]. However, in general, although natural restoration can partially restore aboveground vegetation, it is difficult to restore the area to its pre-subsidence state; rather, it forms a new homeostasis that is suitable for the environment, which is consistent with the conclusion of Liu et al. [35] that vegetation of a coal mining subsidence area can only be restored to a state close to its original state under natural restoration state.

4.2. The Natural Restoration of Soil Physical and Chemical Properties Changes Regulation

Although it has been reported that coal mining causes a decrease in soil water content [36], subsidence instead causes soil water content to be somewhat higher than before subsidence due to the high groundwater table in this study area. This is especially more pronounced in group year 10, but it may be an individual case. As the method of replacing time with space was used in this study when selecting representative areas at different stages of subsidence, it was very difficult to keep all the conditions consistent in field experiments. Subsidence also leads to changes in the physical structure and chemical form of soil, and the proportion of fine particles with a diameter of less than 0.2 mm in the soil increases significantly after subsidence, thereby increasing the soil's water-holding capacity. At the same time, subsidence also caused the organic matter content, total nitrogen, total phosphorus, and total potassium in the soil to rise significantly after subsidence and then gradually decrease, which is consistent with the conclusion of Wu et al.'s [37] research in a coal mining subsidence area of the semi-arid region of Northwest China. This may be because the cracks and slopes caused by subsidence will promote the movement and loss of surface materials. The slope formed by subsidence will cause the gradual downward transfer of nutrients and fine particles in the soil [10], resulting in a high content of fine particles and nutrients in the soil at the beginning of subsidence that, over time, will be transported alluvially to the bottom of the slope or waterlogged areas. The content of fine particles and nutrients will tend to be stable. Song et al. [38] found a similar phenomenon in their study of subsidence soils in coal mining areas in northern Shaanxi.

Relevant studies have shown [39] that the proportion of soil organic carbon and soil clay particles is significantly positively correlated, and the higher the proportion of clay particles, the stronger the stability of soil organic carbon in the soil. In this study, the ratio of soil organic matter to clay particles also showed the same pattern. At the same time, studies have shown [40] that soil organic carbon is mainly regulated by soil pH and plant biomass. Total soil nitrogen first increased and then decreased with the increase in time, nitrate nitrogen content decreased first and then increased, and ammonium nitrogen content gradually decreased with the increase in subsidence time and then flattened. The results of this study also confirm the positive correlation between soil total nitrogen and soil clay particles [38]. However, with the increase in subsidence time, there was no obvious change trend in soil total phosphorus, but the available phosphorus generally increased. Studies have shown that different phosphorus application rates significantly affect plant photosynthetic characteristics and ultimately affect plant biomass accumulation, but excessive phosphorus content can also produce inhibitory effects [41].

4.3. Changes in Soil Microbial Communities in Natural Restoration

Soil microorganisms are sensitive to environmental changes caused by soil subsidence [42,43], and in this study, it was found that soil subsidence significantly impacts the structure and diversity of microbial communities. Although the number of bacterial species and community diversity recovered over time, these metrics were far lower than those in the non-subsidence areas, which may be because subsidence from coal mining altered the physical and chemical characteristics of the soil, which in turn influenced bacterial colonization in the disturbed areas and decreased the diversity of bacterial communities [44,45]. However, for fungi, subsidence increased the number of fungal species and community diversity. Specifically, the number of species decreased in the later stage, but this decrease was not obvious. The diversity of fungal communities increased rapidly in the first 5 years and then decreased significantly. These results indicate that bacteria and fungi are sensitive to changes in the soil environment caused by subsidence, but there are obvious differences in the reactions between the two, which are also related to the different environmental needs of fungi and bacteria. Bacteria and fungi have different nutritional preferences [46], with bacteria preferring to use easily decomposing carbon sources, which are more susceptible in the early stages of disturbance, while fungi can use carbon sources that are difficult for bacteria to decompose, so the effect of disturbing fungi is not very large [47].

From the perspective of community structure, subsidence also had a certain impact on the community composition of soil bacteria and fungi. Many factors affect the structure of microbial communities, such as environmental and vegetation conditions [48]. In this study, although subsidence led to changes in soil physicochemical properties and vegetation conditions, there was no significant change in the soil microbial community's structure after subsidence and during natural restoration. This is consistent with the community composition results; from the phylum level, whether it is fungi or bacteria, the dominant bacteria do not change at different time stages of natural recovery before and after subsidence. However, the proportions of different dominant bacteria in the whole community are slightly different at different stages of subsidence, which is also a result of the influence of environmental and vegetation conditions during the subsidence process [49]. As one of the most widely distributed bacterial phyla in soil, *Actinobacteria* showed a sudden increase in their percentage in the first year just after subsidence in this study, which may be related to the key ecophysiological role of *Actinobacteria* in decomposing plant residues [50]. Aboveground biomass and cover of vegetation in the first year of subsidence were remarkably reduced compared to the pre-subsidence period, suggesting that there was significant vegetation mortality. The residues of the dead vegetation provided abundant food for the *Actinobacteria*, leading to their proliferation. As natural recovery occurred, the vegetation cover and biomass gradually recovered, and the number of *Actinobacteria* gradually returned to the level before subsidence. Our study also found, during the first 2 years after the subsidence, a gradual decrease in the *Firmicutes* phyla with the degradation of the grassland, which is in agreement with the results of the study in the temperate grassland. With natural recovery, the thick-walled phylum gradually recovered to the pre-sinking level. The largest percentage of fungi are *Ascomycetes*, known for their preference for challenging environments and significant involvement in the decomposition of complex organic matter [51].

4.4. The Role of Soil Microbial Communities in Natural Restoration

In this study, the number of aboveground vegetation species, coverage, fine sand content in the soil, TN, NH_4N, and AP are the dominant factors influencing the bacterial community as can be observed from the results of the RDA analysis and heatmap. Sun et al. [52] also showed that aboveground vegetation had a significant impact on the structure of bacterial communities in soil and pointed out that SOM had the greatest impact on the composition and distribution of bacterial communities, and AP played an important role in the distribution of bacteria. Compared with previous studies, they showed that soil total carbon, AP, and AN strongly promoted soil bacterial community diversity. However, there was little correlation between TN content and bacterial community [53], indicating that the bacterial community structure in soil was related to different nitrogen contents. The results showed that the pH value was the dominant factor in the horizontal structure of bacterial communities in the coal mining subsidence area. However, in this study, the pH change was not obvious, and the relationship with bacterial diversity was not significant, which matched the results of Du et al. [18].

However, there are many dominant factors affecting the fungal community, and almost all the factors investigated, including soil physicochemical properties and vegetation conditions, will have a significant impact on the composition of the fungal community, which is also the reason why the fungal community structure changes greatly after subsidence. Compared with bacteria, aboveground vegetation is more pronounced for fungi, which may be because as vegetation biomass increases, vegetation provides resource heterogeneity to soil microbial communities through various factors such as litter decomposition and root exudates, so higher biomass leads to better coexistence of microbial communities [54]. Similarly, fungi are more sensitive to changes in the soil environment. The dominant fungal phyla include *Ascomycetes*, *Basidiomycetes*, and *Mycobacteria*, similar to the findings of other researchers [55]. Among these, *Ascomycetes* and *Basidiomycetes* are considered to be the main fungal decomposers in soils due to their broad-spectrum degradation capacity [56]. *Ascomycetes* dominate soils globally due to their ability to adapt to a variety of environments [57,58], and they are saprophytic fungi that play an important role in soil material cycling. However, *Basidiomycetes* can degrade lignin and other recalcitrant organic substances [59], which may be the reason for the relative abundance of *Basidiomycetes* in grasslands. Therefore, soil organic matter plays an important driving role in fungal community composition.

5. Conclusions

Coal mining subsidence altered the soil's microbiological community, physical and chemical characteristics, and aboveground vegetation community in the affected area. In the process of 15 years of natural restoration, although the microbial communities in the aboveground vegetation and soil were partially restored, it is difficult for them to return to their pre-subsidence state. However, a new homeostasis adapted to the environment was formed. Soil physicochemical properties and vegetation conditions were found to have a significant impact on the composition of bacterial and fungal communities, especially the number of aboveground vegetation species, vegetation coverage, and the content of fine sand particles, TN, NH_4-N, and AP in the soil, which is also an important indicator for assessing the effectiveness of ecological restoration. This study revealed the response of the soil microbial community to the natural restoration of subsidence by studying the changes and influencing factors of microbial diversity and community structure in different natural restoration processes, which has important guiding significance for the efficient ecological restoration of coal mining subsidence areas.

Supplementary Materials: The following supporting information can be downloaded at: https://www.mdpi.com/article/10.3390/microorganisms12010087/s1, Table S1: Main Plants in the Study Area.

Author Contributions: Conceptualization, D.L. and Z.M.; methodology, Z.M.; software, Y.T.; validation, D.L. and Y.T.; formal analysis, D.L.; investigation, B.F.; resources, Y.T.; data curation, Y.T., B.F. and L.X.; writing—original draft preparation, D.L.; writing—review and editing, Z.M.; supervision, Z.M.; project administration, Z.M.; funding acquisition, Z.M. All authors have read and agreed to the published version of the manuscript.

Funding: This research was funded by the National Natural Science Foundation of China, grant number 52004274.

Data Availability Statement: Data is contained within the article. The data presented in this study are available in this manuscript.

Acknowledgments: This study was supported by the National Natural Science Foundation of China (grant number 52004274). The partial data were analyzed on the online platform of Majorbio Cloud Platform (https://cloud.majorbio.com, accessed on 1 March 2023).

Conflicts of Interest: The authors declare no conflicts of interest.

56. Yelle, D.J.; Ralph, J.; Lu, F.; Hammel, K.E. Evidence for cleavage of lignin by a brown rot basidiomycete. *Environ. Microbiol.* **2008**, *10*, 1844–1849. [CrossRef]
57. Egidi, E.; Delgado-Baquerizo, M.; Plett, J.M.; Wang, J.; Eldridge, D.J.; Bardgett, R.D.; Maestre, F.T.; Singh, B.K. A few Ascomycota taxa dominate soil fungal communities worldwide. *Nat. Commun.* **2019**, *10*, 2369. [CrossRef]
58. Tedersoo, L.; Mikryukov, V.; Zizka, A.; Bahram, M.; Hagh-Doust, N.; Anslan, S.; Prylutskyi, O.; Delgado-Baquerizo, M.; Maestre, F.T.; Pärn, J.; et al. Global patterns in endemicity and vulnerability of soil fungi. *Glob. Chang. Biol.* **2022**, *28*, 6696–6710. [CrossRef]
59. Voříšková, J.; Baldrian, P. Fungal community on decomposing leaf litter undergoes rapid successional changes. *ISME J.* **2013**, *7*, 477–486. [CrossRef]

microorganisms

MDPI

Article

The Impact and Determinants of Mountainous Topographical Factors on Soil Microbial Community Characteristics

Jiantao Yu [1], Suyan Li *, Xiangyang Sun *, Wenzhi Zhou, Libing He, Guanyu Zhao, Zhe Chen, Xueting Bai and Jinshuo Zhang

The Key Laboratory for Silviculture and Conservation of Ministry of Education, College of Forestry, Beijing Forestry University, Beijing 100083, China
* Correspondence: lisuyan@bjfu.edu.cn (S.L.); sunxy@bjfu.edu.cn (X.S.)

Abstract: Soil bacterial and fungal community communities play significant ecological functions in mountain ecosystems. However, it is not clear how topographic factors and soil physicochemical properties influence changes in microbial community structure and diversity. This study aims to investigate how altitude and slope orientation affect soil physicochemical properties, soil microbial communities, and their contributing factors. The assessment was conducted using Illumina MiSeq sequencing in various altitude gradients and on slopes with different aspects (shady slopes and sunny slopes) in the subalpine meadow of Dongling Mountain, Beijing. Topographical factors had a significant effect on soil physicochemical properties: the primary factors determining the structure of microbial communities are total potassium (TK), ammonium nitrogen (NH_4^+-N), and soil organic carbon (SOC). There was no significant change in the diversity of the bacterial community, whereas the diversity of the fungal community displayed a single-peaked trend. The effect of slope orientation on microbial communities was not as significant as the effect of elevation on them. The number of bacterial communities with significant differences showed a unimodal trend, while the number of fungal communities showed a decreasing trend. The co-occurrence network of fungal communities exhibits greater intricacy than that of bacterial communities, and bacterial communities are more complex in soils with sunny slopes compared to soils with shady slopes, and the opposite is true for fungal communities. The identification of the main factors that control soil microbial diversity and composition in this study, provided the groundwork for investigating the soil microbial response and adaptation to environmental changes in subalpine meadows.

Keywords: microbial community; diversity; soil characteristics; elevation; illumina MiSeq sequencing

1. Introduction

Mountain ecosystems are one of the typical fragile ecosystems on earth, and the differences in topography and climatic environments make them an open natural platform for studying microbial communities. Subalpine meadow is a special kind of alpine meadow, which is extremely sensitive to the environment, and plays a decisive role in maintaining ecosystem stability and durability [1]. Soil microorganisms are essential constituents of soil ecology, with imperative roles in the material cycle, and the flow of energy within ecosystems [2]. Soil microorganisms are sensible indicators of community composition, ecosystem functioning, and abundance can indicate soil quality, while soil system stability can be reflected by the diversity and structure of soil microorganisms [3]. The soil contains a diverse range of microorganisms, with bacteria and fungi being the most abundant [4]. Studies have shown that in alpine meadows the bacterial community is mainly composed of Proteobacteria and Acidobacteria, and the most numerous fungal communities are Ascomycota and Basidiomycota, and not only with high plasticity and adaptability to the environment, but also, perform a vital function in crucial processes such as plant production, greenhouse gas emissions, and nutrient cycling, which are essential for the maintenance of numerous ecosystem functions [5–8].

Environmental heterogeneity is seen as a universal driver of microbial diversity [9], where topographic factors lead to dramatic changes in temperature and precipitation indirectly affecting changes in plant traits, soil physicochemical properties, and ecosystem functioning [10], thus driving diversity and variability in the distribution of soil microbial communities [11,12]. Studies exploring forest soil microbial diversity patterns across environmental gradients in montane ecosystems are relatively well-studied, however, there are relatively few studies on the combined effects of topographic factors and soil nutrients on soil microbial communities in subalpine meadows. Slope orientation and elevation are the main topographic parameters that generate environmental heterogeneity in microclimate, pedogenic processes, and vegetation patterns, as well as soil microbial variability. Changes in net solar radiation due to slope orientation lead to substantial soil temperature variations [13–15], leading to different soil formations and development of microenvironments, affecting vegetation establishment and there was a decisive impact on the composition of the soil fungal and bacterial community [16]. Consequently, studying fungal and bacterial communities across altitudinal gradients and slope orientations can help to unravel the reply from soil fungal and bacterial to climate change, explore the linkages between topography and microbial communities, and provide insights into their intricate distribution patterns and ecological importance [17–19].

The subalpine meadows of Dongling Mountain are located at the edge of the forest and play a crucial role in maintaining the balance and stability of the Dongling Mountain ecosystem. [20]. We selected Dongling Mountain subalpine meadow as the study area to investigate the structural composition and diversity of soil bacterial and fungal communities in meadows with different slope orientations along the altitudinal gradient (shady and sunny slopes) and the main environmental factors influencing the changes have been analyzed as well as the network of microbial communities in different slope orientations. The purpose of this study is to enhance comprehension of the impact of topography and soil factors on soil microbial communities in meadow ecosystems and to promote ecosystem management protection, and restoration. This study will generate data that supports fundamental ecological predictions on the diversity and constituents of soil bacterial communities and soil fungal communities in subalpine meadow ecosystems, in response to climate change.

2. Materials and Methods

2.1. Description of the Site and the Methodology of the Experiment

Dongling Mountain (39°48′–40°01′ N,115°24′–115°35′ E) is located in Beijing, China, belonging to the remnants of Xiaowutai Mountain in the Taihang Mountain Range, which is the natural ecological environmental protection barrier of Beijing. The highest point of Dongling Mountain is 2303 m above sea level, making it the highest peak in Beijing. It has four distinct seasons, with wet and winter being cold and dry and summer being warm. The average yearly temperature is 5.4 °C, with the highest and lowest temperatures occurring in July and February, respectively. The annual precipitation is 570.3 ± 112.2 mm, with June–August being the peak season every year. The predominant soil types are mountain brown soil and subalpine meadow soil, and their thickness ranges from 90 to 120 cm.

We selected the subalpine meadows with an altitude of 1600~2300 m as the study area in Dongling Mountain Nature Reserve, which had similar slopes and was all free from human interference and selected three altitudinal gradients (1700 m, 1900 m, 2100 m), set two different slope orientations (sunny and shady slopes) at the same altitude, total 6 study areas. And selected three areas of 10 × 10 m were to be sampled using the five-point method, respectively, each of which was more than 100 m apart. Each sampling area was spaced more than 100 m apart, totaling 18 sampling areas (Figure S1). After documenting vegetation information in each sampling area, soil samples were taken from the top

0–20 cm. All samples were collected in sterile plastic bags, placed on dry ice immediately after collection, and transported to the laboratory where they were divided into three portions according to the needs of the study; one portion was stored at −80 °C for DNA high-throughput sequencing analysis, another portion was airdried for physicochemical indexes testing and analysis, and the last portion, collected with a loop knife, was used for investigating and evaluating soil BD and soil moisture(Table 1).

Table 1. Site characteristics of different altitudes.

Altitude (m, above Sea Level.)	Aspect	Coordinates	Soil Type	Dominant Plant Taxa
1700	Sunny	40°2′02″ E, 115°29′38″ N	paramo soil	*Polygonum aviculare* L., *Artemisia carvifolia*, *Sedum aizoon* L., *Vicia cracca* L., *Poa spondylosis Trin.*, *Equisetum arvense* L., *Dendranthema chanetii*, *Bupleurum chinense DC.*, *Halenia corniculata*, *Delphinium grandiflorum*, *Dianthus chinensis* L.
	Shady	40°2′31″ E, 115°29′25″ N		
1900	Sunny	40°2′01″ E, 115°28′56″ N	paramo soil	*Sanguisorba officinalis* L., *Rumex acetosa* L., *Sedum aizoon* L., *Carex* L., *Asparagus cochinchinensis*, *Dendranthema chanetii*, *Heteropappus hispidus (Thunb.)*, *Delphinium grandiflorum*, *Echinops sphaerocephalus* L., *Allium wallichii*, *Dendranthema zawadskii*, *Dracocephalum rupestre Hance*, *Viola verecunda*, *Artemisia sacrorum Ledeb.*, *Arundinella hirta*, *Saussurea japonica*, *Saussurea Chinensis*, *Dianthus chinensis* L.
	Shady	40°2′21″ E, 115°28′57″ N		
2100	Sunny	40°2′04″ E, 115°28′42″ N	paramo soil	*Carex duriuscula C.*, *Potentilla discolor Bge.*, *Sanguisorba officinalis* L., *Euonymus fortunei*, *Carex* L., *Dendranthema zawadskii*, *Stellaria dichotoma* L., *Gentiana macrophylla Pall.*, *Saussurea japonica*, *Saussurea Chinensis*, *Heteropappus hispidus*, *Ctenopteris*
	Shady	40°2′12″ E, 115°28′41″ N		

2.2. Physical and Chemical Properties of Soil

The soil's physical and chemical properties were determined following the guidelines provided in the Soil Analysis Handbook (Springer (Berlin/Heidelberg, Germany), 2006). A digital pH meter (1:2.5 w/v) was used in a soil water sample to determine the pH of the soil. Soil Bulk Density (BD) and moisture content were measured by the drying method, meaning that they were calculated by comparing the weights of soil samples before and after drying in an oven at 105 degrees Celsius for 24 h. Soil-available phosphorus (AP) was determined by the molybdenum antimony anti-colorimetric method with 0.5 M $NaHCO_3$. Soil organic carbon (SOC) content was measured using a total organic carbon (TOC) analyzer (Trace Elemental Instruments (Delft, The Netherlands)—XPERT-TOC/TN_b) Soil's total nitrogen (TN) content was analyzed via the Kjeldahl method (Shanghai Shengsheng Automation Analysis Instrument Co. (Shanghai, China)—K5100). Total phosphorus (TP) and potassium (TK) levels were determined using the NaOH melt method. The available potassium (AK) in the soil was measured through ammonium acetate-flame photometry (Shanghai Jingke Industry Co. (Shanghai, China)—FP6140). Extraction distillation was used to measure ammonium nitrogen (NH_4^+-N), while reduction distillation was employed to measure nitrate nitrogen (NO_3^--N).

2.3. The Extraction of DNA and Its Amplification through PCR

Total microbial genomic DNA was extracted from soil samples using the E.Z.N.A.® soil DNA Kit (Omega Bio-tek, Norcross, GA, USA) according to the manufacturer's instructions. The quality and concentration of DNA were determined by 1.0% agarose gel electrophoresis and a NanoDrop2000 spectrophotometer (Thermo Scientific, Waltham, MA, USA) and kept at −80 °C before further use. The hypervariable region V3-V4 of the bacterial 16S rRNA gene was amplified with primer pairs 338F (5′-ACTCCTACGGGAGGCAGCAG-3′) and 806R (5′-GGACTACHVGGGTWTCTAAT-3′) byT100 Thermal Cycler PCR thermocycler (BIO-RAD, Hercules, CA, USA). The fungal ITS rRNA gene was amplified on a T100 Thermal Cycler PCR thermal cycler (BIO-RAD, Hercules, CA, USA) using primer pairs ITS1F (5′-CTTGGTCATTTAGAGAGGAAGTAA-3′) and ITS2R (5′-GCTGCGTTCTTCATCGATGC-3′). The PCR reaction mixture including 4 μL 5 × Fast Pfu buffer, 2 μL 2.5 mM dNTPs, 0.8 μL each primer (5 μM), 0.4 μL Fast Pfu polymerase, 10 ng of template DNA, and ddH2O to a final volume of 20 μL. PCR amplification cycling conditions were as follows: initial denaturation at 95 °C for 3 min, followed by 27 and 35 cycles of denaturing at 95 °C for 30 s, annealing at 55 °C for 30 s, and extension at 72 °C for 45 s, and single extension at 72 °C for 10 min, and end at 4 °C. The PCR product was extracted from 2% agarose gel and purified using the PCR Clean-Up Kit (YuHua, Shanghai, China) according to the manufacturer's instructions and quantified using Qubit 4.0 (Thermo Fisher Scientific, Waltham, MA, USA).

2.4. Sequencing Data Processing

Purified amplicons were pooled in equimolar amounts and paired-end sequenced on an Illumina PE300 platform (Illumina, San Diego, CA, USA) according to the standard protocols by Majorbio Bio-Pharm Technology Co., Ltd. (Shanghai, China).

2.5. Data Analysis

To analyze the soil's chemical properties and physical properties, as well as the diversity values of soil samples with varying slope directions at different elevations, and to test for statistical significance ($p < 0.05$) using TWO-WAY ANOVA analysis of variance, sorting of data was done using Excel and statistical analysis was carried out using SPSS 27 software. Homogenization of a minimum number of sequences for each soil sample was carried out to minimize the impact of varying sample counts during the analysis of relevant data (34,425 bacterial sequences and 31,691 fungal sequences). Map the microbial communities using R language tools, community Bar mapping was employed, and Circos-0.67–7 was used to plot Circos samples against species. For data analysis of Illumina MiSeq sequencing data, Qiime1.9.1 can generate abundance tables for each taxonomic level and perform beta diversity distance calculations. Mothur1.30.2 can perform alpha diversity analyses to obtain Sobs, Shannon PD, Chao1, and Faith indices. The plotting operation was executed using Origin 2023 software. Normality and variance alignment of the data were assessed using the Levene tests and Shapiro-Wilk, respectively. Beta diversity-based nonmetric multidimensional scaling analysis (NMDS) was conducted to assess bacterial and fungal community similarity, using the Bray-Curtis distance metric. ANOSIM and PERMANOVA were performed using LEfSe for all-to-all comparison (more stringent). We used Redundancy Analysis (RDA) to determine the primary factors influencing the spatial distribution of bacteria and fungi across various gradients and elevations. We calculated the correlation coefficients between environmental factors and selected species using R (version 3.3.1). The resulting matrix was visualized with Heatmap plots.

Analysis of co-occurrence networks was performed using an absolute abundance of dominant OTUs using Networkx 2.8 software. Based on Spearman's correlation analysis, fungal and bacterial community complexity networks with different slope directions at different elevations were constructed and visualized by Gephi v.0.9.2 and Cytoscaoe v.8.2.

3. Results

3.1. Physicochemical Properties of Soils at Different Altitude Gradients

Soil pH, soil BD, TN, TP, and NH_4^+-N ($p < 0.05$, Table S1) were significantly affected by altitude and had a very significant effect on SOC, soil moisture, TK, and AK ($p < 0.001$). pH showed a significant trend of decreasing with increasing altitude, BD exhibited a pattern of initial decrease followed by an increase, and the Soil moisture, SOC, AK, and NH_4^+-N content exhibited an initial increase followed by a decrease. Soil TN and TP were highest at the lowest elevation of 1700 m, and TK was lowest at the lowest elevation of 1700 m. Slope orientation had a conspicuous positive influence ($p < 0.05$) on BD, TK, and AP, and a highly conspicuous positive influence ($p < 0.001$) on Soil moisture and AK. Soil BD and AK were significantly greater on the sunny slope than on the shady side of the north slope, with Soil moisture being greater on the shady than on the sunny side of the slope. TK was greater on the sunny side of the slope than on the shady side, except at 1900 m. It was higher on the shady side of the slope at both 1700 m and 2100 m, while AP was greater on the shady than on the sunny side of the slope at 1900 m, and vice versa at the rest of the elevation.

The interaction of slope orientation and elevation had a significant impact on TK and TP ($p < 0.05$) and a highly significant effect on TK, AP, and AK ($p < 0.001$).

3.2. Characterization of Soil Microbial Communities

The 16S rRNA genes of soil bacteria and the ITS genes of fungi were sequenced on the Illumina MiSeq platform, and the resulting data were pumped flat to obtain a total of 619,650 optimized soil bacterial sequences and 570,438 optimized soil fungal sequences, of which the valid sequences (removing chimeras) were 355,374 soil bacterial sequences and 555,678 soil Fungal sequences. The average sequence length distributions of soil bacteria and fungi were 418 bp and 246 bp, and their 16S and ITS coverage indices were >96% and >99%, respectively. The gene dilution curves of all samples leveled off at 97% OUT sequence similarity (Figure S2), the information that was gathered through this sequencing was sufficient for analyzing the composition and diversity of the bacterial and fungal communities present in every sample.

A total of 3391 soil bacterial OTUs, distributed across 35 phyla, 105 classes, and 611 genera were identified. Actinobacteriota (28.84% of the total OTUs), Proteobacteria (21.74%), and Acidobacteriota (17.08%) accounted for 67.66% of the total 67.66%. Proteobacteria ($p < 0.05$) were significantly affected by altitude, exhibiting an increasing and then decreasing trend, and slope orientation had a greater significant influence on Actinobacteriota ($p < 0.01$) being more shaded than sunny (Figure 1A, Table S2). The relative abundance of Alphaproteobacteria, Actinobacteria, and Thermoleophilia at the class level was 16.70%, 11.53%, and 10.38%, respectively, and together accounted for 38.61% of the total, being the dominant phyla (Figure 1B). Altitude had a conspicuous influence ($p < 0.05$) on Alphaproteobacteria and Actinobacteria, both showing an increasing and then decreasing trend. Slope orientation had a conspicuous effect ($p < 0.05$) on Thermoleophilia, there was a significant increase in shady slopes compared to sunny slopes. Actinobacteria ($p < 0.05$, Table S2) Actinobacteria were significantly influenced by the interplay of altitude and slope orientation. Soil bacterial community at genus level Bacillus (6.82%), Vicinamibacterales (5.89%), and RB41 (4.56%) were the dominant genera accounting for 17.27% of the total bacterial sequences.

A total of 3994 soil fungi OTUs were identified, which were distributed among 15 phyla, 59 classes, and 673 genera. Among them, Ascomycota, Mortierellomycota, and Basidiomycota were the dominating phyla, accounting for 51.19%, 24.28%, and 18.80%, respectively, of the total quantity of obtained bacterial sequences, totaling 94.27%, while the remaining 12 phyla accounted for only 5.73% (Figure 1C). The interaction of elevation and slope orientation had a significant influence ($p < 0.05$) on Ascomycota. The relative abundance of Mortierellomycetes, Sordariomycetes, and Eurotiomycetes at the class level was 24.22%, 16.46%, and 13.37%, respectively, and together they accounted for 54.05% of the total, making them dominant phyla (Figure 1D). Altitude had a highly significant effect on Sordariomycetes ($p < 0.001$), which tended to decrease with increasing altitude, and both altitude and slope orientation had a significant impact on Eurotiomycetes ($p < 0.05$, Table S3), reaching the highest abundance at 1900 m on the sunny slopes, and the interaction between slope orientation and altitude had a major influence on the relative abundance of Sordariomycetes, and Eurotiomycetes had a significant influence ($p < 0.05$). The dominant genera in the soil fungal community were Mortierella (24.26%), Exophiala (5.38%), and Ascomycota (5.02%), which accounted for 34.66% of the total number of fungal sequences.

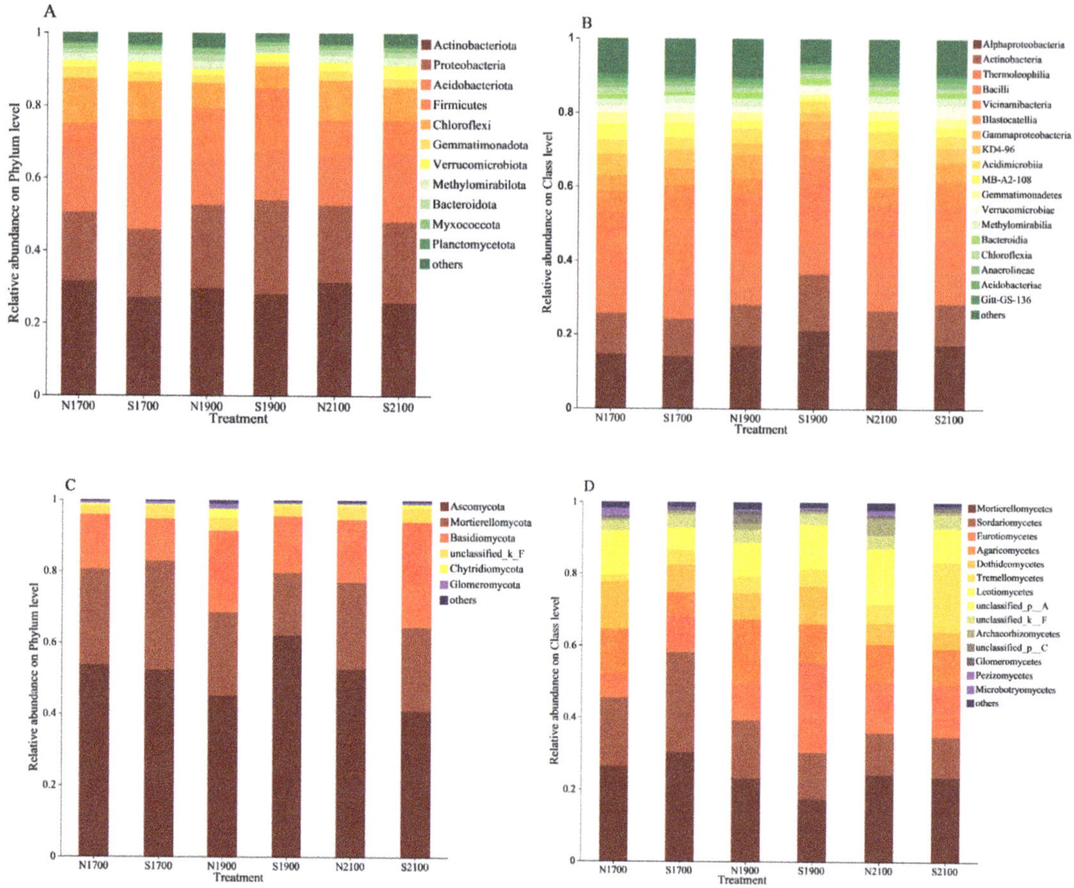

Figure 1. Comparison of the relative abundances of the major Bacterial phyla (**A**) and classes (**B**) found in soil. Fungal phyla (**C**) and classes (**D**) for a variety of slope directions and soils at several different elevations. N for north, shady slopes, S for south, sunny slopes, for example, N1700 means 1700 m above sea level on the north slope.

The dominant flora was similar in the different slope orientations at different altitudes, and the abundance of soil bacterial Actinobacteriota was greater than that in the shaded and sunny slopes of the three altitudinal gradients of 1700 m, 1900 m, and 2100 m in a total of six samples, increasing by 11% in the shaded and sunny slopes at each altitudinal gradient. In terms of the whole collection of Actinobacteriota species, the proportion of shaded slopes was also greater than that of sunny slopes for all three elevation gradients, with a total increase of 6%. For the six samples in this study, the proportion of Proteobacteria abundance increased by 11% on the 1900 m gradient compared to the 1700 m gradient, and for the total number of Proteobacteria species, the 1900 m gradient was also higher than the 1700 m gradient, with an increase of 10% (Figure 2A). For Ascomycota, the dominant soil fungal phylum, the highest percentage of Ascomycota abundance, S1900, was 21% higher than the lowest percentage of abundance, S2100, in the six samples. Of the total Ascomycota species, S1900 was 7% higher than S2100 (Figure 2B).

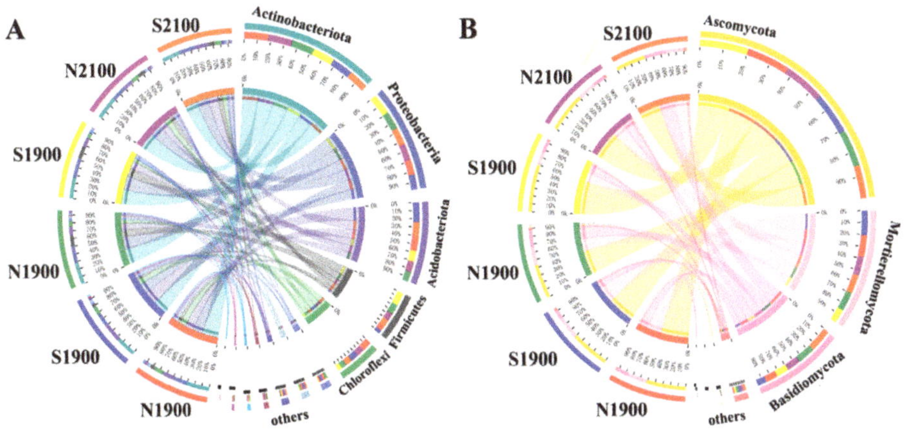

Figure 2. Relative abundances of dominant Bacterial phylum (**A**) and Fungi phylum (**B**) in the soil under three altitude gradients and two slope aspects. The width of the bars from each phylum indicates the relative abundance of that phylum in the sample.

3.3. *Index of Soil Microbial Community Diversity*

Altitude significantly affected both the Shannon and PD diversity indices of the soil bacterial community, displaying a decreasing and then increasing trend. The Shannon index of soil bacteria on sunny slopes at 1900 m was 6.4% lower than that at 1700 m ($p < 0.05$), and the PD index of soil bacteria at 1900 m was 10.9% lower than that at 2100 m ($p < 0.05$). The orientation of the slope did not significantly impact the alpha diversity of soil bacterial communities (Figure 3A–D). The altitude significantly influenced the sobs, Chao 1, and PD diversity indices of soil fungal communities. The total fungal community diversity showed a unimodal pattern of increasing and then decreasing. Moreover, on shaded slopes, fungal sobs and Chao 1 diversity indices were 23.7% and 27.1% higher, respectively, at 1900 m than at 2100 m ($p < 0.05$). At 1900 m, the fungal PD diversity index was 17.1% higher than that at 1700 m ($p < 0.05$). There was no discernible effect that could be attributed to the direction of the slope on the alpha diversity of the soil fungus communities (Figure 3E–H).

The NMDS analyses were performed on soil fungal and bacterial sequencing data collected from different altitudes exhibiting varied slope orientations, based on Bray–Curtis distance and both bacterial and fungal NMDS analyses had STRESS values less than 0.2, indicating that they were representative. The grouping of soil bacterial and soil fungal communities was partially influenced by altitude, and the bacterial and fungal aggregation was firmer and more similar on the 2100 m. It was evident that the bacterial communities and fungal communities at 1700 m and 1900 m, were organized according to slope direction. (Figure 4A,B). According to the ANOSIM and PERMANOVA analyses, there is a significant difference ($p < 0.01$) between the fungal and bacterial community structures at different altitudes. The PERMANOVA analysis of all the samples revealed that altitude had a more demonstrable impact on the structure of the fungal and bacterial communities in the soil than slope. ($p < 0.01$, Table 2).

Discriminant analyses of different taxonomic levels (from phylum to genus) of soil microorganisms at different altitudes and slope orientations using LEfSe analysis revealed statistically significant differences in community types for bacteria and fungi. When the LDA threshold was 3.5, a total of 46 bacterial branches showed statistical differences ($p < 0.05$, Figure 5A) in the bacterial community at different altitudes and slope orientations and different taxonomic levels, and 62 branches in the fungal community showed statistical differences ($p < 0.05$, Figure 5B). At various altitudes and slope orientations, the number of soil bacterial and fungal communities exhibited significant variation. As the elevation increased, the number of bacterial communities with significant differences exhibited a monotonic increase and then decrease. At a gradient of 1700 m, the slopes that are shaded are higher than the slopes, with no significant differences in bacterial branches on the sunny slopes at the 1700 m elevation gradient, and higher on sunny slopes than on shady slopes, and higher on sunny slopes than on the shady slopes at the 1900 m and 2100 m elevation gradients (Figure S3A), with only three bacterial branches significantly enriched on the sunny slopes at the 2100 m elevation gradient and only two bacterial branches significantly enriched on the shaded slope. The number of fungal communities with significant differences with increasing elevation showed a decreasing trend, and the number of fungi on the sunny slopes was greater than that on the shady slopes, except for the 2100 m elevation gradient (Figure S3B), where only three fungal branches were significantly enriched on the sunny slopes of the 2100 m elevation gradient.

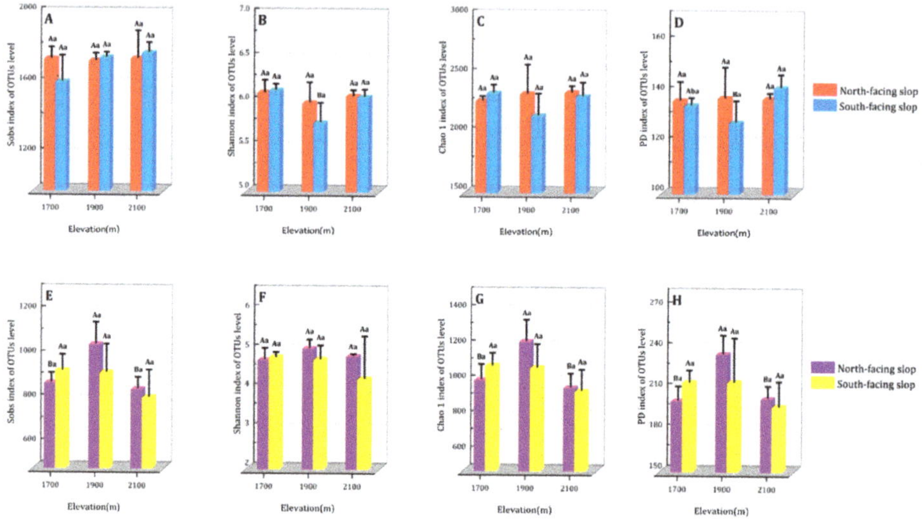

Figure 3. Sobs, PD, Shannon, Chao1 indices of bacterial (**A–D**) and fungal (**E–H**) communities at different altitudes and slops. Differences between shaded and sunny slopes are written in uppercase ($p < 0.05$), and differences across the three elevation gradients are written in lowercase ($p < 0.05$).

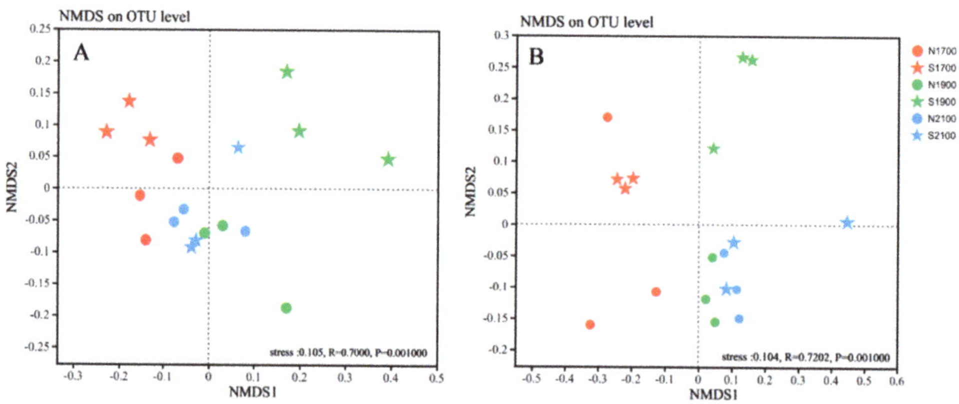

Figure 4. NMDS analyses were conducted on the soil bacterial (**A**), and fungal (**B**) communities.

Table 2. Non-parametric multivariate analysis (PERMANOVA) and ANOSIM soil fungal and bacterial community analysis by altitude and slope aspect.

	ANOSIM		PERMANOVA		
	Statistic	***p***		**R^2**	***p***
Bacteria	0.7000	0.001	Altitude	0.60805	0.001
			Slope aspect	0.11409	0.042
Fungi	0.7202	0.001	Altitude	0.58703	0.001
			Slope aspect	0.09881	0.031

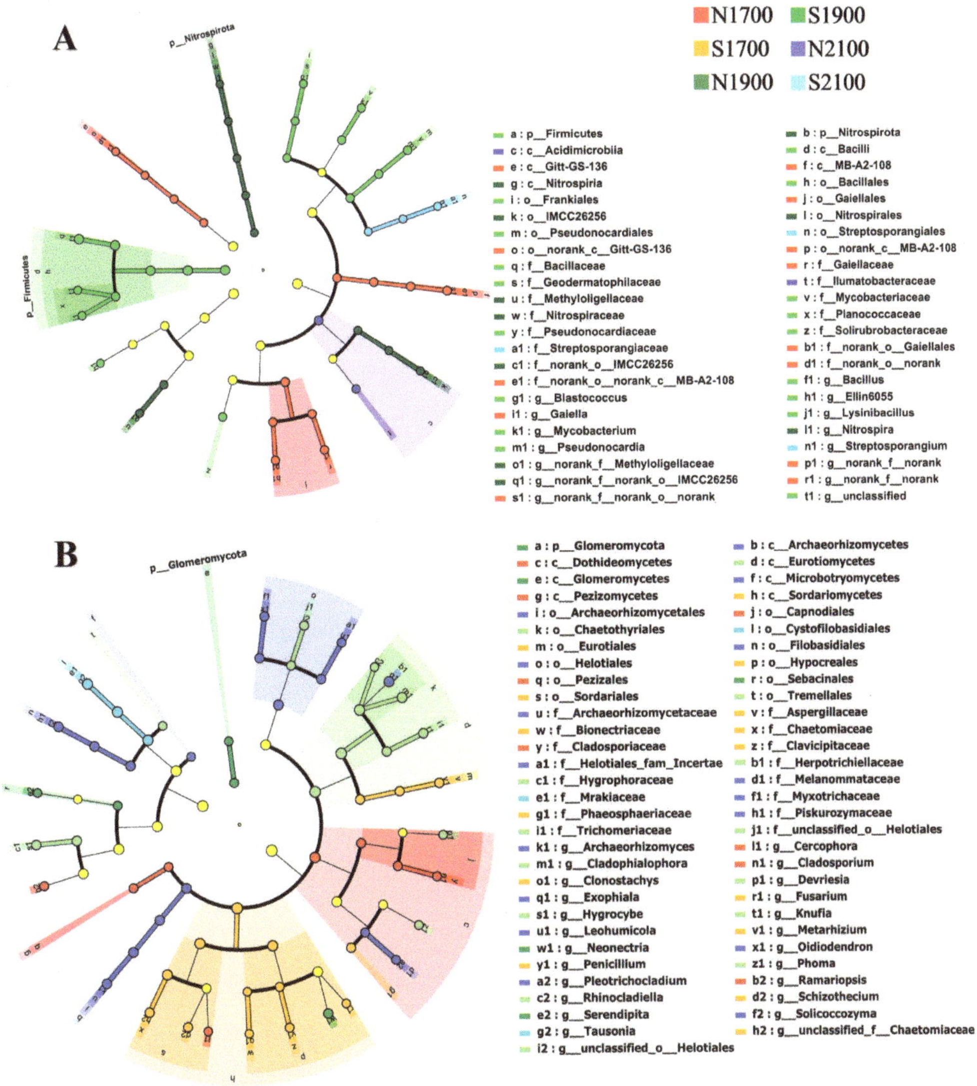

Figure 5. Variations in the community structure of soil bacteria (**A**) and fungi (**B**) at various altitudes and slope aspects in subalpine meadows. Different-colored regions represent different study areas. Circles indicate phylogenetic levels from phylum to genus. The diameter of each circle is proportional to the abundance of the group.

3.4. Association between Soil Physicochemical Properties and Soil Microbial Communities

The article analyzes redundancy and correlation to reflect and evaluate the impact of soil environmental factors on the structure of microbial communities, while also assessing their correlation using heatmap plot analysis. According to the genus-level bacterial community RDA plot, the first two axes of the soil bacterial community explain over 72% of the structural changes (Figure 6A), which was biostatistically significant. Among them, NH_4^+-N, TK, AK, and SOC had a significant effect ($p < 0.01$) with R2 of 0.644, 0.639, 0.63, and 0.629, respectively, followed by TN and soil moisture ($p < 0.05$). For soil fungi, as shown, more than 63% of the structural variation was explained by the first two axes (Figure 6B), which was biologically statistically significant. Among them, TK, NH_4^+-N, SOC, and TN produced highly significant effects ($p < 0.01$) with R2 of 0.689, 0.511, 0.489, and 0.404, followed by AK ($p < 0.05$).

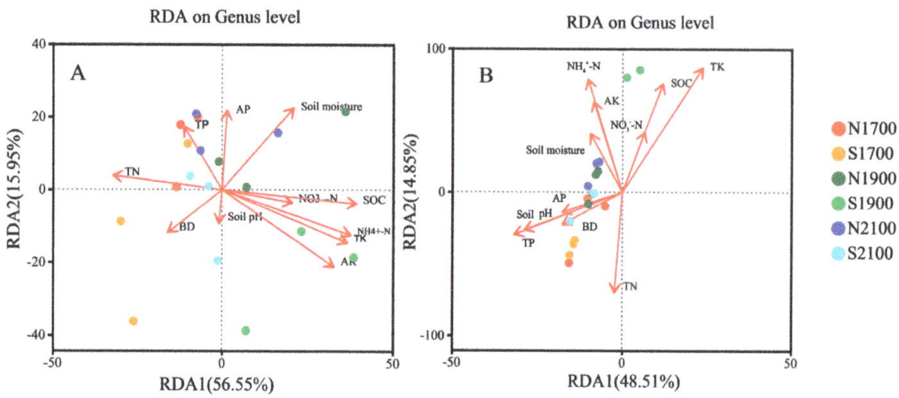

Figure 6. Analysis of soil bacterial genus (**A**), fungal genus (**B**) communities, and factors (red arrows) for Redundancy. BD: Soil bulk density; TN: Total nitrogen; TP: total phosphorus; AK: Available potassium; SOC: soil organic carbon; TK: total potassium AP: Available phosphorus; NO_3^--N: Nitrate nitrogen; NH_4^+-N: Ammonium nitrogen.

According to the compartmentalized soil bacterial communities and fungal communities heatmap (Figure 7A,B), the same environmental factor had different effects on different microbial communities, and in the soil bacterial community, SOC, TK, and NH_4^+-N had highly significant positive correlation ($p < 0.001$) and highly significant negative correlation ($p < 0.001$) correlations with the different bacterial communities, respectively. ($p < 0.001$). Based on their clustering, microbiological communities in the soil can be observed, soil pH, BD, TN, and TP had similar effects on bacteria and fungi, and NH_4^+-N, NO_3^--N, AK, TK, and SOC had a similar influence on bacteria and fungi. Based on the clustering of the bacterial communities and fungal communities, the various responses to environmental factors at the phylum level were categorized into four groups of bacterial communities and four groups of fungal communities.

3.5. Co-Occurrence Network of Soil Microbial Communities in Different Slope Aspects

In the four co-occurrence networks (Figure 8). The shaded slope bacterial community had 312 links (edges) out of 262 OTUs (nodes), of which 302 (96.79%) were positively correlated and 10 (3.21%) were negatively correlated. The sunny slope bacterial community has 663 links (edges) out of 427 OTUs (nodes), of which 643 (96.98%) are positively correlated and 20 (3.02%) are negatively correlated. The shaded slope fungal community has 1333 links (edges) out of 359 OTUs (nodes), of which 1326 (99.47%) are positively correlated and 7 (0.53%) are negatively correlated. The sunny slope fungal community has 990 links (edges) out of 289 OTUs (nodes), of which 983 (99.29%) are positively correlated and 7 (0.71%) are negatively correlated.

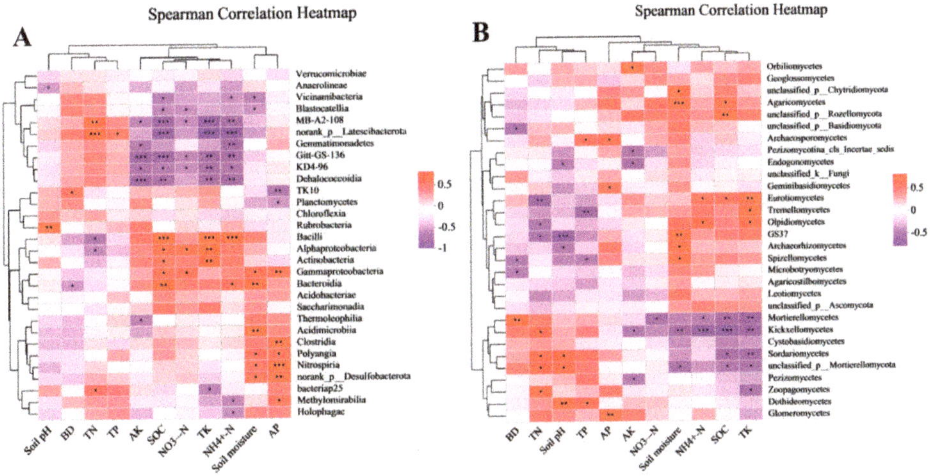

Figure 7. Heatmap depicting the associations between bacterial (**A**) and fungal (**B**) communities and soil variables. BD: Soil bulk density; TN: Total nitrogen; TP: total phosphorus; AK: Available potassium; SOC: soil organic carbon; TK: total potassium AP: Available phosphorus; NO_3^--N: Nitrate nitrogen; NH_4^+-N: Ammonium nitrogen. *, **, *** indicate statistical significance at $p < 0.05$, $p < 0.01$, $p < 0.001$.

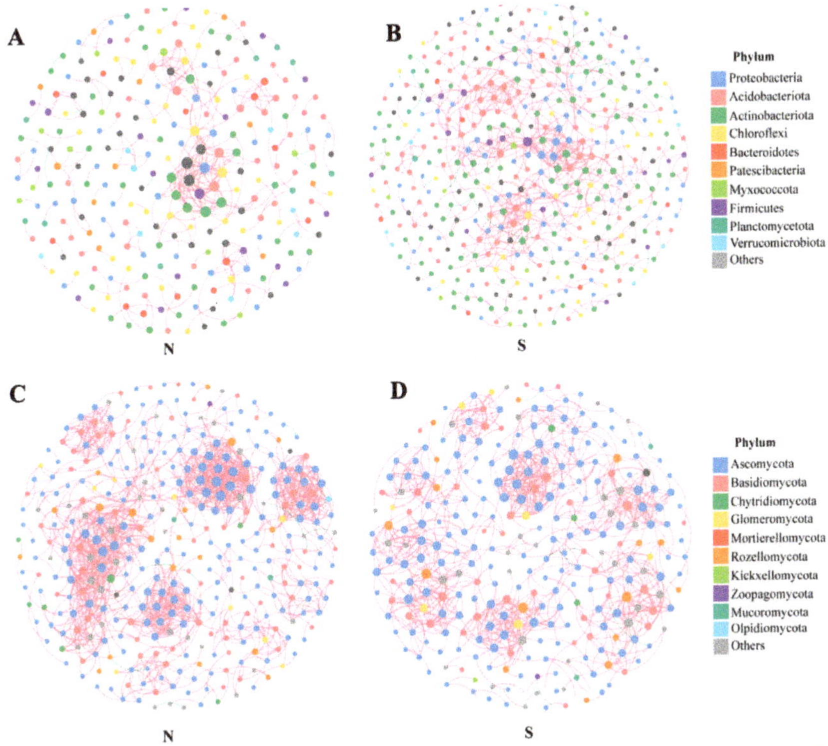

Figure 8. Network co-occurrence analysis of bacteria in the shaded slope (**A**) and fungi in the shaded slope (**B**), and bacteria in the sunny slope (**C**) and fungi in the sunny slope (**D**).

4. Discussion

4.1. Alterations in Physicochemical Properties of Soil along Altitudinal Gradients and Slope Aspects

In this study, there was an increase and subsequent decrease in soil moisture, SOC, TK, AK, NO_3^--N, and NH_4^+-N content as altitude increased. This may be due to vegetation community richness. It is interesting that plant richness also demonstrated the same tendency, which is an increase followed by a decrease with rising altitude. The highest number of plant species was observed at an altitude of 1900 m. Zhou's (2019) research demonstrated that altitude significantly affects plant growth and distribution in subalpine meadows. Hydrothermal conditions vary with altitude and slope direction, leading to an uneven distribution of the environment, causing differences in the composition of plant communities [21], further leading to the uneven distribution of plant communities and changing the nutrient content of the soil. As the altitude increased, grassland vegetation richness and cover at 1900 m reached a maximum, apoplastic material increased, and air temperature and potential evapotranspiration decreased. As a result, the soil moisture content increased and the soil BD decreased. This increased grassroot vigor and productivity, causing root necrosis and secretions that added more carbon and other elements to the soil [13]. This led to a positive feedback cycle that increased some of the soil's nutrient contents. With increasing altitude up to 2100 m and reaching the summit of the main peak of Dongling Mountain, the ambient temperature decreased further. The low temperature was unfavorable for plant growth and led to a decrease in vegetation richness. Evaporation of soil increased, causing a decrease in water content and an increase in soil capacity [22]. This led to a decrease in some soil nutrient contents. Furthermore, at the same altitude in our study, soil moisture, SOC, TP, TN, TK, and AK were greater than or equal to those on sunny slopes at the same altitude. This could be attributed to the significant difference in solar radiation at the same altitude on shady slopes when compared to sunny slopes. It is the main determining factor for soil water evaporation, temperature, carbon, and nitrogen cycling, resulting in significant differences in soil moisture content and bulk density [13].

4.2. Community Composition and Diversity of Soil Microbial Communities

4.2.1. Structural Composition of Soil Microbial Communities at Different Altitudes and Slope Aspects

The microbial abundance of different taxa at different taxonomic levels responded differently to elevation and slope direction, which requires further investigation at finer taxonomic levels. In our study, we found that the dominant phyla of soil bacterial communities were Actinobacteriota, Proteobacteria, and Acidobacteriota, which is in keeping with previous research indicating that these microbes are prevalent in the soils of various ecosystems [23]. Changes in the three altitudinal gradients have a strong influence on the bacterial phylum, which is relatively abundant in soil microorganisms (Proteobacteria, Firmicutes, Chloroflexi, and Gemmatimonadota) and slope orientation had a significant effect on Actinobacteriota and Thermoleophilia in the bacterial community. He's (2023) showed that altitudinal gradients are associated with changes in different environmental factors that trigger different habitat-specific patterns and have different effects on different microbial communities [24–26]. Zhou's (2019) showed that Actinobacteriota, an oligotrophic taxon with strong metabolism and DNA repair mechanisms even at low temperatures, prefers nutrient-poor alkaline environments [21,27]. This is consistent with our study that the total percentage of Actinobacteriota was the highest in the study area and the relative abundance of Actinobacteriota was the lowest in the area of the 2100 m elevation gradient on the sunny slopes with the lowest pH. Proteobacteria were mainly affected by the significant positive correlation between NO_3^--N, TK, SOC, and NH_4^+-N, which showed off the three altitude gradients, the lowest altitude gradient had the lowest levels, which then increased to the highest point before decreasing. Showing a single peak pattern, which is consistent with the previous view that Proteobacteria prefer eutrophic conditions [28]. Proteobacteria, although involved in energy metabolism and playing key roles in phylogenetic, ecological, and pathogenic values, have a symbiotic lifestyle, and are less abundant in harsh and infertile soil environments, which explains why from 1900 m to 2100 m Proteobacteria the relative abundance of Acidobacteriota decreased from 1900 m to 2100 m. Acidobacteriota had a significant negative correlation with soil moisture, NO_3^--N, SOC, and NH_4^+-N. Acidobacteriota plays a significant role in the decomposition of organic matter and nutrient cycling [29].

Ascomycota, Mortierellomycota, and Basidiomycota were the dominant phyla of the soil fungal communities, and the results of this research were comparable to those of Li's (2022) study, and Chen's (2022) study [22,30]. Changes in the three altitudinal gradients in this study had a significant influence on fungal phyla (Sordariomycetes and Eurotiomycetes). Slope direction had a significant effect on Eurotiomycetes in the fungal community. The greatest abundance of Ascomycota in the subalpine meadows of Dongling Mountain can be mainly attributed to its strong reproductive ability, rapid growth, a quantitative advantage in producing a large number of spores, and the fact that Ascomycota is more adaptable to harsh habitats. Mortierellomycota had the highest abundance at an altitude of 2100 m. Fungal saprophytes such as Mortierellomycota and Mucor. are considered high-altitude opportunists due to their tolerance to low temperatures and may be able to colonize and survive on high-altitude peaks [31]. Basidiomycota fungi have a potent capacity for lignocellulose decomposition, and in contrast, in this study, the relative abundance of Basidiomycota in subalpine meadow soils decreased with increasing elevation gradients, possibly because increased nutrient efficiency inhibits its growth [4].

4.2.2. Soil Microbial Community Diversity at Different Altitudes and Slope Aspects

Soil microbial diversity is a direct reflection of changes in its abundance, richness, and uniformity, and the majority of previous studies reporting the pattern of soil microbial diversity at various altitudes using high-throughput sequencing techniques have yielded comparable results. The study revealed a decrease in the diversity of soil fungal and bacterial communities with increasing altitude [32–34]. This partially supports the conclusion that the diversity and structural composition of soil bacterial and fungal communities decrease with an increase in altitude. Generally, the environmental conditions became harsher as the altitude increased, and it was therefore predicted that the abundance of soil microbes and fungi would decrease along the gradient of altitude [35]. Our study found that soil bacterial diversity in subalpine meadows did not exhibit a clear trend with increasing elevation. Instead, it showed a general trend of decreasing until reaching the lowest point and then increasing. This is not entirely consistent with the findings of previous studies. However, consistent with the findings of Fierer's (2011) study that although there was There are no changes that can be seen at a glance in bacterial diversity across the gradient, significantly different compositions of soil bacterial communities in subalpine meadows across three altitudinal gradients [10]. This may be because bacterial community diversity is mainly influenced by soil pH, and there were two altitudinal gradients in the study area where pH was not significantly different [36]. In our study, soil fungal diversity showed a unimodal trend of increasing with altitude and then decreasing after reaching the highest point of elevation. Recent studies have highlighted the inconsistency in the physiological geographical patterns of bacteria and fungi [33]. Distinct vertical distribution patterns of soil bacteria and fungi at altitude arise due to significant differences in the life and evolutionary histories of various taxa. The results of Li's (2022) are in agreement with the unimodal trend observed in fungal communities [22]. This may be the result of elevated soil nutrient concentrations (soil moisture, SOC, AK, NH_4^+-N) at 1900 m, which favored the growth of soil microorganisms. In addition, for the soil bacterial community, the bacterial diversity at three altitudes was relatively flat on the sunny slopes with no significant change, while it showed a decreasing and then increasing trend on the shady slopes. When it came to the fungal community in the soil, the soil fungal diversity index was greater on the sunny side than on the shady side at the same elevation, except at 1700 m. These findings are comparable to those of Li's (2022), who conducted a study on the same topic [22]. It was found that the diversity index of the meadow soil bacterial community was stronger than that of the meadow soil fungal community. Ji's (2022) research in the cold temperate zone supports this finding. Additionally, the study proposes that various microbial taxa in Dongling Mountain have evolved niche differentiation across the altitudinal gradient [30].

The response of the features of the grassland vegetation and the soil nutrient variables to slope and elevation was consistent with the variation in the bacterial and fungal diversity indices found in the soil. Soil physicochemical characteristics may affect soil microbial communities [37]. The beta diversity of soil bacterial communities and soil fungal communities could be distinguished visually at different elevations and slope directions, the properties of soil fungal communities vary considerably more with altitude and slope direction than those of soil bacterial communities in our study. Changes in elevation and slope direction typically lead to variations in climatic factors, soil characteristics, and vegetation, resulting in alterations in the composition of bacterial and fungal communities, and increased species richness and diversity, to facilitate adaptation to altered environmental conditions [38].

4.3. Environmental Factors Drive Soil Microbial Communities

Previous research has demonstrated that soil organic carbon substantially affects the composition and diversity of soil fungal communities in alpine meadows during succession [34]. Wang's (2023) showed that the pH of the soil is an essential factor of bacterial community structure in various soil ecosystems, and it indirectly influences soil bacterial habitats by modulating other influences, thereby influencing the composition of soil bacterial communities [36,39,40]. Nevertheless, the results of our study differ from previous studies, we found that soil physicochemical variables had a high degree of explanatory power for bacterial community composition and that NH_4^+-N, TK, AK, and SOC had a significant impact on soil bacterial communities, with NH_4^+-N having the greatest influence. And TK, TN, NH_4^+-N, SOC had a significant impact on the soil fungal community, with TK having the greatest effect. Soil pH in this study differed significantly across the altitudinal gradient, with the high altitude of 2100 m being lower than other altitudinal gradients, but the effects on soil microbial communities were not significant, as can be seen from the heat map graph, which shows that there was only a slight negative correlation with Basidiomycota of the fungal community. The research of Zheng (2019) demonstrated that certain soil bacterial communities could adapt to variations in soil pH [41]. It is believed that SOC, as a basic substrate and energy source for soil microorganisms, and Soil carbon and soil nitrogen greatly influence microbial distribution patterns at elevation and slope by influencing the metabolism of soil microbial communities [16]. In alpine meadows, Previous research has found that the relative abundance of taxa clades (Actinobacterioidae and Firmicutes) varied with soil organic carbon, indicating that soil carbon efficiency influences soil bacterial diversity and soil community composition [17]. Furthermore, soil NH_4^+-N strongly influenced microbial diversity and the composition of soil fungal and soil bacterial communities in subalpine meadows, supporting the notion that soil conditions influence Soil microbial diversity and soil microbial community composition. Soil-effective nitrogen represents the process of microbial dynamic nitrogen cycling and increases in nitrogen nutrients have a greater significant impact on the constitute of microbial communities by regulating microbial nitrogen energy sources [42].

Bacterial communities are more affected by altitude and slope as compared to fungal communities, perhaps due to the narrower physiological spectrum of soil fungi. For instance, soil bacteria may be photosynthetic autotrophs or chemoautotrophs, but soil fungi are only heterotrophs [43]. Additionally, we confirmed that the bacterial community in the subalpine meadow ecosystem of Dongling Mountain, which is in the warm temperate zone, was more affected by slope and altitude direction than the fungal community, and that elevation had a greater significant impact on the structure of the microbial communities than slope direction. Changes in elevation and slope factors synergistically influence the composition and diversity of soil bacterial and fungal communities in subalpine meadows. The community structure of soil microbes and fungi can be altered in response to environmental stimuli.

4.4. Co-Occurrence Networks of Microbial Communities with Different Characteristics on Shady and Sunny Slopes

Microorganisms in natural ecosystems are not simple assemblages of individual populations, however, these organisms create elaborate ecological networks that are essential for the maintenance of ecosystem processes and services [44]. Co-occurrence networks reveal dominant microorganisms and closely interacting microbial communities that are critical for preserving the stability of the microbial community structure and functioning in the environment [45]. Li's (2023) showed that along the elevation gradient, the distribution patterns of bacterial and fungal community symbiosis networks differed, possibly due to ecological niche differentiation caused by the high variability of the soil environment and plant communities [46]. We speculated that the structure of the co-occurrence network between soil microbial communities would vary depending on whether the slope was shaded or sunny. We found that all four networks showed significantly greater modularity and clustering coefficients compared to the random network. The co-occurrence networks of the soil fungal communities were more interactive than those of the soil bacterial communities, and the increased complexity indicates greater stability and environmental resistance of the fungal communities. This is in line with our assumption. Co-occurrence networks of the microbial communities on the shaded and sunny slopes changed significantly, but the positive correlation ratios were both much larger than the negative correlation ratios, indicating that both had synergistic effects and that there was more cooperation and reciprocity than competition in the microbial communities [47–49]. In the bacterial community, the number of nodes, the number of links, and the positive correlation links were slightly increased in the sunny slope compared to the shaded slope, indicating that the co-occurrence network of the bacterial community was more developed in the sunny slope. The dominant phylum Actinobacteriota, Acidobacteriota, Proteobacteria, and Chloroflexi were more strongly associated with the bacterial community. The total number of nodes, links, and positively correlated links in the fungal community was greater on the shady side of the slope than on the sunny side. Ascomycota were the most significant elements in the fungal community. The distribution and co-occurrence network of fungal and bacterial communities in subalpine meadow ecosystems varied with slope direction.

5. Conclusions

Our study showed that altitude and slope orientation play a significant role in the variation of soil physicochemical properties and have a significant effect on the composition and diversity of soil fungal and bacterial communities. Altitude had a significant influence on soil pH, soil bulk density, total nitrogen, total phosphorus and ammonium nitrogen, soil moisture, soil organic carbon, soil total potassium, and available potassium, and slope orientation had a significant influence on soil bulk density, total potassium and available phosphorus, soil moisture, and available potassium. Significant variations were found between the main phyla and genera found in the soil bacterial and fungal communities of subalpine meadows located at different elevations and along different slopes. The bacterial dominant phyla were Actinobacteriota (28.84%), Proteobacteria (21.74%), and Acidobacteriota (17.08%), while the fungal dominant phyla were Ascomycota (51.19%), Mortierellomycota (24.28%) and Basidiomycota (18.80%). An increase in elevation had a significant effect on Shannon and PD diversity indices of soil bacterial communities, showing a decreasing and then increasing trend. The soil fungal community showed a single-peaked trend of increasing and then decreasing with increasing altitude. Except at 1700 m, the diversity of soil fungi was greater on shady slopes than on sunny slopes at the same altitude. Bacterial diversity was greater than fungal diversity overall. The co-occurrence network of communities of fungi was considerably more intricate than that of communities of bacteria, with the bacterial community network being more complex in soils on sunny slopes and the fungal community network being more developed in soils on shady slopes. Soil total potassium, ammonium nitrogen, and soil organic carbon were the primary factors influencing the community structure of bacteria and fungi. Additional research on soil microbial communities in subalpine meadows would allow us to comprehend the dispersion of microbial communities in the various topographies present in this delicate ecosystem. It would also establish a theoretical foundation to anticipate the reflection, adaptations, and feedback of microbial communities' changes with the environment, which is pivotal in assessing the sustainability of meadow ecosystems and having a comprehensive understanding of microbial ecology.

Supplementary Materials: The following supporting information can be downloaded at: https://www.mdpi.com/article/10.3390/microorganisms11122878/s1, Figure S1: Location of the research area; Figure S2: Rarefaction contours of operational taxonomic units (OTUs) for soil bacterial (A) and fungal (B) communities; Figure S3: Variations in the community structure of soil bacteria (A) and fungi (B) at various altitudes and slope; Table S1: The soil physicochemical properties for South-facing slop and North-facing slop soils in different altitudes; Table S2: Two-way variance analysis of dominant bacterial phyla and classes; Table S3: Two-way variance analysis of dominant fungal phyla and classes.

Author Contributions: Conceptualization, J.Y. and S.L.; Methodology, L.H.; Validation, W.Z.; Formal analysis, X.S.; Investigation, G.Z., Z.C., X.B. and J.Z.; Data curation, J.Y.; Writing—original draft, J.Y.; Writing—review & editing, S.L.; Visualization, J.Y. All authors have read and agreed to the published version of the manuscript.

Funding: This study was funded by the Special Program for Survey of National Basic Scientific and Technological Resources (No. 2021FY00802).

Data Availability Statement: The data presented in this study are available upon request from the corresponding author.

Conflicts of Interest: The authors declare no conflict of interest.

References

1. He, L.; Sun, X.; Li, S.; Zhou, W.; Chen, Z.; Bai, X. The vertical distribution and control factor of microbial biomass and bacterial community at macroecological scales. *Sci. Total Environ.* **2023**, *869*, 161754. [CrossRef] [PubMed]
2. Fernandez-Guisuraga, J.M.; Marcos, E.; Saenz de Miera, L.E.; Ansola, G.; Pinto, R.; Calvo, L. Short-term responses of ecosystem multifunctionality to fire severity are modulated by fire-induced impacts on plant and soil microbial communities. *Sci. Total Environ.* **2023**, *898*, 165477. [CrossRef]
3. García-Carmona, M.; Lepinay, C.; Mataix-Solera, J.; Baldrian, P.; Arcenegui, V.; Cajthaml, T.; García-Orenes, F. Post-fire wood mulch negatively affects the moss biocrust cover and its positive effects on microbial diversity in a semi-arid Mediterranean forest. *Appl. Soil Ecol.* **2023**, *191*, 105026. [CrossRef]
4. Kooch, Y.; Dolat Zarei, F. The effect of different canopy composition of shrublands on soil quality indicators in a semi-arid climate of Iran. *Geoderma Reg.* **2023**, *34*, e00688. [CrossRef]
5. Kooch, Y.; Nouraei, A.; Haghverdi, K.; Kolb, S.; Francaviglia, R. Landfill leachate has multiple negative impacts on soil health indicators in Hyrcanian forest, northern Iran. *Sci. Total Environ.* **2023**, *896*, 166341. [CrossRef] [PubMed]
6. Sapsirisuk, S.; Polburee, P.; Lorliam, W.; Limtong, S. Discovery of Oleaginous Yeast from Mountain Forest Soil in Thailand. *J. Fungi* **2022**, *8*, 1100. [CrossRef]
7. Cong, J.; Cong, W.; Lu, H.; Zhang, Y. Distinct Elevational Patterns and Their Linkages of Soil Bacteria and Plant Community in An Alpine Meadow of the Qinghai–Tibetan Plateau. *Microorganisms* **2022**, *10*, 1049. [CrossRef] [PubMed]
8. Li, Z.; Yang, Y.; Zheng, H.; Hu, B.; Dai, X.; Meng, N.; Yan, D. Environmental changes drive soil microbial community assembly across arid alpine grasslands on the Qinghai-Tibetan Plateau, China. *CATENA* **2023**, *228*, 107175. [CrossRef]
9. Bhople, P.; Samad, A.; Šišić, A.; Antonielli, L.; Sessitsch, A.; Keiblinger, K.; Djukic, I.; Zehetner, F.; Zechmeister-Boltenstern, S.; Joergensen, R.G.; et al. Variations in fungal community structure along elevation gradients in contrasting Austrian Alpine ecosystems. *Appl. Soil Ecol.* **2022**, *177*, 104508. [CrossRef]
10. Fierer, N.; McCain, C.M.; Meir, P.; Zimmermann, M.; Rapp, J.M.; Silman, M.R.; Knight, R. Microbes do not follow the elevational diversity patterns of plants and animals. *Ecology* **2011**, *92*, 797–804. [CrossRef]
11. Li, J.; Li, C.; Kou, Y.; Yao, M.; He, Z.; Li, X. Distinct mechanisms shape soil bacterial and fungal co-occurrence networks in a mountain ecosystem. *FEMS Microbiol. Ecol.* **2020**, *96*, fiaa030. [CrossRef] [PubMed]

12. Pandey, R.; Bargali, S.S.; Bargali, K.; Karki, H.; Kumar, M.; Sahoo, U.K. Fine root dynamics and associated nutrient flux in Sal dominated forest ecosystems of Central Himalaya, India. *Front. For. Glob. Chang.* **2023**, *5*, 1064502. [CrossRef]
13. Adamczyk, M.; Hagedorn, F.; Wipf, S.; Donhauser, J.; Vittoz, P.; Rixen, C.; Frossard, A.; Theurillat, J.P.; Frey, B. The Soil Microbiome of GLORIA Mountain Summits in the Swiss Alps. *Front. Microbiol.* **2019**, *10*, 1080. [CrossRef] [PubMed]
14. Soares, M.; Rousk, J. Microbial growth and carbon use efficiency in soil: Links to fungal-bacterial dominance, SOC-quality and stoichiometry. *Soil Biol. Biochem.* **2019**, *131*, 195–205. [CrossRef]
15. Ramm, E.; Ambus, P.L.; Gschwendtner, S.; Liu, C.; Schloter, M.; Dannenmann, M. Fire intensity regulates the short-term postfire response of the microbiome in Arctic tundra soil. *Geoderma* **2023**, *438*, 116627. [CrossRef]
16. Roy, F.; Ibayev, O.; Arnstadt, T.; Bassler, C.; Borken, W.; Gross, C.; Hoppe, B.; Hossen, S.; Kahl, T.; Moll, J.; et al. Nitrogen addition increases mass loss of gymnosperm but not of angiosperm deadwood without changing microbial communities. *Sci. Total Environ.* **2023**, *900*, 165868. [CrossRef]
17. Runnel, K.; Tamm, H.; Kohv, M.; Pent, M.; Vellak, K.; Lodjak, J.; Lohmus, A. Short-term responses of the soil microbiome and its environment indicate an uncertain future of restored peatland forests. *J. Environ. Manag.* **2023**, *345*, 118879. [CrossRef]
18. Kewessa, G.; Dejene, T.; Alem, D.; Tolera, M.; Martín-Pinto, P. Forest Type and Site Conditions Influence the Diversity and Biomass of Edible Macrofungal Species in Ethiopia. *J. Fungi* **2022**, *8*, 1023. [CrossRef]
19. Łomża, P.; Krucoń, T.; Tabernacka, A. Potential of Microbial Communities to Perform Dehalogenation Processes in Natural and Anthropogenically Modified Environments—A Metagenomic Study. *Microorganisms* **2023**, *11*, 1702. [CrossRef]
20. Craig, M.E.; Geyer, K.M.; Beidler, K.V.; Brzostek, E.R.; Frey, S.D.; Stuart Grandy, A.; Liang, C.; Phillips, R.P. Fast-decaying plant litter enhances soil carbon in temperate forests but not through microbial physiological traits. *Nat. Commun.* **2022**, *13*, 1229. [CrossRef]
21. Zhou, H.; Zhang, D.; Jiang, Z.; Sun, P.; Xiao, H.; Yuxin, W.; Chen, J. Changes in the soil microbial communities of alpine steppe at Qinghai-Tibetan Plateau under different degradation levels. *Sci. Total Environ.* **2019**, *651*, 2281–2291. [CrossRef]
22. Li, Q.; He, G.; Wen, T.; Zhang, D.; Liu, X. Distribution pattern of soil fungi community diversity in alpine meadow in Qilian Mountains of eastern Qinghai-Tibetan Plateau. *Ecol. Indic.* **2022**, *141*, 109054. [CrossRef]
23. Sokołowska, J.; Józefowska, A.; Woźnica, K.; Zaleski, T. Succession from meadow to mature forest: Impacts on soil biological, chemical and physical properties—Evidence from the Pieniny Mountains, Poland. *CATENA* **2020**, *189*, 104503. [CrossRef]
24. He, P.; Ling, N.; Lü, X.-T.; Zhang, H.-Y.; Wang, C.; Wang, R.-Z.; Wei, C.-Z.; Yao, J.; Wang, X.-B.; Han, X.-G.; et al. Contributions of abundant and rare bacteria to soil multifunctionality depend on aridity and elevation. *Appl. Soil Ecol.* **2023**, *188*, 104881. [CrossRef]
25. Dumonceaux, T. Composition and Dynamics of Plant- and Soil-Associated Microbial Communities in Forest and Agricultural Ecosystems. *Microorganisms* **2023**, *11*, 1782. [CrossRef] [PubMed]
26. Yudina, A.; Ovchinnikova, O.; Cheptsov, V.; Fomin, D. Localization of C Cycle Enzymes in Arable and Forest Phaeozems within Levels of Soil Microstructure. *Microorganisms* **2023**, *11*, 1343. [CrossRef] [PubMed]
27. Teague, R.; Dowhower, S. Links of microbial and vegetation communities with soil physical and chemical factors for a broad range of management of tallgrass prairie. *Ecol. Indic.* **2022**, *142*, 109280. [CrossRef]
28. Sushko, S.; Ovsepyan, L.; Gavrichkova, O.; Yevdokimov, I.; Komarova, A.; Zhuravleva, A.; Blagodatsky, S.; Kadulin, M.; Ivashchenko, K. Contribution of microbial activity and vegetation cover to the spatial distribution of soil respiration in mountains. *Front. Microbiol.* **2023**, *14*, 1165045. [CrossRef] [PubMed]
29. Zald HS, J.; Callahan, C.C.; Hurteau, M.D.; Goodwin, M.J.; North, M.P. Tree growth responses to extreme drought after mechanical thinning and prescribed fire in a Sierra Nevada mixed-conifer forest, USA. *For. Ecol. Manag.* **2022**, *510*, 120107. [CrossRef]
30. Chen, J.; Shi, Z.; Liu, S.; Zhang, M.; Cao, X.; Chen, M.; Xu, G.; Xing, H.; Li, F.; Feng, Q. Altitudinal Variation Influences Soil Fungal Community Composition and Diversity in Alpine-Gorge Region on the Eastern Qinghai-Tibetan Plateau. *J. Fungi* **2022**, *8*, 807. [CrossRef]
31. Frey, B.; Rime, T.; Phillips, M.; Stierli, B.; Hajdas, I.; Widmer, F.; Hartmann, M. Microbial diversity in European alpine permafrost and active layers. *FEMS Microbiol. Ecol.* **2016**, *92*, fiw018. [CrossRef] [PubMed]
32. Ji, L.; Shen, F.; Liu, Y.; Yang, Y.; Wang, J.; Purahong, W.; Yang, L. Contrasting altitudinal patterns and co-occurrence networks of soil bacterial and fungal communities along soil depths in the cold-temperate montane forests of China. *CATENA* **2022**, *209*, 105844. [CrossRef]
33. Peay, K.G.; von Sperber, C.; Cardarelli, E.; Toju, H.; Francis, C.A.; Chadwick, O.A.; Vitousek, P.M. Convergence and contrast in the community structure of Bacteria, Fungi and Archaea along a tropical elevation-climate gradient. *FEMS Microbiol. Ecol.* **2017**, *93*, fix045. [CrossRef] [PubMed]
34. Shankar, A.; Garkoti, S.C. Influence of forest types on soil physicochemical and biological characteristics of associated agroecosystems in the central Himalaya. *Sci. Total Environ.* **2023**, *906*, 167731. [CrossRef] [PubMed]
35. Margesin, R.; Jud, M.; Tscherko, D.; Schinner, F. Microbial communities and activities in alpine and subalpine soils. *FEMS Microbiol. Ecol.* **2009**, *67*, 208–218. [CrossRef]
36. Cui, Y.; Bing, H.; Fang, L.; Wu, Y.; Yu, J.; Shen, G.; Jiang, M.; Wang, X.; Zhang, X. Diversity patterns of the rhizosphere and bulk soil microbial communities along an altitudinal gradient in an alpine ecosystem of the eastern Tibetan Plateau. *Geoderma* **2019**, *338*, 118–127. [CrossRef]

37. Ai, S.; Chen, J.; Gao, D.; Ai, Y. Distribution patterns and drivers of artificial soil bacterial community on cut-slopes in alpine mountain area of southwest China. *CATENA* **2020**, *194*, 104695. [CrossRef]
38. Labouyrie, M.; Ballabio, C.; Romero, F.; Panagos, P.; Jones, A.; Schmid, M.W.; Mikryukov, V.; Dulya, O.; Tedersoo, L.; Bahram, M.; et al. Patterns in soil microbial diversity across Europe. *Nat. Commun.* **2023**, *14*, 3311. [CrossRef]
39. Wang, C.; Guo, L.; Shen, R.F. Rare microbial communities drive ecosystem multifunctionality in acidic soils of southern China. *Appl. Soil Ecol.* **2023**, *189*, 104895. [CrossRef]
40. Wang, Q.; Liu, K.; Tao, K.; Hou, T. Biogeographical patterns and drivers of bacterial community in the Qinghai-Tibetan Plateau. *Appl. Soil Ecol.* **2023**, *183*, 104757. [CrossRef]
41. Zheng, Q.; Hu, Y.; Zhang, S.; Noll, L.; Bockle, T.; Dietrich, M.; Herbold, C.W.; Eichorst, S.A.; Woebken, D.; Richter, A.; et al. Soil multifunctionality is affected by the soil environment and by microbial community composition and diversity. *Soil Biol. Biochem.* **2019**, *136*, 107521. [CrossRef]
42. Przemieniecki, S.W.; Kosewska, A.; Purwin, C.; Zapałowska, A.; Mastalerz, J.; Kotlarz, K.; Kolaczek, K. Biometric, chemical, and microbiological evaluation of common wheat (*Triticum aestivum* L.) seedlings fertilized with mealworm (*Tenebrio molitor* L.) larvae meal. *Appl. Soil Ecol.* **2021**, *167*, 104037. [CrossRef]
43. Llado, S.; Lopez-Mondejar, R.; Baldrian, P. Forest Soil Bacteria: Diversity, Involvement in Ecosystem Processes, and Response to Global Change. *Microbiol. Mol. Biol. Rev.* **2017**, *81*, e00063-16. [CrossRef]
44. Terrer, C.; Phillips, R.P.; Hungate, B.A.; Rosende, J.; Pett-Ridge, J.; Craig, M.E.; van Groenigen, K.J.; Keenan, T.F.; Sulman, B.N.; Stocker, B.D.; et al. A trade-off between plant and soil carbon storage under elevated CO(2). *Nature* **2021**, *591*, 599–603. [CrossRef] [PubMed]
45. Schaffer, N.; Mariot, R.F.; Gimenes, G.H.; Carlos, F.S.; Roesch, L.F.W.; Garrido, L.R.; de Melo, G.W.B.; de Sá, E.S.; Camargo, F.A.d.O. Vineyard renewal reduces copper and zinc bioavailability and increases microbial diversity in southern Brazil. *Ecol. Eng.* **2023**, *194*, 106995. [CrossRef]
46. Li, J.; Wang, X.; Wu, J.H.; Sun, Y.X.; Zhang, Y.Y.; Zhao, Y.F.; Huang, Z.; Duan, W.H. Climate and geochemistry at different altitudes influence soil fungal community aggregation patterns in alpine grasslands. *Sci. Total Environ.* **2023**, *881*, 163375. [CrossRef]
47. D'Alo, F.; Baldrian, P.; Odriozola, I.; Morais, D.; Vetrovsky, T.; Zucconi, L.; Ripa, C.; Cannone, N.; Malfasi, F.; Onofri, S. Composition and functioning of the soil microbiome in the highest altitudes of the Italian Alps and potential effects of climate change. *FEMS Microbiol. Ecol.* **2022**, *98*, fiac025. [CrossRef] [PubMed]
48. Ruiz, L.; Carrión-Paladines, V.; Vega, M.; López, F.; Benítez, Á. Biological Crust Diversity Related to Elevation and Soil Properties at Local Scale in a Montane Scrub of Ecuador. *J. Fungi* **2023**, *9*, 386. [CrossRef]
49. Leite, M.F.A.; van den Broek, S.W.E.B.; Kuramae, E.E. Current Challenges and Pitfalls in Soil Metagenomics. *Microorganisms* **2022**, *10*, 1900. [CrossRef]

microorganisms

MDPI

Article

The Salinity Survival Strategy of *Chenopodium quinoa*: Investigating Microbial Community Shifts and Nitrogen Cycling in Saline Soils

Xuli Zhao [1], Tianzhu Meng [1,*], Shenghan Jin [1], Kaixing Ren [1], Zhe Cai [1], Bo Cai [1] and Saibao Li [2]

[1] College of Agricultural Science and Engineering, Hohai University, No. 8 Focheng West Road, Nanjing 211100, China

[2] College of Water Resources and Civil Engineering, Tibet Agricultural and Animal Husbandry University, No. 8 Xueyuan Road, Linzhi 860000, China

* Correspondence: tianzhumeng@hhu.edu.cn; Tel.: +86-138-5198-8983

Abstract: Quinoa is extensively cultivated for its nutritional value, and its exceptional capacity to endure elevated salt levels presents a promising resolution to the agricultural quandaries posed by salinity stress. However, limited research has been dedicated to elucidating the correlation between alterations in the salinity soil microbial community and nitrogen transformations. To scrutinize the underlying mechanisms behind quinoa's salt tolerance, we assessed the changes in microbial community structure and the abundance of nitrogen transformation genes across three distinct salinity thresholds ($1\ g \cdot kg^{-1}$, $3\ g \cdot kg^{-1}$, and $6\ g \cdot kg^{-1}$) at two distinct time points (35 and 70 days). The results showed the positive effect of quinoa on the soil microbial community structure, including changes in key populations and its regulatory role in soil nitrogen cycling under salt stress. *Choroflexi*, *Acidobacteriota*, and *Myxococcota* were inhibited by increased salinity, while the relative abundance of *Bacteroidota* increased. *Proteobacteria* and *Actinobacteria* showed relatively stable abundances across time and salinity levels. Quinoa possesses the ability to synthesize or modify the composition of keystone species or promote the establishment of highly complex microbial networks (modularity index > 0.4) to cope with fluctuations in external salt stress environments. Furthermore, quinoa exhibited nitrogen (N) cycling by downregulating denitrification genes (*nirS*, *nosZ*), upregulating nitrification genes (Archaeal *amoA* (AOA), Bacterial *amoA* (AOB)), and stabilizing nitrogen fixation genes (*nifH*) to absorb nitrate–nitrogen (NO_3^-_N). This study paves the way for future research on regulating quinoa, promoting soil microbial communities, and nitrogen transformation in saline environments.

Keywords: *Chenopodium quinoa*; saline soils; microbial community; nitrogen transformations; salt tolerance

1. Introduction

Soil salinization is a widespread environmental stressor that adversely affects plant growth and soil health [1]. It is estimated that salt-affected soil covers approximately 8.7% of Earth's land area [2]. The low permeability of saline soils directly inhibits the structure and metabolic activity of the soil microbial community [3,4]. It greatly impacts microbial-mediated soil ecological functions and a range of processes related to soil fertility [5]. Additionally, salinization slows down or inhibits various nitrification processes in the soil, leading to the accumulation of nitrite or accelerated loss of ammonia–nitrogen (NH_4^+) through volatilization [6]. This disruption disrupts the normal nitrogen (N) transformation in the soil and affects plant uptake and the utilization of N, as well as soil microbial activity [7].

Quinoa (*Chenopodium quinoa*), renowned for its ability to grow in extreme environments and high nutritional value [8], serves as an ideal crop model for studying salt stress tolerance through interactions among microorganisms, plants, and soil [9–11]. Studies reveal that quinoa can enhance soil microbial diversity and functionality in saline ecosystems [11–13]. Therefore, quinoa can be utilized to rebuild the composition and functionality of microbial communities in saline soils, aiming to enhance and safeguard the essential physicochemical and biological properties of salt-affected soils [14].

Keystones have been demonstrated to have the potential to promote nutrient transformation and plant growth [15]. Furthermore, the biodiversity of keystones largely determines the functional capability of soil microbial communities [16]. Hence, focusing on keystones offers the potential to organize information regarding microbial interactions more effectively, simplify microbiome analysis, and identify the critical components of individual genera within microbial communities [17,18]. Microbial communities function as keystones that regulate the interactions and functional aspects of microbial community dynamics [19,20]. For instance, the nutrient cycling of nitrogen relies heavily on various nitrogen transformation reactions performed by diverse and versatile microorganisms [21]. This includes functional genes involved in nitrogen transformations such as nitrogen fixation (*nifH*), nitrification (Archaeal *amoA* (AOA), Bacterial *amoA* (AOB)), and denitrification (*nirK*, *nirS*, *nosZ*). These nitrogen transformations are strongly inhibited in saline environments due to the detrimental effects of salt on functional microorganisms [22]. However, there have been limited studies examining the mechanisms of salt tolerance in quinoa from the perspective of soil microbial communities and nitrogen transformations.

So, to investigate how quinoa enhances microbial resistance to salt stress and influences the expression of nitrogen transformation genes in various saline environments, this study primarily focuses on observing the seedling stage of quinoa (the critical adaptation of quinoa to adverse environmental conditions). Accordingly, this study aims to address the following questions: (1) How does quinoa influence the structure of soil microbial communities under different salinity levels? (2) What are the characteristics of key microbial taxa associated with quinoa in the soil exposed to varying salt stress conditions? (3) What are the effects of quinoa, under salt stress, on microbial nitrogen transformation processes? Overall, this study aims to explore the impact of quinoa on soil microbial communities, keystones, and microbial N transformations under salt stress across three salinity levels. Understanding the succession of soil microbial communities in quinoa is crucial for scientific cultivation and salinization management.

2. Material and Methods

2.1. Experimental Design and Soil Processing

This study was conducted at the Water-Saving Experimental Station of Hohai University in the Jiangning district (31°91′ N, 118°79′ E), Nanjing, Jiangsu Province, China from October to December 2022. The region is characterized by its subtropical humid monsoon climate. Figure 1 illustrates the average monthly temperature values and fluctuations observed in the greenhouse.

The quinoa seeds utilized in this study were obtained from the Seed Base in Suqian, Jiangsu Province, China. The saline soil used in the experiment was sourced from the demonstration field of Jianfeng Agricultural Industry Co., Ltd., located in the coastal mudflats of Yancheng, Jiangsu Province, China. The original physicochemical properties of the soil are shown in Table 1. A two-factor randomized complete block design with three replications was established, examining six different treatments that varied in the number of cultivation days and salt levels (Figure 2). To alleviate the "pulse effect", a two-week dark treatment was conducted after accurately introducing sodium chloride (NaCl) to establish controlled salinity levels of 1 $g \cdot kg^{-1}$, 3 $g \cdot kg^{-1}$, and 6 $g \cdot kg^{-1}$, respectively. On 24 October, two quinoa seedlings were planted in each pot. Destructive sampling was conducted at 35 and 70 days of cultivation, respectively. During the cultivation period, the plants were irrigated moderately based on the greenhouse's humidity and soil conditions. They were nourished with Hogland nutrition solution twice a week, with a dilution ratio of $1/2\times$.

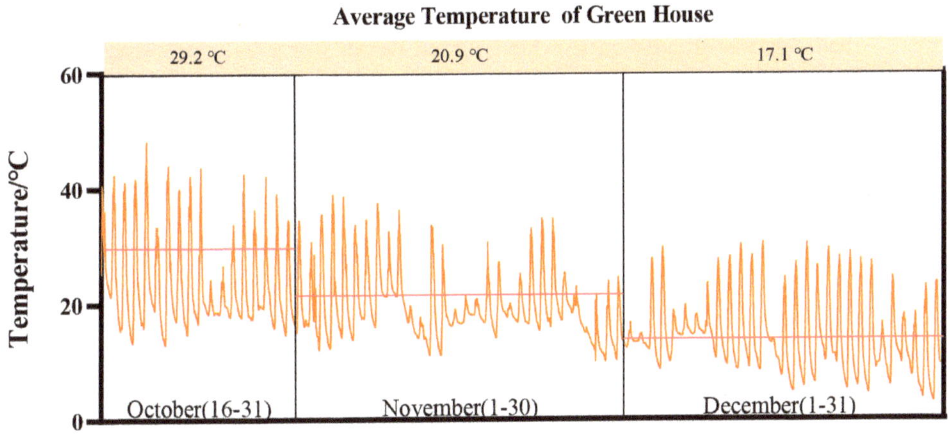

Figure 1. The average temperature of the greenhouse.

Table 1. The initial rationalization properties of the tested soil samples.

pH	EC (ms·cm^{-1})	Salinity (g·kg^{-1})	TOC (%)	TN (g·kg^{-1})	AP (mg·g^{-1})	SWC (%)
7.35 ± 0.03	0.28 ± 0.01	0.72 ± 0.10	1.47 ± 0.03	0.94 ± 0.03	23.89 ± 0.20	27.87

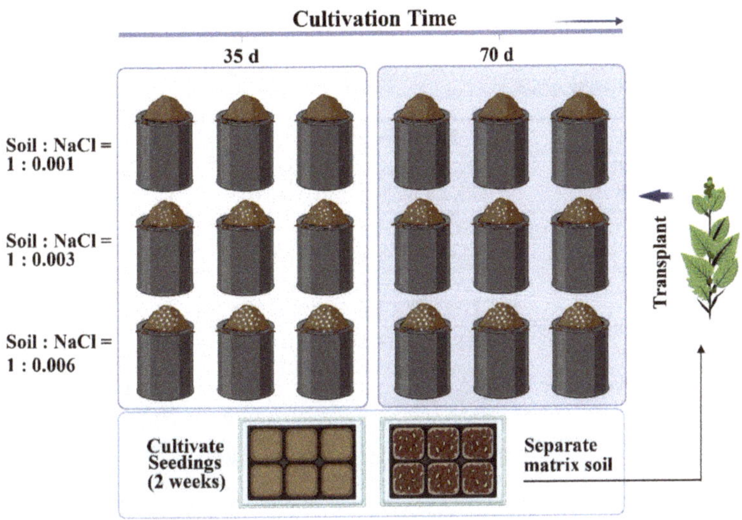

Figure 2. Experiential design. (Untreated and artificially treated soil was filled in each experimental pot; height 22 cm; diameter 18 cm).

After the harvest, all samples were divided into two parts: one part was stored at 4 °C for a physicochemical analysis, and the other part was stored at −80 °C for subsequent DNA extraction. The measured soil physicochemical properties include pH, EC (electrical conductivity), the amounts of water-soluble salts in the soil, ammonia–nitrogen (NH_4^+_N), and nitrate–nitrogen (NO_3^-_N).

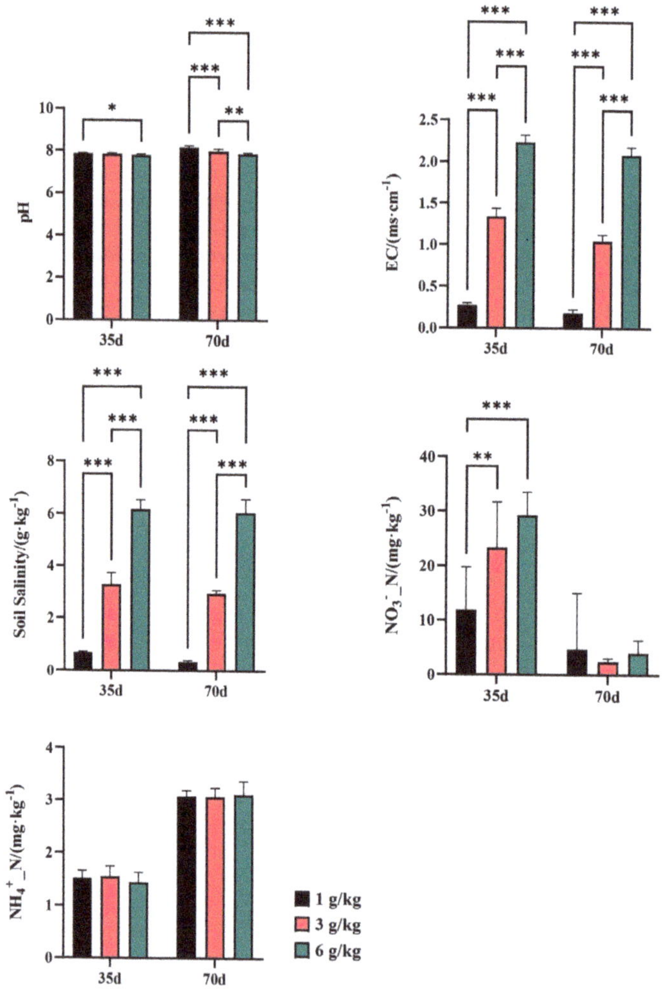

Figure 4. Basic physical and chemical properties of soil; '*' for $p < 0.05$; '**' for $p < 0.01$; '***' for $p < 0.001$.

The NMDS ordination showed the differences between the bacterial and fungal communities in terms of diversity. Soil salt concentration and cultivation period are two main factors that significantly distinguish bacteria and fungi (Figure 3b). A permutational multivariate analysis of variance (PREMANOVA) further corroborated that the variations in the bacterial communities were mainly driven by soil water-soluble salt conversion (Table 4). However, both the cultivation period and the salinity gradient did not exhibit a significant influence on the alterations in fungal communities. This indicates that fungi are more stable than bacteria under salt stress, and Rath et al. [36] have also observed the same conclusion. It can be explained that variations in bacterial communities under salt stress can affect the penetration of quinoa. Additionally, Otlewska et al. [37] have reported that microbial communities directly/indirectly participate in the osmotic adjustment of plants.

Table 4. The effects of soil salinity and incubation time on the differentiation of soil microbial communities based on permutational multivariate analysis of variance (PERMANOVA).

Microbial Community	Phylum	Salinity	Day
Bacteria	R^2	0.236	0.129
	P	0.002	0.031
	Sig.	**	*
Fungi	R^2	0.195	0.086
	P	0.085	0.243
	Sig.	-	-

Note: '*' for $p < 0.05$, '**' for $p < 0.01$.

Quinoa benefits from being associated with different microbiological communities, and saline soil is a crucial environmental filter that can promote quinoa to choose soil microbial communities with highly specific traits (Qin et al., 2016 [38]). Figure 3c represents the reduction in bacterial-specific and fungal-specific OUTs in different treatment groups. This means that the functions of quinoa may be altered due to changes in microbial diversity [39]. Numerous studies have highlighted the significance of microbiome diversity in aiding plant adaptation to saline environments [11,40,41]. Toubali and Meddich [11] found that microbial communities promoted the growth of quinoa plants in saline environments and improved soil chemical quality by enhancing enzyme-driven antioxidant and osmotic regulation systems. Different bacterial communities exhibit varying responses to high salt concentrations. Salt stress exerts inhibitory effects on microorganisms such as *Choroflexi*, *Acidobacteriota*, and *Myxococcota*. However, as salt concentrations increase, there is an increasing trend in the abundance of *Bacteroidota*, indicating their ability to thrive in high-salt environments (Figure 5). This discovery aligns with the findings conducted by Rath et al. [42]. These microorganisms exhibiting distinct thresholds in response to salt stress may be selectively favored or recruited by quinoa, leading to the maintenance of a relatively stable population. Similar findings have been observed in studies on rice [43] and *Arabidopsis thaliana* [44]. *Proteobacteria*, *Chloroflexi*, and *Actinobacteria* are the predominant components of soil bacterial communities, and their relative proportions remain relatively constant over time intervals (Figure 6). Additionally, the abundance of Proteobacteria and Actinobacteria is considered a reliable indicator of soil health [45]. In the treatments of CQL1, CQH1, and CQH2, the average relative abundance of *Acidobacteriota*, *Bacteroidota*, and *Gemmatimonadota* was significantly higher, indicating their enrichment in the soil microbial communities (Figure 5). Quinoa and microorganisms engage in mutual interactions that lead to the establishment of specific associations between the plant and certain bacteria at different stages of growth in a salty environment [46]. Importantly, halophytes possess a natural reservoir of salt-resistant microorganisms, which facilitates their growth and development under unfavorable conditions [47]. Moreover, prolonged fluctuations in salinity can contribute to the adaptation of microbial communities [47]. The finding that the average relative abundance of *Planctomycetota* and *Gemmatimonadota* increases with rising salt concentrations and longer growth periods also supports this perspective (Figure 5).

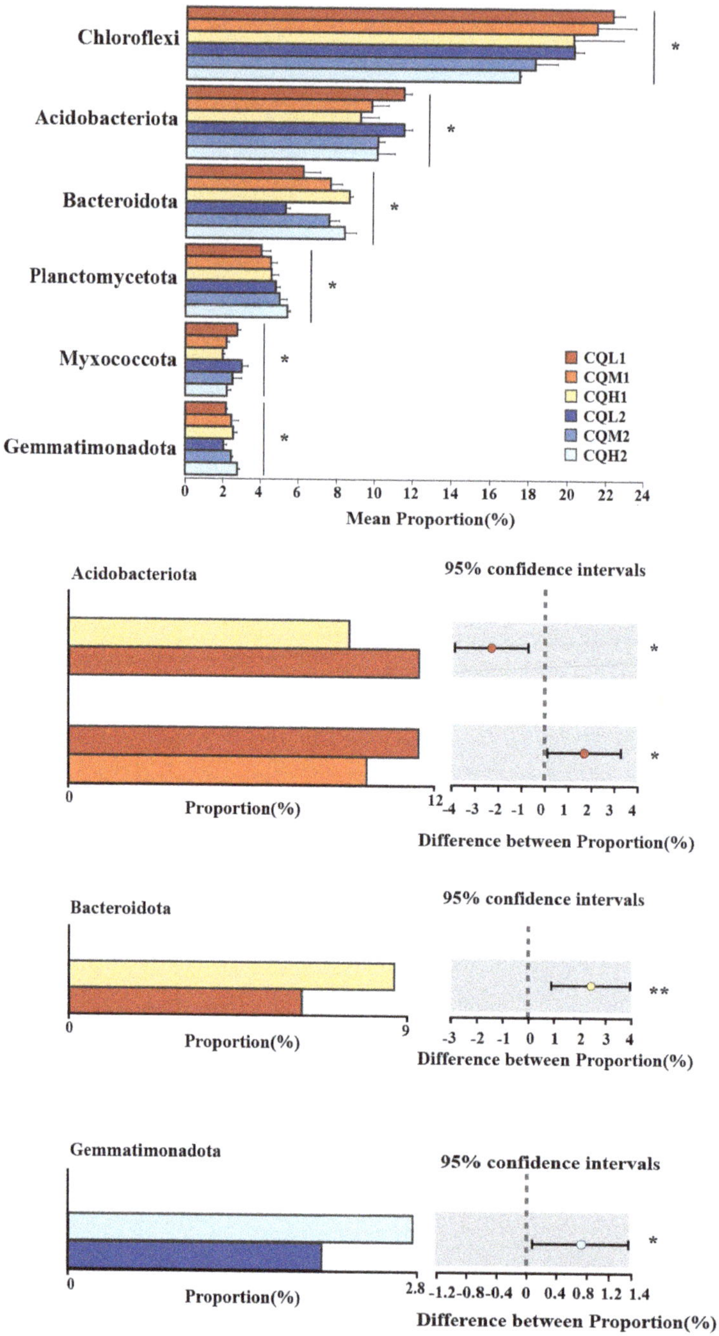

Figure 5. Differentially abundant microbiome at the phylum level; '*' for $p < 0.05$; '**' for $p < 0.01$.

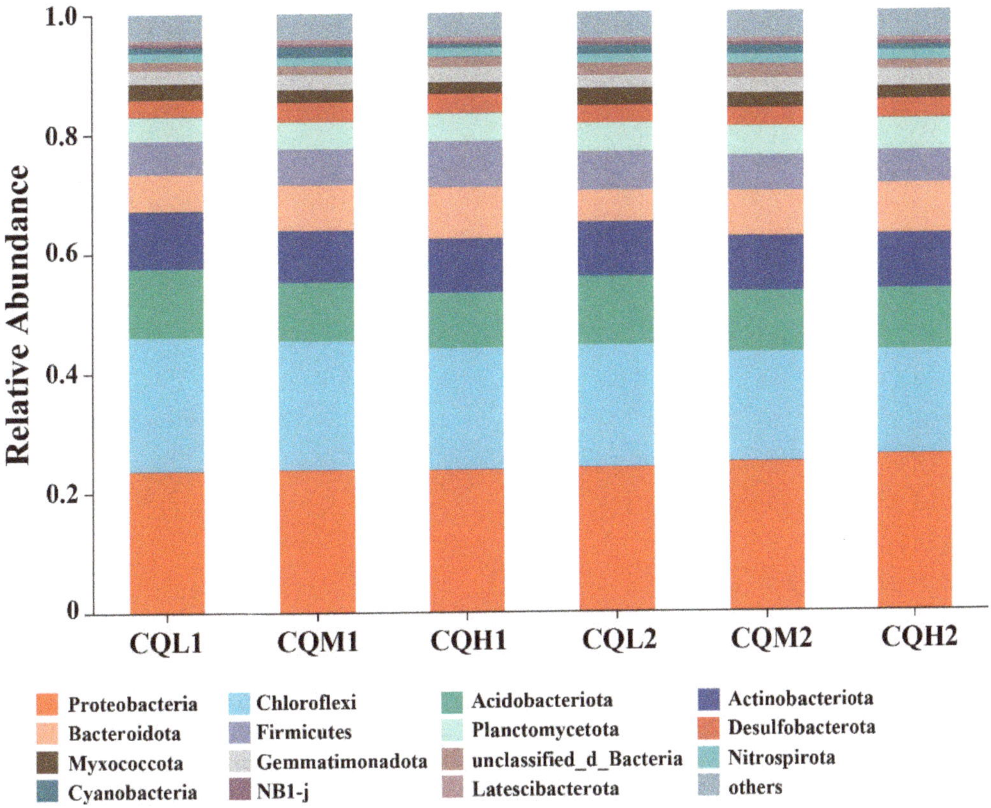

Figure 6. Relative abundance of microbial communities at the phylum level in six treatments.

3.2. Soil Microbial Co-Occurrence and Keystone Taxa

Co-occurrence networks are frequently used to explore potential connections between microbial populations [22,48,49]. At the genus level, co-occurrence networks were constructed for the bacterial and fungal communities, and keystones were identified to thoroughly investigate the impacts of soil salinity and cultivation days on the soil microbial communities. All genus occurred in the network of the six treatments. Keystones are selected as benchmarks based on the 1% betweenness centrality distribution, considering the top five abundance ratings among these genera [20]. The bacteria keystone taxa in CQL1, CQM1, and CQH1 are seven, nine, and six, respectively, and the bacteria keystone taxa for CQL2, CQM2, and CQH2 are five (Figure 7a). The fungi keystone taxa for CQL1, CQM1, and CQH1 are 10, 12, and 10, respectively, and the fungi Keystone taxa for CQL2, CQM2, and CQH2 are all 10 (Figure 7b). The results demonstrate the consistent response of microbial communities to diverse environmental factors. These genera, through their shared environmental preferences and dispersal capabilities, indicate important classification units that are interconnected and hold significant value [50].

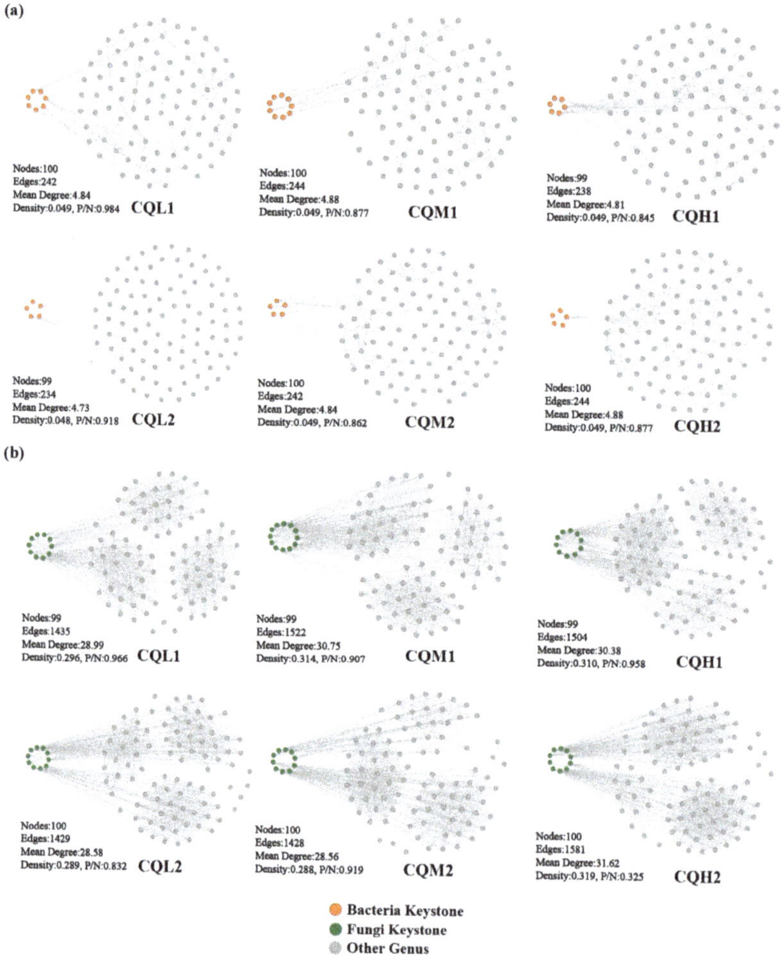

Figure 7. Co-occurrence networks of microbial communities in different treatment groups at the genus level. Topological properties of co-occurrence networks were indicated under each network, including the number of nodes, number of edges, mean degree, density, and the ratio of positive to negative interactions (P/N) of the whole network. (**a**) Networks of bacterial communities, (**b**) Networks of fungal communities.

The modularity index of each treatment exceeds 0.4, indicating a characteristic modular structure in the microbial networks of each process (Figure 8). In our study, as the degree of salt stress increased, there were no significant changes observed in the nodes, edges, and network topological indices of both the bacterial and fungal networks (Figure 7). This indicates that salt stress appears to have no negative impact on the interactions among microbial communities. This finding contradicts the conclusion proposed by Xu et al. [51] that salt stress weakens the co-occurrence network of rhizosphere microorganisms. So, we analyzed the sampling abundance of these keystones and found that the bacterial communities are grouped into CQM1 and CQH1 and CQM2 and CQH2, separately from CQL1 and CQL2. Although the layer clustering of fungal communities at the phylum level did not follow a consistent pattern, there was a noticeable fluctuation in sample abundance (Figure 9). This suggests that quinoa may, under salt stress, either promote the emergence of new keystone taxa or alter the abundance of existing keystones to maintain a stable microbial structure [52]. It is also possible that quinoa facilitates the establishment of highly connected and complex networks among soil microorganisms during the early stages to cope with fluctuations in external salt stress conditions [53].

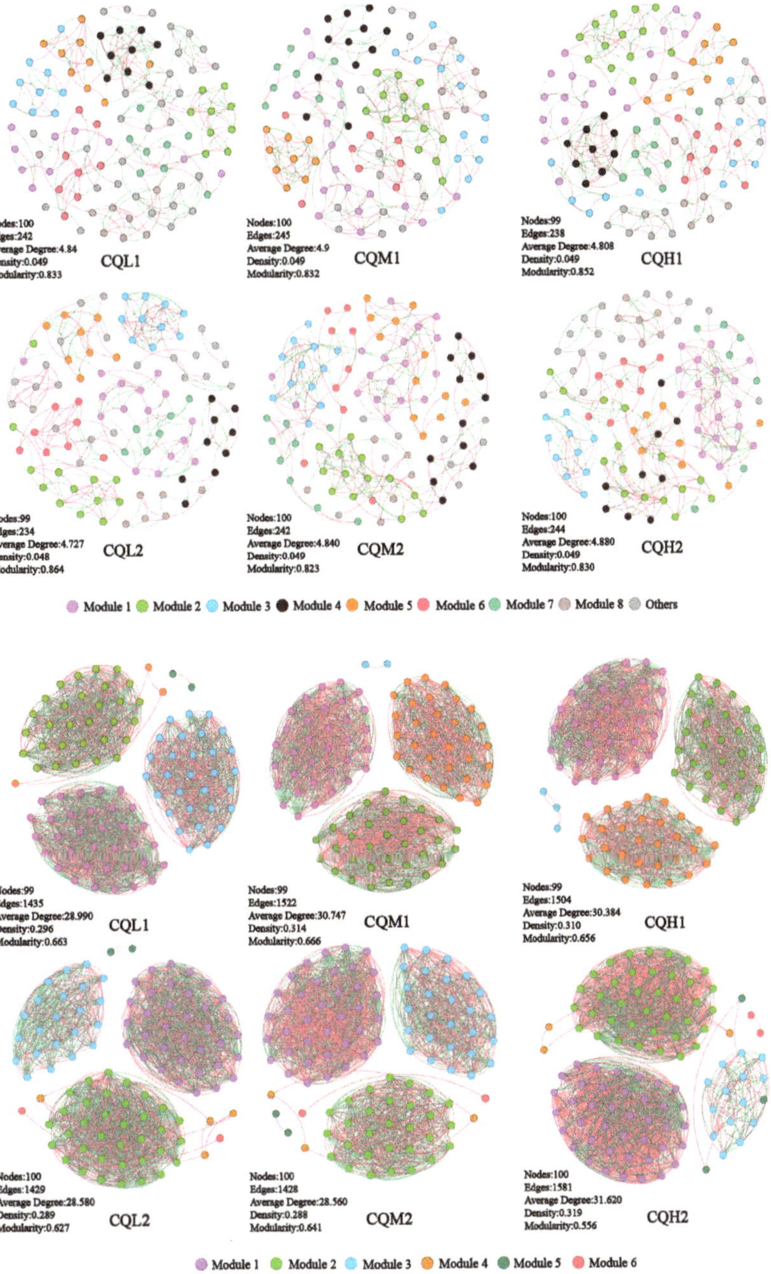

Figure 8. Co-occurrence networks of the microbial communities in the different treatment groups.

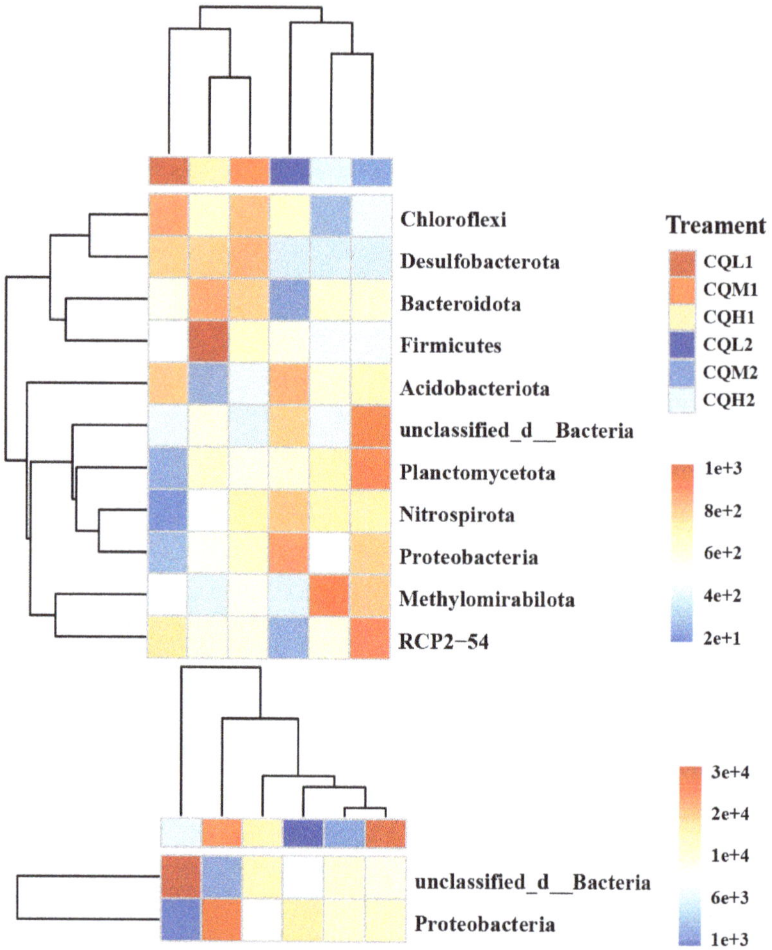

Figure 9. Significant responses of keystone taxa. These keystone taxa are further annotated at the phylum level in the heatmap. Calculating the distance between samples using Euclidean Distance, average linkage clustering determines the relationship between the samples.

In addition, keystones have been demonstrated to have the proficiency to enhance nutrient conversion and promote plant growth [54]. Keystones, at the phylum level, such as *Proteobacteria* and *Acidobacteria*, are often regarded as copiotrophic microorganisms [19]. The abundant bacterial clusters of *Bacteroidetes* and *Firmicutes* can be regarded as highly salt-resistant species. When the soil is treated with appropriate salinity levels, its relative abundance increases compared to that of untreated soil conditions. Our study yielded similar conclusions to [49] (Figure 9). Changes in community composition at the phylum level may significantly impact the functional characteristics of these communities. Previous studies have found that phylogenetic groups belonging to *Firmicutes* can withstand environmental stress through the formation of their Gram-positive cell walls and spores [55]. Pérez Castro et al. [16] discovered a correlation between *Chloroflexi* and genes related to the metabolism of amino sugars, sugar alcohols, and simple carbohydrates. Therefore, we propose that quinoa enhances its stress resilience by directly influencing microbial abundance and the synthesis of keystones or indirectly participating in microbially mediated nutrient transformations.

3.3. Correlations among the Environmental Variables, Bacterial Community, and Nitrogen Functional Genes

A redundancy analysis (RDA) is employed to assess the relationship between the bacterial communities in the different treatment groups and environmental factors, examine the impact of soil properties on community composition at the phylum level, and determine the significance of the association between soil traits and bacterial abundance. The first two axes of the RDA represent 36.96% and 12.92%, respectively, of the total data variation (Figure 10). In bacterial communities, the highest total salinity variation is followed by NH_4^+_N (Table 5). In CQL1, *Chloroflexi* predominate, followed by *Acidobacteriota* at CQL2, *Bacteroidota* at CQM1 and CQH1, and *Proteobacteria* at CQM2 and CQH2. In contrast to *Acidobacteriota* and *Proteobacteria*, which are significantly positively correlated with NH_4^+_N ($p < 0.05$), *Bacteroidota* and *Chloroflexi* have a positive association with NO_3^-_N. Thus, *Bacteroidota* and *Chloroflexi* may have contributed to the conversion and release of NO_3^-_N in the quinoa culture for up to 35 days, while *Acidobacteriota* and *Proteobacteria* may have played a more important role in the transformation and discharge of NH_4^+_N at 70 days, as confirmed by our measurement of the concentrations of NO_3^-_N and NH_4^+_N in the soil (Figure 4). Generally, quinoa has the potential to influence the abundance of bacteria during different stages of cultivation, consequently impacting the conversion of nitrogen by different salt sub-processing groups.

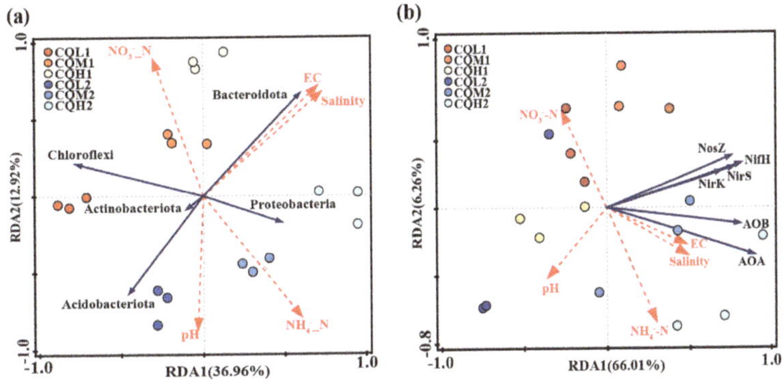

Figure 10. Redundancy analysis based on the relationships between environmental variables, top 5 phyla (**a**), and nitrogen functional genes (**b**) during the overall experimental process. The red arrow represents environmental factors, while the blue arrows respectively represent phyla and nitrogen functional genes.

Table 5. Significant physicochemical parameters and explanations of RDA of (a) top 5 phyla and (b) nitrogen functional genes.

Variable	Explains (%)	F	P
Salinity	24.8	5.3	0.006 **
NH4+_N	22.4	6.4	0.014 *
	(a)		
NH4+_N	41.6	12.8	0.002 **
EC	13.8	6.6	0.018 *
	(b)		

Note: '*' for $p < 0.05$, '**' for $p < 0.01$.

The salinity level plays a crucial role in determining the response of nitrogen metabolism to saline stress [1]. We performed the RDA to examine the relationship between the nitrogen transformation genes of the different processing groups and environmental variables (Figure 10b). The results showed that environmental factors accounted for 75.3% of the total variation in nitrogen conversion genes. The interpretation of NH_4^+_N and EC changes in the composition of nitrogen-transforming genes were identified as the two primary environmental factors, accounting for 41.6% and 13.8%, respectively, of the variation in the six nitrogen transformer genes (Table 5). The limitations imposed by salinity stress on nitrogen transformation in soil biological systems can be understood by considering the prevalence of nitrogen-transforming genes. On the 35th day of cultivation, the abundance of denitrification genes (*nirS* and *nosZ*) decreased with increasing salt content, while the abundance of nitrification genes (AOA and AOB) showed minimal changes (Figure 11). These factors may be the primary reasons for the increase in NO_3^-_N content in the soil (Figure 4). On the 70th day of cultivation, the abundance of denitrification genes (*nirS*, *nosZ*, and *nirK*) showed a positive correlation with salt concentration, while the abundance of nitrification genes (AOA and AOB) exhibited a similar trend (Figure 11). However, there was no significant difference in the NO_3^-_N content of the soil, and it exhibited a sharp decline compared to the content at 35 days (Figure 4). We suggest that this could be explained by quinoa actively absorbing Cl^- to cope with the prolonged salt stress. Simultaneously, the membrane transport proteins that mediate the transport of Cl^- and NO_3^- simultaneously could indirectly facilitate the uptake of NO_3^- from the soil by quinoa [9]. Another possibility is that soil microorganisms are converting NO_3^- into NH_4^+ [50]. Due to the relatively low and stable abundance of the nitrogen fixation gene (*nifH*), microorganisms maintain stable ammonium assimilation throughout different cultivation periods (Figure 11). This makes quinoa more inclined to utilize nitrate–nitrogen as a nitrogen source. Miranda-Apodaca et al. [1] have revealed that quinoa employs precise regulation of NO_3^-_N and Cl^- under salt stress to maintain an optimal root N concentration, enabling effective osmotic regulation. Huang et al. [56] and Zhang et al. [22] also observed similar patterns.

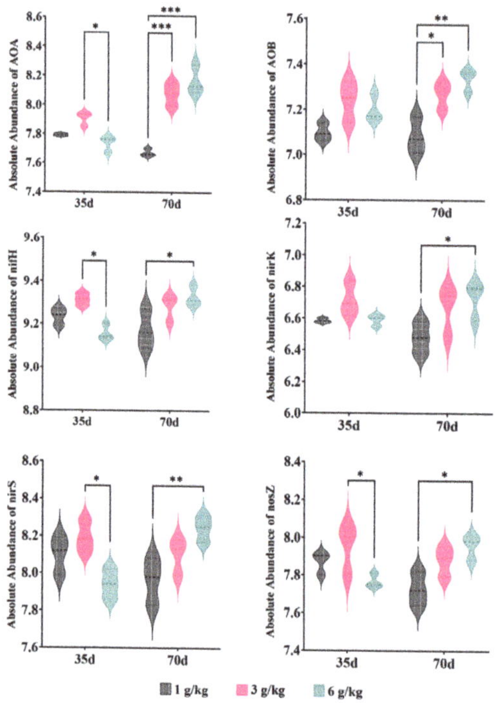

Figure 11. Relative abundance of nitrogen genes; '*' for $p < 0.05$; '**' for $p < 0.01$; '***' for $p < 0.001$.

4. Conclusions

In conclusion, this study explored the effects of quinoa on soil microbial community structure, changes in key populations, and the regulation of soil nitrogen cycling. The results indicate that soil bacterial communities are more susceptible to salt stress than fungal communities. Quinoa mitigates the adverse effects of salt on bacterial communities by absorbing or accumulating inorganic ions, thereby slightly reducing the soil salinity. Microbial diversity plays an essential role in facilitating quinoa's adaptation to saline environments. Certain microbial groups, such as *Bacteroidota*, show increased relative abundance under high salt conditions.

It was observed that different environmental factors consistently influence soil microbial community composition, and salt stress does not seem to have a negative impact on microbial interactions. Quinoa may maintain a stable microbial structure by promoting the emergence of new keystones or altering the abundance of existing keystones. Changes at the different phylum levels of these keystones can significantly affect the functional characteristics of microbial communities. Quinoa may enhance its stress tolerance by directly influencing the microbial abundance and the synthesis of keystones or indirectly participating in microbe-mediated nutrient transformations.

There is a correlation between soil bacterial communities and environmental factors in different treatment groups during the period of cultivation. Soil properties significantly influence bacterial community composition at the phylum level. During quinoa cultivation, different salt treatment groups have varying effects on the conversion of NH_4^+_N and NO_3^-_N. Salt content also influences the abundance of nitrogen transformation genes in the soil. The long-term salt stress adaptation of quinoa may involve the passive absorption of Cl^- and the conversion of NO_3^- by soil microorganisms. These results suggest that quinoa adapts and maintains normal growth under salt stress by regulating nitrogen metabolism and ion absorption.

Author Contributions: Conceptualization, T.M.; methodology, X.Z., S.J., K.R., Z.C., B.C. and S.L.; software, X.Z.; validation, X.Z.; formal analysis, X.Z.; investigation, X.Z. and S.J.; resources, T.M.; data curation, X.Z. and K.R.; writing—original draft preparation, X.Z.; writing—review and editing, X.Z.; visualization, X.Z.; supervision, T.M.; project administration, T.M.; funding acquisition, T.M. All authors have read and agreed to the published version of the manuscript.

Funding: This work was financially supported by the National Key R&D Program of China (No. 2020YFD0900705).

Data Availability Statement: Data are contained within the article.

Conflicts of Interest: The authors declare no conflict of interest.

References

1. Miranda-Apodaca, J.; Agirresarobe, A.; Martínez-Goñi, X.S.; Yoldi-Achalandabaso, A.; Pérez-López, U. N Metabolism Performance in Chenopodium Quinoa Subjected to Drought or Salt Stress Conditions. *Plant Physiol. Biochem.* **2020**, *155*, 725–734. [CrossRef] [PubMed]
2. FAO. *The Impact of Disasters and Crises on Agriculture and Food Security*; Report; FAO: Rome, Italy, 2018.
3. Rath, K.M.; Rousk, J. Salt Effects on the Soil Microbial Decomposer Community and Their Role in Organic Carbon Cycling: A Review. *Soil Biol. Biochem.* **2015**, *81*, 108–123. [CrossRef]
4. Zhang, L.; Tang, C.; Yang, J.; Yao, R.; Wang, X.; Xie, W.; Ge, A.-H. Salinity-Dependent Potential Soil Fungal Decomposers under Straw Amendment. *Sci. Total Environ.* **2023**, *891*, 164569. [CrossRef]
5. Yuan, B.-C.; Li, Z.-Z.; Liu, H.; Gao, M.; Zhang, Y.-Y. Microbial Biomass and Activity in Salt Affected Soils under Arid Conditions. *Appl. Soil Ecol.* **2007**, *35*, 319–328. [CrossRef]
6. Zhang, S.; Rasool, G.; Wang, S.; Zhang, Y.; Guo, X.; Wei, Z.; Zhang, X.; Yang, X.; Wang, T. Biochar and Chlorella Increase Rice Yield by Improving Saline-Alkali Soil Physicochemical Properties and Regulating Bacteria under Aquaculture Wastewater Irrigation. *Chemosphere* **2023**, *340*, 139850. [CrossRef]
7. Akhtar, M.; Hussain, F.; Ashraf, M.Y.; Qureshi, T.M.; Akhter, J.; Awan, A.R. Influence of Salinity on Nitrogen Transformations in Soil. *Commun. Soil Sci. Plant Anal.* **2012**, *43*, 1674–1683. [CrossRef]
8. Jacobsen, S.-E.; Mujica, A.; Jensen, C. The Resistance of Quinoa (Chenopodium quinoaWilld.) to Adverse Abiotic Factors. *Food Rev. Int.-Food Rev. Int.* **2003**, *19*, 99–109. [CrossRef]

9. Böhm, J.; Messerer, M.; Müller, H.M.; Scholz-Starke, J.; Gradogna, A.; Scherzer, S.; Maierhofer, T.; Bazihizina, N.; Zhang, H.; Stigloher, C.; et al. Understanding the Molecular Basis of Salt Sequestration in Epidermal Bladder Cells of Chenopodium Quinoa. *Curr. Biol.* **2018**, *28*, 3075–3085.e7. [CrossRef] [PubMed]
10. Estrada, R.; Cosme, R.; Porras, T.; Reynoso, A.; Calderon, C.; Arbizu, C.I.; Arone, G.J. Changes in Bulk and Rhizosphere Soil Microbial Diversity Communities of Native Quinoa Due to the Monocropping in the Peruvian Central Andes. *Microorganisms* **2023**, *11*, 1926. [CrossRef]
11. Toubali, S.; Meddich, A. Role of Combined Use of Mycorrhizae Fungi and Plant Growth Promoting Rhizobacteria in the Tolerance of Quinoa Plants Under Salt Stress. *Gesunde Pflanz.* **2023**, *75*, 1855–1869. [CrossRef]
12. Hu, H.; Shao, T.; Gao, X.; Long, X.; Rengel, Z. Effects of Planting Quinoa on Soil Properties and Microbiome in Saline Soil. *Land Degrad. Dev.* **2022**, *33*, 2689–2698. [CrossRef]
13. Hui, X.; Chen, H.; Shen, S.; Zhi, H.; Li, W. Establishment of Residual Methods for Matrine in Quinoa Plants and Soil and the Effect on Soil Bacterial Community and Composition. *Foods* **2023**, *12*, 1337. [CrossRef] [PubMed]
14. Zhang, H.; Chang, D.; Zhu, Z.; Meng, C.; Wang, K. Soil Priming Effects and Involved Microbial Community along Salt Gradients. *Biogeosci. Discuss.* **2023**, *114*, 1–29. [CrossRef]
15. Fan, K.; Delgado-Baquerizo, M.; Guo, X.; Wang, D.; Wu, Y.; Zhu, M.; Yu, W.; Yao, H.; Zhu, Y.; Chu, H. Suppressed N Fixation and Diazotrophs after Four Decades of Fertilization. *Microbiome* **2019**, *7*, 143. [CrossRef]
16. Pérez Castro, S.; Cleland, E.E.; Wagner, R.; Sawad, R.A.; Lipson, D.A. Soil Microbial Responses to Drought and Exotic Plants Shift Carbon Metabolism. *ISME J.* **2019**, *13*, 1776–1787. [CrossRef]
17. Agler, M.T.; Ruhe, J.; Kroll, S.; Morhenn, C.; Kim, S.-T.; Weigel, D.; Kemen, E.M. Microbial Hub Taxa Link Host and Abiotic Factors to Plant Microbiome Variation. *PLoS Biol.* **2016**, *14*, e1002352. [CrossRef]
18. Herren, C.M.; McMahon, K.D. Keystone Taxa Predict Compositional Change in Microbial Communities. *Environ. Microbiol.* **2018**, *20*, 2207–2217. [CrossRef]
19. Jia, L.; Wang, Z.; Ji, L.; De Neve, S.; Struik, P.C.; Yao, Y.; Lv, J.; Zhou, T.; Jin, K. Keystone Microbiome in the Rhizosphere Soil Reveals the Effect of Long-Term Conservation Tillage on Crop Growth in the Chinese Loess Plateau. *Plant Soil* **2022**, *473*, 457–472. [CrossRef]
20. Zhang, Z.; Han, X.; Yan, J.; Zou, W.; Wang, E.; Lu, X.; Chen, X. Keystone Microbiomes Revealed by 14 Years of Field Restoration of the Degraded Agricultural Soil Under Distinct Vegetation Scenarios. *Front. Microbiol.* **2020**, *11*, 1915. [CrossRef]
21. Kuypers, M.M.M.; Marchant, H.K.; Kartal, B. The Microbial Nitrogen-Cycling Network. *Nat. Rev. Microbiol.* **2018**, *16*, 263–276. [CrossRef] [PubMed]
22. Zhang, M.; Han, F.; Liu, Z.; Han, Y.; Li, Y.; Zhou, W. Ammonium-Assimilating Microbiome: A Halophilic Biosystem Rationally Optimized by Carbon to Nitrogen Ratios with Stable Nitrogen Conversion and Microbial Structure. *Bioresour. Technol.* **2022**, *350*, 126911. [CrossRef]
23. Quince, C.; Lanzen, A.; Davenport, R.J.; Turnbaugh, P.J. Removing Noise From Pyrosequenced Amplicons. *BMC Bioinformatics* **2011**, *12*, 38. [CrossRef]
24. Lane, D.; Stackebrandt, E.; Goodfellow, M. Nucleic Acid Techniques in Bacterial Systematics. In *Nucleic Acid Techniques in Bacterial Systematics*, 1st ed.; Wiley: Hoboken, NJ, USA, 1991; Volume 125–175.
25. Gurr, S. PCR Protocols-A Guide to Methods and Applications. *Biochem. Educ.* **1991**, *19*, 45. [CrossRef]
26. Edgar, R.C. UPARSE: Highly Accurate OTU Sequences from Microbial Amplicon Reads. *Nat. Methods* **2013**, *10*, 996–998. [CrossRef] [PubMed]
27. Tourna, M.; Freitag, T.E.; Nicol, G.W.; Prosser, J.I. Growth, Activity and Temperature Responses of Ammonia-Oxidizing Archaea and Bacteria in Soil Microcosms. *Environ. Microbiol.* **2008**, *10*, 1357–1364. [CrossRef] [PubMed]
28. Francis, C.A.; Roberts, K.J.; Beman, J.M.; Santoro, A.E.; Oakley, B.B. Ubiquity and Diversity of Ammonia-Oxidizing Archaea in Water Columns and Sediments of the Ocean. *Proc. Natl. Acad. Sci. USA* **2005**, *102*, 14683–14688. [CrossRef] [PubMed]
29. Rotthauwe, J.H.; Witzel, K.P.; Liesack, W. The Ammonia Monooxygenase Structural Gene amoA as a Functional Marker: Molecular Fine-Scale Analysis of Natural Ammonia-Oxidizing Populations. *Appl. Environ. Microbiol.* **1997**, *63*, 4704–4712. [CrossRef]
30. Rösch, C.; Bothe, H. Improved Assessment of Denitrifying, N2-Fixing, and Total-Community Bacteria by Terminal Restriction Fragment Length Polymorphism Analysis Using Multiple Restriction Enzymes. *Appl. Environ. Microbiol.* **2005**, *71*, 2026–2035. [CrossRef]
31. Michotey, V.; Méjean, V.; Bonin, P. Comparison of Methods for Quantification of Cytochrome Cd1-Denitrifying Bacteria in Environmental Marine Samples. *Appl. Environ. Microbiol.* **2000**, *66*, 1564–1571. [CrossRef]
32. Kandeler, E.; Deiglmayr, K.; Tscherko, D.; Bru, D.; Philippot, L. Abundance of narG, nirS, nirK, and nosZ Genes of Denitrifying Bacteria during Primary Successions of a Glacier Foreland. *Appl. Environ. Microbiol.* **2006**, *72*, 5957–5962. [CrossRef] [PubMed]
33. Braker, G.; Fesefeldt, A.; Witzel, K.-P. Development of PCR Primer Systems for Amplification of Nitrite Reductase Genes (nirK and nirS) To Detect Denitrifying Bacteria in Environmental Samples. *Appl. Environ. Microbiol.* **1998**, *64*, 3769–3775. [CrossRef] [PubMed]
34. Henry, S.; Bru, D.; Stres, B.; Hallet, S.; Philippot, L. Quantitative Detection of the nosZ Gene, Encoding Nitrous Oxide Reductase, and Comparison of the Abundances of 16S rRNA, narG, nirK, and nosZ Genes in Soils. *Appl. Environ. Microbiol.* **2006**, *72*, 5181–5189. [CrossRef] [PubMed]

35. Hariadi, Y.; Marandon, K.; Tian, Y.; Jacobsen, S.-E.; Shabala, S. Ionic and Osmotic Relations in Quinoa (Chenopodium Quinoa Willd.) Plants Grown at Various Salinity Levels. *J. Exp. Bot.* **2011**, *62*, 185–193. [CrossRef] [PubMed]
36. Rath, K.M.; Murphy, D.N.; Rousk, J. The Microbial Community Size, Structure, and Process Rates along Natural Gradients of Soil Salinity. *Soil Biol. Biochem.* **2019**, *138*, 107607. [CrossRef]
37. Otlewska, A.; Migliore, M.; Dybka-Stępień, K.; Manfredini, A.; Struszczyk-Świta, K.; Napoli, R.; Białkowska, A.; Canfora, L.; Pinzari, F. When Salt Meddles Between Plant, Soil, and Microorganisms. *Front. Plant Sci.* **2020**, *11*, 1429. [CrossRef] [PubMed]
38. Qin, Y.; Druzhinina, I.S.; Pan, X.; Yuan, Z. Microbially Mediated Plant Salt Tolerance and Microbiome-Based Solutions for Saline Agriculture. *Biotechnol. Adv.* **2016**, *34*, 1245–1259. [CrossRef] [PubMed]
39. Zhou, Y.; Bastida, F.; Zhou, B.; Sun, Y.; Gu, T.; Li, S.; Li, Y. Soil Fertility and Crop Production Are Fostered by Micro-Nano Bubble Irrigation with Associated Changes in Soil Bacterial Community. *Soil Biol. Biochem.* **2020**, *141*, 107663. [CrossRef]
40. Khalmuratova, I.; Choi, D.-H.; Kim, J.-G.; Lee, I.-S. Endophytic Fungi of Salt-Tolerant Plants: Diversity and Ability to Promote Plant Growth. *J. Microbiol. Biotechnol.* **2021**, *31*, 1526–1532. [CrossRef]
41. Liang, W.; Ma, X.; Wan, P.; Liu, L. Plant Salt-Tolerance Mechanism: A Review. *Biochem. Biophys. Res. Commun.* **2018**, *495*, 286–291. [CrossRef]
42. Rath, K.M.; Fierer, N.; Murphy, D.V.; Rousk, J. Linking Bacterial Community Composition to Soil Salinity along Environmental Gradients. *ISME J.* **2019**, *13*, 836–846. [CrossRef]
43. Edwards, J.; Johnson, C.; Santos-Medellín, C.; Lurie, E.; Podishetty, N.K.; Bhatnagar, S.; Eisen, J.A.; Sundaresan, V. Structure, Variation, and Assembly of the Root-Associated Microbiomes of Rice. *Proc. Natl. Acad. Sci. USA* **2015**, *112*, E911–E920. [CrossRef]
44. Yuan, J.; Chaparro, J.M.; Manter, D.K.; Zhang, R.; Vivanco, J.M.; Shen, Q. Roots from Distinct Plant Developmental Stages Are Capable of Rapidly Selecting Their Own Microbiome without the Influence of Environmental and Soil Edaphic Factors. *Soil Biol. Biochem.* **2015**, *89*, 206–209. [CrossRef]
45. Raiger Iustman, L.J.; Almasqué, F.J.; Vullo, D.L. Microbiota Diversity Change as Quality Indicator of Soils Exposed to Intensive Periurban Agriculture. *Curr. Microbiol.* **2021**, *78*, 338–346. [CrossRef] [PubMed]
46. Roy, S.; Chakraborty, A.P.; Chakraborty, R. Understanding the Potential of Root Microbiome Influencing Salt-Tolerance in Plants and Mechanisms Involved at the Transcriptional and Translational Level. *Physiol. Plant.* **2021**, *173*, 1657–1681. [CrossRef] [PubMed]
47. Chaudhary, D.R.; Rathore, A.P.; Kumar, R.; Jha, B. Spatial and Halophyte-Associated Microbial Communities in Intertidal Coastal Region of India. *Int. J. Phytoremediation* **2017**, *19*, 478–489. [CrossRef] [PubMed]
48. Ren, Y.; Shao, Q.; Ge, W.; Li, X.; Wang, H.; Dong, C.; Zhang, Y.; Deshmukh, S.K.; Han, Y. Assembly Processes and Biogeographical Characteristics of Soil Bacterial Sub-Communities of Different Habitats in Urban Green Spaces. *Curr. Microbiol.* **2023**, *80*, 309. [CrossRef]
49. Zhang, G.; Bai, J.; Jia, J.; Wang, W.; Wang, D.; Zhao, Q.; Wang, C.; Chen, G. Soil Microbial Communities Regulate the Threshold Effect of Salinity Stress on SOM Decomposition in Coastal Salt Marshes. *Fundam. Res.* **2023**. [CrossRef]
50. Zheng, Y.; Cao, X.; Zhou, Y.; Li, Z.; Yang, Y.; Zhao, D.; Li, Y.; Xu, Z.; Zhang, C.-S. Effect of Planting Salt-Tolerant Legumes on Coastal Saline Soil Nutrient Availability and Microbial Communities. *J. Environ. Manag.* **2023**, *345*, 118574. [CrossRef] [PubMed]
51. Xu, Y.; Zhang, G.; Ding, H.; Ci, D.; Dai, L.; Zhang, Z. Influence of Salt Stress on the Rhizosphere Soil Bacterial Community Structure and Growth Performance of Groundnut (*Arachis Hypogaea* L.). *Int. Microbiol.* **2020**, *23*, 453–465. [CrossRef]
52. Chen, Z.; Zheng, Y.; Ding, C.; Ren, X.; Yuan, J.; Sun, F.; Li, Y. Integrated Metagenomics and Molecular Ecological Network Analysis of Bacterial Community Composition during the Phytoremediation of Cadmium-Contaminated Soils by Bioenergy Crops. *Ecotoxicol. Environ. Saf.* **2017**, *145*, 111–118. [CrossRef]
53. Santolini, M.; Barabási, A.-L. Predicting Perturbation Patterns from the Topology of Biological Networks. *Proc. Natl. Acad. Sci. USA* **2018**, *115*, E6375–E6383. [CrossRef] [PubMed]
54. Fan, K.; Delgado-Baquerizo, M.; Guo, X.; Wang, D.; Zhu, Y.; Chu, H. Biodiversity of Key-Stone Phylotypes Determines Crop Production in a 4-Decade Fertilization Experiment. *ISME J.* **2021**, *15*, 550–561. [CrossRef] [PubMed]
55. Van Horn, D.J.; Okie, J.G.; Buelow, H.N.; Gooseff, M.N.; Barrett, J.E.; Takacs-Vesbach, C.D. Soil Microbial Responses to Increased Moisture and Organic Resources along a Salinity Gradient in a Polar Desert. *Appl. Environ. Microbiol.* **2014**, *80*, 3034–3043. [CrossRef] [PubMed]
56. Huang, G.; Sun, Y.; Zhang, X.; Rodríguez, L.G.; Luo, J.; Chen, Z.; Ou, Y.; Gao, Y.; Ghaffari, H.; Yao, Y. Adaptation to Low Nitrogen and Salt Stresses in the Desert Poplar by Effective Regulation of Nitrogen Assimilation and Ion Balance. *Plant Physiol. Biochem.* **2022**, *193*, 14–24. [CrossRef] [PubMed]

MDPI AG
Grosspeteranlage 5
4052 Basel
Switzerland
Tel.: +41 61 683 77 34

Microorganisms Editorial Office
E-mail: microorganisms@mdpi.com
www.mdpi.com/journal/microorganisms

www.ingramcontent.com/pod-product-compliance
Lightning Source LLC
Chambersburg PA
CBHW041948260926
51850CB00018B/1968
9783725844722